Lecture Notes in Mathematics

Edited by A. Dold and

888

Padé Approximation and its Applications Amsterdam 1980

Proceedings of a Conference Held in Amsterdam, The Netherlands, October 29–31, 1980

Edited by M.G. de Bruin and H. van Rossum

Springer-Verlag
Berlin Heidelberg New York 1981

Editors

Marcel G. de Bruin
Herman van Rossum
Institut voor Propedeutische Wiskunde, University of Amsterdam
Box 20239, 1000 HE Amsterdam, The Netherlands

AMS Subject Classifications (1980): 30E05, 41A21, 42A16, 42C05, 65B05, 65D15

ISBN 3-540-11154-9 Springer-Verlag Berlin Heidelberg New York
ISBN 0-387-11154-9 Springer-Verlag New York Heidelberg Berlin

Printing and binding: Beltz Offsetdruck, Hemsbach/Bergstr.
2141/3140-543210

PREFACE

As already mentioned by Luc Wuytack (in the preface to the proceedings of the conference held in Antwerp 1979, LNM 765) the interest in Padé approximation, convergence acceleration and related subjects is still growing: the conference on "Padé and Rational Approximation, theory and applications" held at the Instituut voor Propedeutische Wiskunde of the University of Amsterdam on October 29-31, 1980, was attended by approximately fifty mathematicians and physicists.

In this publication the reader can find the papers that were presented during the Amsterdam conference. Two of the speakers, however, had their papers accepted for publication elsewhere (A. Sidi & D. Levin: Rational Approximations from the d-Transformation; J. Fleischer & M. Pindor: On an application of the operator Padé Approximants to Quantum Scattering Problems), these talks are not included. Altogether there were thirty lectures: four invited one-hour talks (part I) and twenty six half hour communications (of which 24 appear in part II).

The conference in Amsterdam already was the sixth in a series of conferences on this subject in Europe (Canterbury '72, Toulon '75, Lille '77 and '78, Antwerp '79) and it turned out that the participants would welcome a continuation along this line; some people even offered to organize a conference in 1982 and in 1983.

Finally we would like to express our gratitude to the "Faculteit der Wiskunde en Natuurwetenschappen" of the University of Amsterdam for the financial support and of course, last but not least, we thank the editors of the Lecture Notes and Springer-Verlag for their help and speedy publication of this volume.

Amsterdam, July 29, 1981

M.G. de Bruin

H. van Rossum

CONTENTS

PART I: Invited speakers

PART II: Short communications

* - in the cases of two authors the talk has been delivered by the person with an asterisk -

THE LONG HISTORY OF CONTINUED FRACTIONS

AND

PADÉ APPROXIMANTS

Claude BREZINSKI

- ★ -

U.E.R I.E.E.A - Informatique
University of Lille
59655 Villeneuve d'Ascq Cédex
FRANCE

- ★ -

INTRODUCTION.

Continued fractions and Padé approximants have played a quite important rôle in the development of pure and applied mathematics and they are still widely used as this congress shows. Thus I think that it is of interest for specialists in this field to have an idea of their history. The reason is not only a cultural one but some old works can also be the starting point of new researches.

The history of continued fractions is rather long since it begins with Euclid's algorithm for the g.c.d. 300 years B.C. Their history also involves most of the well known mathematicians of all ages. Thus a complete history will be too long for these proceedings and I shall only give a brief account of it. The complete history with references will be published later.

I would like to thank Bernard Rouxel from the University of Lille for his autorized advice on the history of mathematics. Thanks are also due to Herman Van Rossum and Marcel de Bruin for interesting discussions and for having accepted this long paper for publication.

THE EARLY AGES.

As it is often the case in sciences, continued fractions have been used a long time before their real discovery. It seems that their first use goes back to the algorithm of EUCLID (c. 306 B.C. - c. 283 B.C.) for computing the g. c. d. of two integers which leads to a terminating continued fraction. Let a and b be two positive integers with a > b. We set $r_0 = a$, $r_1 = b$ and we define the sequence (r_k) by

$$r_k = r_{k+1} q_k + r_{k+2} \qquad k = 0,1,\ldots$$

with $0 \le r_{k+1} < r_k$ and $q_k \in \mathbb{N}$. It can be proved that an index n exists such that $r_{n+2} = 0$. Thus $r_n = r_{n+1} q_n$ and r_{n+1} is the g.c.d. of a and b. Moreover we have

$$r_k/r_{k+1} = q_k + 1 / (r_{k+1}/r_{k+2})$$

and consequently

$$\frac{a}{b} = q_0 + \frac{1|}{|q_1} + \ldots + \frac{1|}{|q_n} .$$

Of course Euclid did not present his algorithm in that way but he used geometrical considerations on the measurement of a segment by another one.

Euclid's algorithm is related to the approximate simplification of ratios as it was practiced by ARCHIMEDES (287 B.C. - 212 B.C.) and ARISTARCHUS OF SAMOS (c. 310 B.C. - c. 230 B.C.). Continued fractions were also implicitely used by greek mathematicians, such as THEON OF ALEXANDRIA (c. 365), in methods for computing the side of a square with a given area. Theon's method for extracting the square root is in fact the beginning of the continued fraction

$$\sqrt{A} = \sqrt{a^2+r} = a + \frac{r|}{|2a} + \frac{r|}{|2a} + \ldots$$

where a is the greatest integer such that $a^2 \le A$. Archimedes also gave the bounds

$$\frac{265}{153} = \frac{1}{3}\left(5 + \frac{1|}{|5} + \frac{1|}{|10}\right) < \sqrt{3} < \frac{1351}{780} = \frac{1}{3}\left(5 + \frac{1|}{|5} + \frac{1|}{|10} + \frac{1|}{|5}\right).$$

Other attempts to approximate square roots have been made among the centuries. Though none of them is directly related to continued fractions they opened the way that will be followed later by those who really created their theory.

Another very ancient problem which also leads to the early use of continued fractions is the problem of what is now called diophantine equations in honour to DIOPHANTUS (c. 250 A.D.) who found a rational solution of

$$ax \pm by = c$$

where a, b and c are given positive integers. Such equations arose from astronomy and their solutions are connected with theappearance of some constellations. This problem has been completely solved by the indian mathematician ĀRYABHATA (475-550). Let us consider

$$ax + by = c$$

where a and b are relatively prime and $a > b > 0$. Then by Euclid's algorithm

$$\frac{a}{b} = q_0 + \frac{1|}{|q_1} + \ldots + \frac{1|}{|q_n}.$$

Let us set

$$\frac{p}{q} = q_0 + \frac{1|}{|q_1} + \ldots + \frac{1|}{|q_{n-1}}.$$

Then $aq - bp = \pm 1$. If we consider the case with the positive sign then

$$ax + by = c(aq-bp)$$

and thus

$$(cq - x)/b = (y + cp)/a = t.$$

Then

$$x = cq - bt \qquad y = at - cp.$$

Giving integral values to t we obtain all the integral solutions of the diophantine equation. This solution was also given by BRAHMAGUPTA (c. 598). In Europe the same method will be rediscovered by Claude Gaspard BACHET DE MEZIRIAC (1581-1638) in 1612. The same method will also be used by Nicholas SAUNDERSON (1682-1739) and by Joseph Louis LAGRANGE (1736-1813) who wrote down explicitely in 1767 the continued fraction for a/b but 1300 years after the indian mathematicians !

One of the most important indian mathematicians is probably BHASCARA who is born in Bidur in 1115 and worked at Ujjayni. Around 1150 he wrote a book "*Līlāvatī*" where he treated the equation $ax - by = c$. He proved that the solution can be obtained from the continued fraction for a/b. He showed that the convergents $C_k = A_k/B_k$ of this continued fraction satisfy

$$A_k = q_k A_{k-1} + A_{k-2}$$
$$B_k = q_k B_{k-1} + B_{k-2}$$
$$A_k B_{k-1} - A_{k-1} B_k = (-1)^{k-1}.$$

Then the solution is given by

$$x = \mp c\,B_{n-1} + bt \qquad y = \mp c\,A_{n-1} + at$$

according as $a\,B_{n-1} - b\,A_{n-1} = \pm 1$.

We must anticipate to say that the recurrence relationship of continued fractions will only be known in Europ 500 years later. Thus it is important to notice that the first english translation of Bhascara's book appeared in 1816 and that his work was probably not known earlier in Europe.(except, maybe, by some latin translation of its arabic translation).

Remark. The notation

$$b_0 + \frac{a_1|}{|b_1} + \frac{a_2|}{|b_2} + \dots$$

has been introduced in 1898 by Alfred PRINGSHEIM (1850-1941) while the notation

$$b_0 + \frac{a_1}{b_1+}\ \frac{a_2}{b_2+} \dots$$

has been introduced in 1820 by Sir John William HERSCHEL (1792-1871).

THE FIRST STEPS.

In Europe the birth place of continued fractions is obviously the north of Italy. The first attempt for a general definition of a continued fraction was made by Leonardo FIBONACCI (c. 1170 - c. 1250) also called Leonardo of Pisa. He was a merchant who traveled quite widely in the East and was in contact with the Arabic mathematical writings. In his very celebrated book "*Liber Abaci*" (written in 1202, revised in 1228 but only published in 1857) he introduced a kind of ascending continued fraction which is not of very great interest.

The first mathematician who really used our modern infinite continued fractions was Rafael BOMBELLI (1526 - 1572) the discoverer of imaginary numbers. Little is known about his life and career but he published a book "*L'Algebra parte maggiore dell' arimetica divisa in tre libri*" in Bologna in 1572 followed by a second edition in 1579 with the title "*L'Algebra Opera*". In this second edition he gave a recursive algorithm for extracting the square root of 13 which is completely equivalent to the infinite continued fraction

$$\sqrt{13} = 3 + \frac{4|}{|6} + \frac{4|}{|6} + \dots$$

Bombelli gives not hint for the success of his method nor how he discovered it but, of course, it consists in writing

$$A = a^2 + r$$

where a is the greatest integer such that $a^2 < A$. Otherwise we have

$$(\sqrt{A} + a)(\sqrt{A} - a) = r$$

and thus

$$\sqrt{A} = a + r/(a + \sqrt{A}).$$

Replacing in the denominator, $\sqrt{A}$ by its expression and repeating indefinitely this process we get

$$\sqrt{A} = a + \frac{r|}{|2a} + \frac{r|}{|2a} + \dots$$

Bombelli admits that the first version of his book was based on AL-KHOWARIZMI, the great muslim mathematician who lived in Bagdad around 830.

This method could also be attributed to AL-HAYYĀM (c. 1048 - c. 1131).

The next and most important contribution to the theory of continued fractions is by Pietro Antonio CATALDI (1548-1626) who can be considered as the real founder of the theory. In his book "*Trattato del modo brevissimo di trovare la radice quadra delli numeri* ..." published in Bologna in 1613, he followed the same method as Bombelli for extracting the square root but he was the first to introduce a symbolism for continued fractions. He computed the continued fraction for $\sqrt{18}$ up to the 15 th convergent and proved that the convergents are alternately greater and smaller than $\sqrt{18}$ and that they converge to it.

At the same period continued fractions (or, more precisely, a device related to them) were still used to find approximate values for ratios and to simplify fractions. About this problem we must mention the contributions of Daniel SCHWENTER (1585-1636), Frans von SCHOOTEN (1615-1660) and Albert GIRARD (1595-1632).

THE BEGINNING OF THE THEORY.

We must now emigrate to England for the next major step in the development of the theory.

In 1655, John WALLIS (1616-1703) in his book "*Arithmetica Infinitorum*" obtained an infinite product for $4/\pi$

$$\frac{4}{\pi} = \frac{2.3.5.5.7.7\ \ldots.}{2.4.4.6.6.8\ \ldots.}$$

Then he says about Lord William BROUNCKER (1620-1684) one of the founders and the first President of the Royal Society

> "*that Most Noble Man, after having considered this matter, saw fit to bring this quantity by a method of infinitesimals peculiar to him, to a form which can thus be conveniently written* (in our modern notation)

$$\frac{4}{\pi} = 1 + \frac{1|}{|2} + \frac{9|}{|2} + \frac{25|}{|2} + \frac{49|}{|2} + \frac{81|}{|2} + \ldots \text{"}$$

and Wallis continues

> "*Nempe si unitati adjungatur fractio, quae denominatorem habeat continue fractum*"

Thus the words "*continued fractions*" were invented.

Wallis' most important contribution arises ten pages farther when he writes

> "*Nos inde hanc colligimus regulam, cujus ope a principio reductionem inchoemur quousq ; libet continuandam*

$$\begin{array}{ccccc} P & Q & P & Q & Q \\ \hline N3 \times & \frac{N1}{D1} & :+\ D3 \times & \frac{N2}{D2} & := \frac{N3}{D3} \end{array} \text{"}$$

This is our modern recurrence relationship for the convergents of a continued fraction. Wallis also made the first step to a proof of convergence when he pointed out that the convergents of Brouncker's continued fraction are successively larger and smaller than $4/\pi$ and when he claimed that the process converges : "*ad numerum justum acceditur*".

We must also mention the Dutch mathematician and astronomer Christiaan HUYGENS (1629-1695) who built, in 1682, an automatic planetarium. He used continued fractions for this purpose as described in his book "*Descriptio automati planetarii*" published after his death. In one year Saturn covers 12° 13' 34" 18"' and the earth 359° 45' 40" 30"' which gives the ratio 2640858/77708431. For finding the smallest integers whose ratio is close to this ratio (which will give him the number of teeth of the wheels of his planetarium) he divided the greastest number by the smallest one, then the smallest by the remainder and so on. He thus obtained, for his ratio, the continued fraction

$$29 + \frac{1|}{|2} + \frac{1|}{|2} + \frac{1|}{|1} + \frac{1|}{|5} + \frac{1|}{|1} + \frac{1|}{|4} + \dots$$

Huygens was also interested by the solution of the diophantine equation $py - qx = \pm 1$. He developed p/q into a continued fraction and noted that the convergents converge to p/q and are alternately smaller and greater than p/q. x and y are respectively given by the numerator and the denominator of the convergent immediately preceding p/q. Then py - qx = +1 or -1 according as x/y is smaller or greater than p/q. This method was in fact Āryabhata's method and it was used by the Englishman Nicholas SAUNDERSON to solve ax - by = c where c is the g.c.d. of a and b. Saunderson proved some additional results on the method such as optimality properties for the convergents of the continued fraction for a/b.

This subject was also studied by Roger COTES (1682-1716), Gottfried Wilhelm LEIBNIZ (1646-1716), Robert SIMSON (1687-1768) and some others.

To end this section we must not forget Johann BERNOULLI I (1667-1748) who used continued fractions to simplify the ratio of large integers and who was Euler's professor.

THE GOLDEN AGE.

The eighteenth century is the golden age of continued fractions. It has been marked by three outstanding mathematicians : Euler, Lambert and Lagrange all of whom belonging to the Academy of Sciences in Berlin.

Obviously the major contribution to the theory of continued fractions is due to Leonhard EULER (1707-1783). In his first paper on the subject, dated 1737, he proved that every rational number can be developed into a terminating continued fraction, that an irrational number gives rise to an infinite continued fraction and that a periodic continued fraction is the root of a quadratic equation. He also gave the continued fractions for e, (e+1)/(e-1) and (e-1)/2 by integrating the Riccati's equation by two different methods. It must be noticed that, apart from the convergence of these continued fractions which he did not treated, Euler proved the irrationality of e and e^2.

The first extensive and systematic exposition of the theory of continued fractions was given by Euler in 1748 in his celebrated book "*Introductio in analysis infinitorum*". He first gives the recurrence relationship for the convergents $C_k = A_k/B_k$ of the continued fraction $b_0 + \frac{a_1|}{|b_1} + \frac{a_2|}{|b_2} + \ldots$ and then shows how to transform a continued fraction into a series

$$C_n - C_{n-1} = (-1)^{n-1} \frac{a_1 \ldots a_n}{B_{n-1} B_n}$$

which leads to the relation

$$C = b_o + \sum_{n=1}^{\infty} (-1)^{n-1} \frac{a_1 \ldots a_n}{B_{n-1} B_n} .$$

Reciprocally Euler shows that an infinite series can be transformed into a continued fraction

$$\sum_{n=1}^{\infty} (-1)^{n-1} c_n = \frac{c_1|}{|1} + \frac{c_2|}{|c_1 - c_2} + \ldots + \frac{c_{n-2} c_n|}{|c_{n-1} - c_n} + \ldots$$

After some examples he treats the case of a power series.

Then he comes to the problem of convergence showing how to compute the value of the periodic continued fraction $C = \frac{1|}{|2} + \frac{1|}{|2} + \ldots$ by writing $C = 1/(2+C)$ which gives $C^2 + 2C = 1$ and thus $C = \sqrt{2} - 1$. From this example he derives Bombelli's method for the continued fraction expansion of the square root and a general method for the solution of a quadratic equation. The chapter ends with Euclid's algorithm and the simplification of fractions with examples.

Euler published some papers where he applied continued fractions to the solution of Riccati's differential equation and to the calculation of integrals. He also showed that certain continued fractions derived from power series can converge outside the domain of convergence of the series. This is, in particular, the case for the divergent series $x - 1!x^2 + 2!x^3 - 3!x^4 + \ldots$. He proved that this series formally satisfies the differential equation $x^2y'' + y = x$ and he got the solution

$$y(x) = \int_0^{\infty} \frac{x\, e^{-t}}{1 + xt} dt.$$

He thus obtained a method for summing a divergent series. Then he converted the preceding series into the continued fraction

$$\frac{x|}{|1} + \frac{x|}{|1} + \frac{x|}{|1} + \frac{2x|}{|1} + \frac{2x|}{|1} + \frac{3x|}{|1} + \frac{3x|}{|1} + \ldots$$

and he used it to compute the "*value*" of the divergent series $1! - 2! + 3! - \dots$ Euler's ideas on the subject will be extended later by Laguerre and Stieltjes.

In a letter dated 1743 and in a paper published in 1762, Euler investigated the problem of finding the integers a for which $a^2 + 1$ is divisible by a given prime of the form $4n + 1 = p^2 + q^2$. Its solution involves the penultimate convergent of the continued fraction for p/q.

In 1765, Euler studied the Pellian equation $x^2 = Dy^2 + 1$. He developed $\sqrt{D}$ into a continued fraction $\sqrt{D} = v + \frac{1|}{|a} + \frac{1|}{|b} + \frac{1|}{|c} + \dots$. He denoted the successive convergents of this continued fraction by

$$\frac{1}{0} \quad \frac{(v)}{1} \quad \frac{(v,a)}{(a)} \quad \frac{(v,a,b)}{(a,b)} \quad \frac{(v,a,b,c)}{(a,b,c,)}$$

He stated several equalities and proved that

$$(v)^2 - D.1^2 = -\alpha \ ; \ (v,a)^2 - D(a)^2 = \beta \ ; \ (v,a,b)^2 - D(a,b)^2 = -\gamma \ ; \ \dots$$

where α, β, γ, ... can be obtained from v, a, b, c, ...

The study of the numerators and denominators of the convergents as functions of the partial denominators was first seriously undertaken by Euler around the same time. Denoting by (a), (a,b)/b, (a,b,c)/(b,c), ... the convergents of $a + \frac{1|}{|b} + \frac{1|}{|c} + \dots$ he proved a long list of identities such as

$$(a,b,c,d,\dots) = a(b,c,d,\dots) + (c,d,\dots)$$
$$(a,b,c,\dots,q) = (q,\dots,c,b,a)$$
$$(a,b)(b,c) - (b)(a,b,c) = 1$$
$$(a,b,c)(d,e,f) - (a,b,\dots,f) = -(a,b)(e,f), \text{ etc} \dots$$

In 1771, Euler applied continued fractions to the approximate determination of the geometric mean of two numbers whose ratio is as $1/x$. The method can be used to get approximate values of $x^{p/q}$.

In 1773, Euler used continued fractions to find x and y making $mx^2 - ny^2$ minimum and in 1780 for seeking f and g such that $fr^2 - gs^2 = x$.

In 1783, Euler proved that the value of the continued fraction $\frac{m+1|}{|2} + \frac{m+2|}{|3} + \dots$ is a rational number when m is an integer not smaller than 2.

Thus Euler was the first mathematician not only to give a clear exposition of continued fractions but also to use them extensively to solve various problems. It is obvious that his influence is prominent in the development of the subject.

In 1775, Daniel BERNOULLI (1700-1782) solved the problem of finding a continued fraction with a given sequence of convergents.

By 1750, the number π had been expressed as infinite series, infinite products and infinite continued fractions but the problem of the quadrature of the circle still remained unsolved. A first step to the negative answer to this problem was done by Johan Heinrich LAMBERT (1728-1777). Using Euler's work on continued fractions he got in 1766 the development

$$\tan x = \frac{x|}{|1} - \frac{x^2|}{|3} - \frac{x^2|}{|5} - \ldots$$

Then he proved that tan x cannot be rational if x is a rational non zero number. Since $\tan \pi/4 = 1$ it follows that neither $\pi/4$ nor π are rational. Then, from an analogy between hyperbolic and trigonometric functions, he proved, from the continued fraction for e^x+1, that $e^n (n \in \mathbb{N})$ is irrational and that all the rational numbers have irrational natural logarithms. Lambert also proved the convergence of the continued fraction for tan x and he ended his work with the conjecture that "*no circular or logarithmic transcendental quantity into which no other transcendental quantity enters can be expressed by any irrational radical quantity*".

Lambert gave some examples of divergent series whose continued fraction converges and he obtained the continued fractions for Arctan x, Log(1+x), $(e^x-1)/(e^x+1)$ and π. It is very much remarkable for that time that Lambert gave a complete theoretical justification of these expansions, although a little bit complicated but perfectly rigourous.

The next fundamental contributions to the theory of continued fractions are due to Joseph Louis LAGRANGE (1736-1813). In 1766 he gave the first proof that $x^2 = Dy^2 + 1$ has integral solutions with $y \neq 0$ if D is a given integer not a square. The proof makes use of the continued fraction for $\sqrt{D}$.

In 1767, Lagrange published a "*Mémoire sur la résolution des équations numériques*" where he gave a method for approximating the real roots of an equation by continued fractions. One year later he wrote an "*Addition*" to the preceding "*Mémoire*" where he proved the converse of Euler's result :

> *"Now I claim that the continued fraction which expresses the value of* x [the real positive zero of a quadratic equation] *will always be necessarily periodic,"*

He showed that the continued fraction for $\sqrt{D}$ is periodic and that the period can only take two different forms which he exhibited. He related his results to the solution of $x^2 = Dy^2 \pm 1$. In the same paper he extended Huygens' and Saunderson's method for solving $py - qx = r$. He noted that the method is *"essentially the same as Bachet's, as are also all methods proposed by other mathematicians"* and that it is equivalent to the usual one of converting p/q into a continued fraction.

An interesting problem treated by Lagrange in 1772 is the solution of linear difference equations with constant coefficients. Let (c_n) be an infinite sequence of numbers ; Lagrange first defines what he calls the generating function of the sequence (c_n)

$$f_o(x) = c_o + c_1 x + c_2 x^2 + \ldots$$

He then shows that if

$$f_o(x) = (a_o + a_1 x + \ldots + a_{k-1} x^{k-1})/(b_o + b_1 x + \ldots + b_k x^k)$$

then the sequence (c_n) is recurrent, i.e. $\forall n \geq k$

$$c_n = -(b_1 c_{n-1} + \ldots + b_k c_{n-k})/b_o.$$

Conversely if (c_n) is recurrent its generating function is a rational function. Lagrange was also interested by the inverse problem of searching hidden periodicities in a sequence. It is equivalent to show that the generating function f_o of the sequence is a rational function. He constructed the sequence (f_k) of functions by

$$f_{-1}(x) = 1$$

$$f_{k-1}(x)/f_k(x) = p_k + q_k x + x^2 \; f_{k+1}(x)/f_k(x).$$

Thus

$$f_o(x) = \frac{1 |}{|p_o+q_o x} + \frac{x^2 |}{|p_1+q_1 x} + \frac{x^2 |}{|p_2+q_2 x} + \ldots$$

If for some k, $f_{k-1}(x) \equiv 0$ then f_o is the ratio of a polynomial of degree k-1 by a polynomial of degree k and thus (c_n) satisfies a difference equation of order k.

In 1774, in an addition to Euler's Algebra, Lagrange proved that if a is a given positive real number then relatively prime integers p and q can be found such that $p-qa < r-sa$ for $r < p$ and $s < q$ by taking p/q as any convergent of the continued fraction for a in which all the terms are positive. He also gave a method, using continued fractions, to solve $Ay^2 - 2Byz + Cz^2 = 1$ in integers and he proved that Pell's equation cannot be solved by use of a continued fraction for $\sqrt{D}$ in which the signs of the partial denominators are arbitrarily chosen.

In 1776, Lagrange published a paper on the use of continued fractions in integral calculus where he developed a general method to obtain the continued fraction expansion of the solution of a differential equation. He then gave some examples and reduced the continued fractions thus obtained to ordinary fractions (by computing their convergents) whose power series expansions in ascending powers of the variable agree with those of the functions *"jusqu'à la puissance de* x *inclusivement qui est la somme des deux plus hautes puissances de* x *dans le numérateur et dans le dénominateur"*.

This is really the birth -certificate of Padé approximants ! In a letter to d'Alembert he says about the volume containing this paper

> *"Il y a comme de raison quelque chose de moi, mais rien qui puisse mériter votre attention ... ".*

This was not a prophetic view !

To end this section let us mention that Lagrange's method for the diophantine equation was used in 1772 by Johann BERNOULLI II (1710-1790) to find the least integer u giving an integral solution for $A = Bt - Cu$ where B and C are relatively prime. If C is odd and $A = (C + 1)/2$ then $u = (B + s - 1)/2$ and $t = (C + r)/2$ where r/s is the penultimate convergent of the continued fraction for C/B.

During the eighteenth century continued fractions were also used and developed by Japanese mathematicians in connection with the expansion of a quadratic surd, i.e. a number of the form $(a \pm \sqrt{b})/c$ where a, b and c are integers and b not a square.

THE NINETEENTH CENTURY.

The theory of continued fractions is now ready to be extensively developed and used and this will be the case during the nineteenth century. Most of the prominent mathematicians of that period have made contributions to the theory or have used it in their proofs.

The first of them is Pierre Simon LAPLACE (1749-1827) who showed that the solution of a difference equation of the first degree and second order can always be expressed as a continued fraction. He gave the continued fraction

$$\int_0^x e^{-t^2} dt = \frac{\sqrt{\pi}}{2} - \frac{e^{-x^2}/2|}{|x} + \frac{1|}{|2x} + \frac{2|}{|x} + \frac{3|}{|2x} + \frac{4|}{|x} + \ldots$$

In 1803, Basilius VISKOVATOFF (1778-1812) proposed a method for transforming the quotient of two power series into a continued fraction and in 1818 Jean Jacques BRET (1781-1819) obtained some continued fraction expansions by Lambert's method for tan x. The study of such questions will directly lead to Padé approximants since it is possible to develop the series $c_0 + c_1x + \ldots$ into a continued fraction of the form $b_0 + \frac{x|}{|b_1} + \frac{x|}{|b_2} + \ldots$. This problem was investigated by Maritz Abraham STERN (1807-1894) in 1833 who found a recurrence relationship for the b_i's and by O. HEILERMANN who gave them explicitely in 1846. Heilermann also studied the continued fraction

$$b_0 - \frac{a_1|}{|x+b_1} - \frac{a_2|}{|x+b_2} - \ldots$$

for the ratio of two series in x^{-1}. These two developments are unique and the connection between them was established by Heinrich Eduard HEINE (1821-1881) in 1878.

The continued fraction for Log (1+x)/(1-x) was obtained by Carl Friedrich GAUSS (1777-1855) who used it in his very celebrated paper on gaussian quadrature methods presented to the Göttingen Society in 1814. Gauss also gave some other continued fraction expansions such as that of the ratio of two hypergeometric series.

In 1833, M.A. STERN studied the transformation of an infinite product into an equivalent continued fraction and reciprocally.

The operations of contraction and extension were introduced by Philipp Ludwig von SEIDEL (1821-1896) in 1855 although special cases were already treated by Lagrange in 1774 and 1776. The relationship between the corresponding and the associated continued fractions was considered by Johann Bernhard H. HEILERMAN (1820-1899) in 1860. Determinantal formulas for the coefficients of these continued fractions have been obtained by J.B.H. HEILERMAN in 1845, Hermann HANKEL (1839-1873) in 1862, G. BAUER in 1872, Thomas MUIR (1844-1934) in 1875, Georg Ferdinand FROBENIUS (1849-1917) in 1886 and Thomas Jan STIELTJES (1856-1894) in 1894. Of course it is possible to transform an infinite continued fraction into infinitely many equivalent continued fractions where the successive convergents are only partially conserved. In 1855, L. SEIDEL showed that a converging continued fraction can be thus transformed into a diverging one and conversely. A cautious application of equivalent transformations can be used to accelerate the convergence, as showed by August Ferdinand MÖBIUS (1790-1868) in 1830, or to determine the value of the continued fraction, as showed by M.A. STERN in 1834.

Adrien Marie LEGENDRE (1752-1833) widely used continued fractions in his book on number theory. He proved, by a modification of Lambert's proof for π, that π^2, e and e^2 are irrational numbers.

Carl Gustav Jacob JACOBI (1804-1851) made numerous contributions to the theory of continued fractions, the most well known of them being his determinantal formula for Padé approximants dated 1845. In a paper published in 1868, Jacobi showed that cubic irrationals have some properties in common with quadratic irrationals since they lead to algorithms similar to the algorithm of periodic continued fractions. In another paper, published in 1869, Jacobi proposed a solution to the problem of simultaneous diophantine approximations. It consists in finding two sequences of rational approximations with the same denominators, (A_n/C_n) and (B_n/C_n), of two real numbers. Jacobi was leaded to a difference equation of the third order whose solutions were (A_n), (B_n) and (C_n). This algorithm, which is a generalization of the continued fraction algorithm, will be extensively developed later by Oskar PERRON and is known as the Jacobi-Perron algorithm.

A determinantal formula for the numerators and denominators of the convergents of a continued fraction has been given in 1853 by James Joseph SYLVESTER (1814-1897) and in 1856 by William SPOTTISWOODE (1825-1883) in the case where $a_n = \pm 1$. The solution of the three terms recurrence relation satisfied by the numerators and denominators of the convergents has been obtained by L. PAINVIN in 1858 and the result was applied for the first time to continued fractions by S. GÜNTHER in 1872. Reciprocally every three terms recurrence relationship gives rise to a continued fraction ; this remark has been used by G. BAUER in 1859, H.E. HEINE in 1860 and W. SCHEIBNER in 1864 to prove some results on continued fractions.

Many important results in arithmetic and in number theory have been obtained by continued fractions. We already saw some of them. In 1851, Joseph LIOUVILLE (1809-1882) proved, by means of continued fractions, the existence of transcendental numbers (that is numbers which don't satisfy any algebraic equation with integral coefficients). He proved that there exist infinitely many such numbers and he used continued fractions for approximating them. He also proved that neither e nor e^2 are quadratic irrationals. Using some particular continued fractions, Adolf HURWITZ (1859-1919) showed in 1896 that e cannot be a zero of a cubic equation with integral coefficients. An important contribution to number theory is due to Charles HERMITE (1822-1901) who proved in 1873, always by continued fractions, that e is a transcendental number. Hermite's fundamental idea is as follows : let $n_1,\ldots,n_m$ be arbitrary positive integers, $N = n_1+\ldots+n_m$ and let $k_1,\ldots,k_m$ be m distinct real or complex numbers. Then it is possible to construct polynomials $p_1,\ldots,p_m$ such that p_i has the exact degree $N-n_i$ and that

$$e^{k_i x}\, p_\ell(x) - e^{k_\ell x}\, p_i(x) = O(x^{N+1}) \qquad i,\ \ell = 1,\ldots,m$$

as x tends to zero. The differences between the exponentials and their approximate values can be represented by definite integrals. If x = 1 and if the k_i are integers, it can be proved that e cannot satisfy an algebraic equation with integral coefficients and thus is tnancendental.

Hermite was very much occupied by this idea. In a letter, dated 1873, he was interested by finding the polynomials p_i of degree n_i-1 such that

$$\sum_{i=1}^{m} e^{k_i x}\, p_i(x) = O(x^{N-1}).$$

This type of approximation is now called Padé-Hermite approximation when the functions $e^{k_i x}$ are replaced by arbitrary formal power series. Let us mention that Hermite was Padé's advisor and that Padé also worked on this subject after his thesis.

After his memoir of 1873 on the exponential function Hermite continued his researches on algebraic continued fractions. Since Gauss the rôle played by Legendre polynomials in the development of Log (x-1)/(x+1) in a continued fraction was known. H.E. HEINE and Elwin Bruno CHRISTOFFEL (1829-1900) have related the theory of continued fractions to some linear differential equations of the second order. Hermite extended all these results, showing how a certain linear differential equation of order m+1 is related to his simultaneous approximations. He applied the result to Log $(x-x_i)/(x-x_o)$ generalizing Legendre's polynomials.

Using the same method as Hermite for e, Carl Louis Ferdinand LINDEMANN (1852-1939) gave in 1882 the first proof that π is transcendental, ending by a negative result the controversy on the quadrature of the circle which was an open problem for more than 2000 years.

In 1828, at the age of 17, Evariste GALOIS (1811-1832) proved that if

$$x = a_o + \frac{1|}{|a_1} + \ldots + \frac{1|}{|a_n} + \frac{1|}{|a_o} + \ldots + \frac{1|}{|a_n} + \frac{1|}{|a_o} + \ldots$$

is a zero of a polynomial of arbitrary degree, then

$$y = -\frac{1|}{|a_n} + \frac{1|}{|a_{n-1}} + \ldots + \frac{1|}{|a_o} + \frac{1|}{|a_n} + \ldots + \frac{1|}{|a_o} + \frac{1|}{|a_n} + \ldots$$

is also a zero of the same polynomial. This result was implicitely contained in the earlier work of Lagrange and it was the lecture of Lagrange's "*Mémoire*" on algebraic equations which introduced Galois to the subject.

Coming back to Euler's work, Edmond Nicolas LAGUERRE (1834-1886) studied the differential equation $P(x)y'(x) = Q(x)y(x)+R(x)$ where P, Q and R are polynomials in x. He developed y into a continued fraction for some particular cases. He studied in details

$$y(x) = \int_o^\infty \frac{x\,e^{-t}}{1+xt}\,dt = \frac{x|}{|1} + \frac{x|}{|1} + \frac{x|}{|1} + \frac{2x|}{|1} + \frac{2x|}{|1} + \frac{3x|}{|1} + \frac{3x|}{|1} + \ldots$$

which was already obtained by Euler.

In 1881, Leopold KRONECKER (1823-1891) considered the problem of finding polynomials p and q such that the degree of (gq-fp) q is less than n, where f and g are given polynomials of degrees n and n-m respectively. He treated the special case of finding a rational function p/q having the same derivatives that a given function g at a given point. This is exactly the Padé approximation problem. Kronecker gave two techniques for constructing the solution ; the first one is the Euclidean division algorithm for finding a continued fraction expansion of f/g. Polynomials $f_o = f$, $f_1 = g$, $f_2,\ldots,g_1, g_2,\ldots$ are defined such that $f_{k-1}-g_kf_k+f_{k+1} = 0$ for $k = 1,2,\ldots$. It follows that

$$\frac{g}{f} = \frac{1|}{|g_1} - \frac{1|}{|g_2} - \frac{1|}{|g_3} - \ldots$$

The convergents of this continued fraction are the solutions of the problem. The second method given by Kronecker is to solve the system of linear equations obtained by requiring that the first coefficients of the power series expansion of gq-fp are zero.

In the same year, G. FROBENIUS gave relations between the numerators and denominators of three adjacent approximants in the Padé table. Some of these identies, known as Frobenius identities, are connected with Jacobi's determinantal formulas. They can also be used to obtain explicit formulas for the coefficients of the continued fraction $a_0 + \frac{x|}{|a_1} + \frac{x|}{|a_2} + \ldots$ whose successive convergents form the main diagonal of the Padé table. A recursive method for computing the a_i's is also given by Frobenius. In fact Frobenius gave the first systematic study of Padé approximants and placed their theory on a rigourous basis.

Padé approximants were also considered by Pafnoutiy Lvovitch CHEBYSHEV (1821-1894) in 1885. He used the development

$$\int_a^b \frac{f(x)}{z-x}\,dx = \frac{A_o}{z} + \frac{A_1}{z^2} + \ldots$$

where $A_k = \int_a^b x^k f(x)\,dx$. He then transformed this series into the continued fraction.

$$\frac{1|}{|\alpha_1 z+\beta_1} - \frac{1|}{|\alpha_2 z+\beta_2} - \ldots$$

The kth convergent of this continued fraction agrees with the initial series up to the term $1/z^{2k}$ inclusively.

The must important contribution to the theory of continued fractions during the nineteenth century is certainly due to T.J. STIELTJES who is really the founder of our modern analytic theory of continued fractions. Stieltjes' first paper on the subject appeared in 1884. It was concerned with the gaussian quadrature formula

$$\int_{-1}^1 f(x)dx \simeq \sum_{i=1}^n A_i\, f(x_i)$$

and with the continued fraction

$$C = \frac{2|}{|x} - \frac{1.1/1.3|}{|x} - \frac{3.3/5.7|}{|x} - \frac{4.4/7.9|}{|x} - \ldots$$

Stieltjes proved that if P_n/Q_n is the nth convergent of C then $x_1,\ldots,x_n$ in the quadrature formula are the zeros of Q_n (which is a polynomial of degree n in x) an that

$$\frac{P_n}{Q_n} = \frac{A_1}{x-x_1} + \ldots + \frac{A_n}{x-x_n}\ .$$

If $x \notin [-1,1]$ the continued fraction converges to $\int_{-1}^1 \frac{dt}{x-t}$. The same result also holds for $\int_a^b \frac{f(t)}{x-t}\,dt$ with f non negative on [a,b]. Let us mention that, the same year, Stieljes proved the convergence of Gaussian quadrature formulas.

By these results Stieltjes lays the foundations of the theory of orthogonal polynomials (see below) whose study historically originated from certain type of continued fractions. Stieltjes was in fact very much excited by the analogy between Gaussian quadrature methods and some kind of continued fractions. During ten years he worked very hard on this subject, stimulated by the correspondence and friendship with Hermite, and he finally produced, only a few months before his death, his celebrated paper of 1894 "*Recherches sur les fractions continues*". He starts with the continued fraction

$$\frac{1|}{|a_1x} + \frac{1|}{|a_2} + \frac{1|}{|a_3x} + \frac{1|}{|a_4} + \ldots$$

where the a_i's are positive real numbers and x is a complex variable. He shows that if the series $\sum_{i=1}^{\infty} a_i$ diverges then the continued fraction converges to a function F which is analytic in the complex plane except along the negative real axis and at the origin and that

$$F(x) = \int_0^{\infty} \frac{d\alpha(t)}{t+x} .$$

If the series converges the even and odd parts of the continued fraction converge to distinct limits F_1 and F_2 with

$$F_1(x) = \int_0^{\infty} \frac{d\alpha_1(t)}{t+x} \qquad F_2(x) = \int_0^{\infty} \frac{d\alpha_2(t)}{t+x} .$$

It was known that the preceding continued fraction could be formally developed into the series $c_0 x^{-1} - c_1x^{-2} + c_2x^{-3} - \ldots$ with $c_i > 0$. Stieltjes shows how to obtain the c_i's from the a_i's and that the ratio c_i/c_{i-1} is increasing if the series $\sum a_i$ diverges. If this ratio has a limit λ, then the series converges for $|x| > \lambda$ and it diverges for all x if the ratio has no limit. Thus, although the continued fraction converges if the series does, the converse is not true. When the series diverges one must distinguish two cases according as $\sum a_i$ converges or not. When $\sum a_i$ converges, two different functions are obtained from the even and odd parts of the continued fraction. Stieltjes' result indicates a division of divergent series into two classes : the series which are the expansion of one single function F and those for which there are at least two functions whose expansions are the series. The continued fraction is only an intermediate between the series and the integral.

In the same paper Stieltjes treats the moment problem : let $0 \leq x_1 < x_2 < \ldots$ and $c_k > 0$ be given. The moment problem consists in finding under which conditions there exist $m_i > 0$ such that

$$c_k = \sum_{i=1}^{\infty} m_i x_i^k \qquad k = 0,1,\ldots$$

Stieltjes proved that this problem has a solution if and only if the series $\sum_{i=0}^{\infty} (-1)^i c_i/x^{i+1}$ has a continued fraction expansion of the form

$$\frac{1|}{|a_1x} + \frac{1|}{|a_2} + \frac{1|}{|a_3x} + \frac{1|}{|a_4} + \ldots$$

with all the a_i's positive.

I also would like to mention that Stieltjes showed, in 1889, how to transform the series $c_0 x^{-1} - c_1 x^{-2} + c_2 x^{-3} - \ldots$ into the continued fractions

$$\frac{a_0|}{|x+a_1} + \frac{1+a_2|}{|x+a_3} + \ldots$$

and

$$\frac{a_0|}{|x+a_1} - \frac{a_1a_2|}{|x+a_2+a_3} - \frac{a_3a_4|}{|x+a_4+a_5} - \ldots$$

which, in fact, contains Rustishauser's qd-algorithm.

We already said that the theory of orthogonal polynomials originated from certain type of continued fractions. Special cases of such continued fractions were studied by several authors. For example, P.L. CHEBYSHEV showed that the Legendre's polynomial of degree n is the denominator of the nth convergent of the continued fraction for $\frac{1}{2}$ Log $\frac{x+1}{x-1}$. The connection between continued fractions and orthogonal polynomials is as follows : let $\{P_n\}$ be a family of orthogonal polynomials on $[a,b]$ with respect to $d\alpha$. These polynomials satisfy the usual recurrence relationship

$$P_n(x) = (A_n x + B_n)\, P_{n-1}(x) - C_n P_{n-2}(x).$$

Let us consider the continued fraction

$$\frac{1|}{|A_1x+B_1} - \frac{C_2|}{|A_2x+B_2} - \frac{C_3|}{|A_3x+B_3} - \ldots$$

and let $R_n(x)/S_n(x)$ be its nth convergent. Then $S_n(x) = \sqrt{c_0}\, P_n(x)$ where $c_k = \int_a^b x^k d\alpha(x)$. It has been proved by Andrei Andreevitch MARKOV (1856-1922) in 1896 that

$$\lim_{n\to\infty} R_n(x)/S_n(x) = c_0^{-2}(c_0 c - c_1)^{1/2} \int_a^b d\alpha(t)/(x-t)$$

if x is an arbitrary point in the complex plane cut along the segment $[a,b]$ and that the convergence is uniform on every closed set of the complex plane having no points in common with [a,b].

The connection between orthogonal polynomials and continued fractions was also investigated by E.B. CHRISTOFFEL in 1877. He showed that for every polynomial p of degree not exceeding 2n-1

$$\int_{-1}^{1} p(x)d\alpha(x) = \sum_{i=1}^{n} p(x_i) \frac{Q_n(x_i)}{P_n'(x_i)}$$

where the x_i's are the roots of P_n and where

$$Q_n(t) = \int_{-1}^{1} \frac{P_n(x)-P_n(t)}{x - t} d\alpha(x).$$

He proved the so-called "*Christoffel-Darboux identity*" and gave the continued fraction expansion of the function

$$f(t) = \int_{-1}^{1} \frac{d\alpha(x)}{t-x}$$

which, as we previously saw, is closely related with the family of orthogonal polynomials $\{P_n\}$.

One of the major subjects of interest of the nineteenth century mathematicians was the theory of numbers. It is a field where continued fractions were an essential tool. We already mentioned the work of Liouville, Hermite and Lindemann on transcendental numbers. One of the main chapters where continued fractions were used is diophantine equations and extensions with Peter Gustav Lejeune DIRICHLET (1805-1859) who solved ax-by = 1 by continued fractions using a method due to Euler, A. PLESKOT who studied ax+by+cz = d in 1893 and Wilhelm Franz MEYER (1856-1934) who used recurring sequences obtained by simplifying and extending Jacobi's generalized continued fraction algorithm for solving $a_1x_1 + \dots + a_nx_n = a$.

Another point of interest was the representation of numbers as the sum of two, three or four squares where we find the names of Christian Friedrich KAUSLER (1760-1825), A.M. LEGENDRE, F. ARNDT, C. HERMITE, Joseph Alfred SERRET (1819-1885) and others. Euler's theorem (1752 : every divisor of the sum of two relatively prime squares is itself the sum of two squares) received proofs using continued fractions by V. EUGENIO in 1870 and François Edouard Anatole LUCAS (1842-1891) in 1891. Bachet's theorem (1575 : any number is either a square or the sum of two, three or four squares) also received the same kind of proof by Henry John Stephen SMITH (1826-1883) in 1881. Constructions of integers a and b such that $a^2 + b^2$ is a prime of the form 4n + 1 (stated by Fermat and proved by Euler) were given by A.M. LEGENDRE in 1808,

C.F. GAUSS in 1825 and J.A. SERRET in 1848.

Very many authors used continued fractions in connection with the solution of Pell's equation $ax^2 + bx + c = d^2$ or for the problem of binary quadratic forms. Continued fractions were also used for binomial congruences, the formation of the convergents to a fraction by Farey's sequences, the factorization of numbers and recurring sequences.

It is well known that any real positive number x can be developed into a continued fraction $x = a_o + \frac{1|}{|a_1} + \frac{1|}{|a_2} + \ldots$. Let $C_k = P_k/Q_k$ be the convergents of this continued fraction. H.J.S. SMITH proved that $|x - C_{k+1}| < |x - C_k|$. He also proved that the fraction P_k/Q_k with $Q_k > 1$ is the best approximation to x in the sense that it represents x more accurately than any other fraction with a smaller denominator. This result was already realised by Huygens.

In 1808, A.M. LEGENDRE gave a criterion such that p/q, with $q > 0$ and p/q irreductible, be a convergent of the continued fraction expansion of a given number x. If, $\forall i \quad o < a_i \leq b_i$ and a_i, b_i are integers then the continued fraction $\frac{a_1|}{|b_1} + \frac{a_2|}{|b_2} + \ldots$ converges. Legendre proved that its value is an irrational number smaller than one.

In 1895, K. Th. VAHLEN proved that from two successive convergents P_k/Q_k and P_{k+1}/Q_{k+1} $(k > 1)$ at least one satisfies

$$|x - p/q| < 1/2q^2.$$

Such results are very much useful for proving the irrationality of a given number x.

If, in the continued fraction $b_o + \frac{a_1|}{|b_1} + \frac{a_2|}{|b_2} + \ldots$, a_i and b_i are positive for $i \geq 1$ the sequences (P_k) and (Q_k) are increasing, the sequence $(C_{2k-1}=P_{2k-1}/Q_{2k-1})$ is decreasing and $(C_{2k} = P_{2k}/Q_{2k})$ is increasing. Moreover

$$C_{2k-1} > C_{2k+1} > C_n > C_{2k+2} > C_{2k} \qquad 1 \leq k < n/2-1.$$

Geometrical interpretations of these inequalities have been given by Louis POINSOT (1777-1859) in 1845, J. LIEBLEIN in 1867, Oskar Xavier SCHLÖMILCH (1823-1901) in 1873, M. KOPPE in 1887 and Christian Felix KLEIN (1849-1925) in 1895. These inequalities were used by M. KOPPE in 1887, A. HURWITZ in 1891 and Hermann MINKOWSKI (1864-1909) in 1896 to obtain bounds on the error. The case where $a_1 > 0$, $a_i < 0$ for $i \geq 2$ and $b_i \geq |a_i| + 1$ was treated by M.A. STERN in 1833. The case where the continued fraction is periodic has been investigated by Otto STOLZ (1842-1905) in 1884 and the case where the a_i's and b_i's are rational was treated by Georg LANDSBERG (1865-1912) in 1892.

In 1875, E.B. CHRISTOFFEL introduced the notion of "*characteristic*" of a rational number x in the range]0,1[. The characteristic of a number is a sequence whose elements are zero or one. The characteristic of a rational number is finite and it determines x uniquely. Christoffel gave the explicit relationship between the characteristic of x and its continued fraction. In 1888 he extended his results to irrational numbers.

Before going to a new subject let us mention that a practical rule for the computation of Euler's symbol $(b_o,\ldots,b_k)$ for the denominator of the continued fraction $b_o + \frac{1|}{|b_1} + \frac{1|}{|b_2} + \ldots$ was given by V. SCHLEGEL in 1877 and that J.J. SYLVESTER proved, in 1853, the relation

$$(b_1,\ldots,b_{n+m}) = (b_1,\ldots,b_m)(b_{m+1},\ldots,b_{n+m})+(b_1,\ldots,b_{m-1})(b_{m+2},\ldots,b_{n+m})$$

which generalizes similar identities given by Euler.

A major subject of interest that was much developed during the nineteenth century is the convergence theory of continued fractions. The precise definition of the convergence and divergence of a continued fraction was first given by L. SEIDEL in 1846 and M.A. STERN in 1848. These definitions will be refined by Alfred PRINGSHEIM (1850-1941) in 1898. Seidel and Stern also proved independently that the divergence of the series $\sum_{i=1}^{\infty} a_i$ $(a_i > 0)$ is a necessary and sufficient condition for the convergence of $a_o + \frac{1|}{|a_1} + \frac{1|}{|a_2} + \ldots$. They showed that the divergence of at least one of the series

$$\sum_{n=1}^{\infty} \frac{a_1a_3 \ldots a_{2n-1}}{a_2a_4 \ldots a_{2n}} b_{2n} \qquad \sum_{n=1}^{\infty} \frac{a_2a_4 \ldots a_{2n}}{a_2a_3 \ldots a_{2n+1}} b_{2n+1}$$

is a necessary and sufficient condition for the convergence of $b_o + \frac{a_1|}{|b_1} + \frac{a_2|}{|b_2} + \ldots$ when a_i, $b_i > 0$. The result has been extended by O. STOLZ to the case when the b_i's have arbitrary signs.

In 1855, L. SEIDEL showed that among the continued fractions of the form $-\frac{1|}{|q_1} - \frac{1|}{|q_2} - \ldots$ with $q_n < 2$ and $\lim_{n\to\infty} q_n = 2$, some are convergent and some are divergent.

A. PRINGSHEIM proved that the condition $|b_n| \geq |a_n| + 1$ is sufficient for the convergence of $\frac{a_1|}{|b_1} + \frac{a_2|}{|b_2} + \ldots$ and that its modulus is not greater than one.

When the b_i's are positive, O. STOLZ showed that $\frac{1|}{|b_1} + \frac{1|}{|b_2} + \ldots$ converges if the series $b_1b_2 + b_2b_3 + \ldots$ diverges. L. SAALSCHÜTZ proved in 1899 that this continued fraction converges if the series $\sqrt{b_1b_2} + \sqrt{b_2b_3} + \ldots$ diverges. These conditions are only sufficient.

In manuscripts found after his death, Georg Friedrich Bernhard RIEMANN (1826-1866) gave a proof of the convergence of the continued fraction given by Gauss for the ratio of two hypergeometric series. According to Padé, this is the first proof of convergence for Padé approximants. The convergence of the same continued fraction was also studied by Ludwig Wilhelm THOME in 1866.

In 1883, Ludwig Henrik Ferdinand OPPERMAN (1817-1883) proved that the divergence of the infinite product $\prod_{i=1}^{\infty}(1+a_i)$ is a necessary and sufficient condition for the convergence of $a_o + \frac{1|}{|a_1} + \frac{1|}{|a_2} + \ldots$

It was showed by Julius Daniel Theodor WORPITZKY (1835-1895) in 1865 and I.V. SLESHINSKII in 1889 that the condition $|c_n| \leq 1/4$ for $n = 2,3,\ldots$ is sufficient for the convergence of $\frac{c_1|}{|1} + \frac{c_2|}{|1} + \ldots$.

T.J. STIELTJES proved in 1894 that the conditions $a_i < 0$, a_{2i+1} not all zero and the series $\sum a_i$ divergent are sufficient for the uniform convergence of $\frac{x|}{|a_1} + \frac{x|}{|a_2} + \ldots$ in any finite domain of the complex plane cut along the negative real axis.

The first investigation of the convergence of continued fractions derived from power series (i.e. concerning classes of functions and not a special function whose continued fraction expansion is given in close form) was done by A. MARKOV in 1895 using an earlier work of P.L. CHEBYSHEV dated 1860.

About the convergence theory the name of Helge von KOCH (1870-1924) must also be mentioned. He proved that $\frac{1|}{|a_1} + \frac{1|}{|a_2} + \ldots$ diverges if $\sum |a_i|$ converges.

Miscellaneous results on or by continued fractions were obtained by various authors. For example they were used by Georg CANTOR (1845-1918) to prove that $\mathbb{R}$ and $\mathbb{R}^2$ have the same power and by A. MARKOV and P.L. CHEBYSHEV in their researches on probability theory.

The last contribution (but, of course, not the least) I would like to discuss is that of Henri Eugène PADE (1863-1953). Everybody knows his thesis "*Sur la repré-*

sentation approchée d'une fonction par des fractions rationnelles" which was presented at the Sorbonne in Paris on June 21, 1892 with the jury : C. HERMITE (chairman and advisor), Paul APPELL (1855-1930) and Emile PICARD.

In his introduction, Padé says :

> "*Nous avons été amené à nous occuper de cette question par une parole de M. Hermite, recueillie dans une de ses leçons, et par laquelle il laissait entrevoir les richesses que cachait sans doute encore cette théorie*" .

In this thesis, Padé gave a systematical study of the "*Padé approximants*". He classified them, arranged them in the "*Padé table*" and investigated the different types of continued fractions whose convergents form a descending staircase or a diagonal in the table. He studied the exponential function in details and showed that its Padé approximants are identical with the rational approximants obtained by Gaston Jean DARBOUX (1842-1917) in 1876 for the same function. He showed that

$$[n+k/m]_f\ (t) = \sum_{i=0}^{k-1} c_i t^i + t^k [n/m]_g\ (t)$$

where $f(t) = c_o + c_1 t + \ldots$ and $g(t) = c_k + c_{k+1} t + \ldots$ and studied the connection between the two halves of the table. Padé also investigated quite carefully what is now called the block structure of the Padé table.

THE FIRST PART OF THE TWENTIETH CENTURY.

The researches on continued fractions during the first part of the twentieth century (up to 1938) are mostly devoted to their analytic theory.

Using a result given by Jacques HADAMARD (1865-1963) in his thesis, Robert Fernand Bernard, Viscount de MONTESSUS DE BALLORE (1870-1937) gave, in 1902, his celebrated result on the convergence of the sequence $([n/k]_f)$ when n goes to infinity where f is a series having k poles and no other singularities in a given circle C.
Hadamard's results were extended in 1905 by Paul DIENES (1882-1952). This allowed R. WILSON to investigate in 1927 the behaviour of $([n/k]_f)$ upon the circle C and at the included poles.

In 1906, Thorvald Nicolai THIELE (1838-1910) proved that

$$f(x) = f(x_o) + \frac{x-x_o|}{|\rho_1^{(o)}-\rho_{-1}^{(o)}} + \frac{x-x_1|}{|\rho_2^{(o)}-\rho_o^{(o)}} + \ldots$$

where the $\rho_k^{(n)}$ are the so-called reciprocal differences which are calculated by

$$\rho_{-1}^{(n)} = 0 \qquad \rho_0^{(n)} = f(x_n) \qquad n = 0,1,\ldots$$

$$\rho_{k+1}^{(n)} = \rho_{k-1}^{(n+1)} + (x_{n+k+1}-x_n)/(\rho_k^{(n+1)}-\rho_k^{(n)}) \qquad n,k = 0,1,\ldots$$

The convergent C_{2k+1} of this continued fraction is the ratio of a polynomial of degree k+1 by a polynomial of degree k. C_{2k} is the ratio of two polynomials of degree k and $C_k(x_i) = f(x_i)$ for $i = 0,\ldots,k$. If we set $x_n = t + nh$, $x = t + h$ and if we let h tends to zero we obtain Thiele's expansion formula

$$f(t+h) = f(t) + \frac{h|}{|\rho_1(t) - \rho_{-1}(t)} + \frac{h|}{|\rho_2(g) - \rho_0(t)} + \ldots$$

with

$$\rho_{-1}(t) = 0 \qquad \rho_0(t) = f(t)$$

$$\rho_{k+1}(t) = \rho_{k-1}(t) + (k+1) / \rho_k'(t) \qquad k = 0,1,\ldots$$

This formula, which gives the continued fraction expansion of f, can be compared with the Taylor's formula which gives the series expansion of f. Thiele's formula terminates when f is a rational function of degrees k+1 by k or k by k just as Taylor's formula terminates when f is a polynomial.

Let us now describe the development of the analytic theory of continued fractions after Stieltjes' researches.

In 1903, Edward Burr VAN VLECK (1863-1943) undertook to extend Stieltjes' theory to continued fractions of the form $\frac{1|}{|x+b_1} - \frac{a_1|}{|x+b_2} - \frac{a_2|}{|x+b_3} - \ldots$ where the a_k's are arbitrary positive numbers and the b_k's are arbitrary real numbers. He connected, in certain cases, these continued fractions with Stieltjes' type definite integrals with the range of integration taken over the entire real axis. He also extended Stieltjes' theory to the Padé table.

In 1906, David HILBERT (1862-1943) gave his famous theory of infinite matrices and bounded quadratic forms in infinitely many variables. In 1914 Hilbert's theory was used by Ernst HELLINGER (1883-1950) and Otto TOEPLITZ (1881-1940) to connect integrals of the form $\int_a^b d\alpha(t)/(x-t)$, a and b finite, with the continued fractions considered by Van Vleck. The same year J. GROMMER extended these results to more general cases where the range of integration is the entire real axis. The complete theory was obtained by Hellinger in 1922 using Hilbert's theory of infinite linear systems. The same goal was reacked by several other mathematicians at about the same time by different methods : Rolf Hermann NEVANLINNA (1895-1980) in 1933 by methods of function theory and asymptotic series, Torsten CARLEMAN (1892-1949) in 1923 by integral equations and Marcel RIESZ in 1921 and 1923 by successive approximations.

Using the results by Van Vleck, Hubert Stanley WALL (1902-1971), in his thesis dated 1927 under Van Vleck's direction, gave a complete analysis of the convergence behaviour of the forward diagonal sequences of the Padé table derived from a Stieltjes series, i.e. whose coefficients are given by $c_i = \int_0^\infty t^i d\alpha(t)$ with α bounded and nondecreasing in $[0,\infty)$.

In 1931 and 1932 he extended these results to the cases where the range of integration is $[a,b]$ with $-\infty \leq a < b \leq \infty$ or with $-\infty < a < b < \infty$.

Another kind of investigation had been going on in the meantime. Around 1900, PRINGSHEIM and VAN VLECK studied the convergence of continued fractions with complex elements $\frac{1|}{|1} + \frac{c_1|}{|1} + \frac{c_2|}{|1} + \ldots$ and $\frac{1|}{|b_1} + \frac{1|}{|b_2} + \ldots$ Pringsheim proved that the first continued fraction converges if $|c_i| \leq (1-g_{i-1})\, g_i$, where $0 < g_i < 1$ $i = 0,1,\ldots$ and Van Vleck arrived to the same conclusion with $g_0 = 0$ and the condition that the series

$$1 + \sum_{i=1}^{\infty} \frac{g_1 \ldots g_i}{(1-g_1) \ldots (1-g_i)}$$

converges. Both these results include Worpitzky's result.

O. STOLZ proved that $b_0 + \frac{1|}{|b_1} + \frac{1|}{|b_2} + \ldots$ where the b_i's are complex numbers, diverges if the series $|b_1| + |b_2| + \ldots$ converges. In 1901, VAN VLECK proved that this condition is also sufficient if $\forall n > N$, $\mathrm{Re}(b_n)$ have the same sign and $\mathrm{Im}(b_n)$ are alternately positive and negative. This condition is still sufficient if $\forall n > N$, $\mathrm{Re}(b_n)$ have the same sign and $|\mathrm{Im}(b_n)/\mathrm{Re}(b_n)| < K$ or if $\forall n > N$, $\mathrm{Im}(b_n)$ have alternate signs and $|\mathrm{Re}(b_n)/\mathrm{Im}(b_n)| < K$ where K is a given number. Van Vleck also proved that if $c_i = a_i x$, $\lim_{i\to\infty} a_i = a$ then the first continued fraction converges except at certain isolated points and except when x is on the rectilinear cut from $-1/4a$ to infinity in the direction of the vector from the origin to $-1/4a$.

In 1905, A. PRINGSHEIM proved that the continued fraction $\frac{a_1|}{|b_1} + \frac{a_2|}{|b_2} + \ldots$ is perfectly converging if $\forall n \geq 2$

$$\left|\frac{a_n}{b_{n-1}\, b_n}\right| \leq \frac{P_n - 1}{P_{n-1}\, P_n}$$

where (P_n) is a given sequence of numbers with $P_1 \geq 1$. This result generalizes the result given by von Koch in 1895 and that of Van Vleck. Pringsheim's result also applies for complex continued fractions. Many other convergence results had been previously obtained by Pringsheim most of them being particular cases of this last condition.

In 1903, Josef Anton GMEINER (1862-1927) obtained convergence results for $b_o - \frac{a_1|}{|b_1} + \frac{a_2|}{|b_2} - \frac{a_3|}{|b_3} + \dots$ which are not particular cases of Pringsheim's criterion.

Some other results on the convergence of $\frac{1|}{|1} + \frac{c_1|}{|1} + \frac{c_2|}{|1} + \dots$ were obtained by Otto SZASZ (1884-1952) between 1912 and1916. These results extend those of Worpitzky, von Kock and Perron. In 1915 he wrote a paper on the continued fraction of the Stieltjes-Markov type answering a conjecture by Perron. With Felix BERNSTEIN he worked on the irrational character of certain continued fractions in 1915.

The inequality found by Pringsheim and Van Vleck restrict the c_i's to lie in the neighborhood of the origin. This inequality has been replaced by H.S. WALL (and some of his students) by inequalities restricting the c_i's to lie in domains bounded by parobolas with foci at the origin. Wall also developed the theory of positive definite continued fractions thus extending Stieltjes' theory to complex J-continued fractions. This theory is closely related to that of tri-diagonal matrices. Wall's other contributions deal with the application of continued fractions to function theoretic problems such as a characterization of the Hausdorff moment problem, the extension and unification of a major part of the convergence theory and a work on continuous continued fractions and harmonic matrices. But, in Wall's opinion, his most important contributions were his 62 doctoral students.

During the first part of the twentieth century number theory still was a domain where continued fractions were extensively used mostly for diophantine and Pell's equations, factorization of numbers and transcendental numbers. Among the contributors are many outstanding mathematicians such as P. BACHMANN, F. and S. BERNSTEIN, H. HAMBURGER, F. HAUSDORFF, G. POLYA, S. RAMANUJAN, J. SHOHAT, W. SIERPINSKI, A. TAUBER, W. TURNBULL, E.T. WHITTAKER and others.

As a conclusion it can be said that there is only one other subject in mathematics with such a long history and influence : NUMBERS !

*

* *

EFFICIENT RELIABLE RATIONAL INTERPOLATION

P.R. Graves-Morris,
Mathematical Institute,
University of Kent,
Canterbury, England.

Summary It is shown that Thiele fractions and Thiele-Werner fractions always provide representations for the solution of a given soluble, rational interpolation problem. A strategy which guarantees the accuracy of construction of Thiele-Werner interpolants is reviewed. Some difficulties in the selection of best library algorithms for rational interpolation are considered.

§1 Introduction

In certain instances, there is good reason to suppose that interpolation by rational fractions is likely to yield better approximations than interpolation by polynomials. If the function to be interpolated is itself rational, or if it is meromorphic (e.g. tan(x)), then rational interpolation is expected to have a wider range of accurate approximation than polynomial interpolation to corresponding order.

Example 1 We consider interpolation of $\Gamma(x)$ with a [2/1] type rational interpolant and a cubic polynomial. We use the data

x_i	4	3	2	1
$\Gamma(x_i)$	6	2	1	1

We find the Thiele interpolant to be

$$r(x) = 6 + \frac{x-4}{1/4} + \frac{x-3}{-20/3} + \frac{x-2}{-21/20} . \tag{1.1}$$

We find the Newton interpolating polynomial to be

$$\pi_3(x) = 6 + 4(x-4) + \tfrac{3}{2}(x-4)(x-3) + \tfrac{1}{3}(x-4)(x-3)(x-2).$$

At the mid-point of the range, we compare the exact value of the gamma function with its approximations:

$$\Gamma(2.5) = 1.33 , \quad r(2.5) = 1.31 , \quad \pi_3(2.5) = 1.25.$$

We note that we have one case in which we expect rational interpolation to be superior to polynomial interpolation for approximation purposes, and it is. One example of a more serious demonstration of the use of rational functions as approximations to the gamma function is given by Werner, [1963].

Successive value transformations The Thiele fraction (1.1) was computed by the method of successive value transformations. We seek the result in the form

$$r_0(x) = b_0 + \frac{x-4}{b_1} + \frac{x-3}{b_2} + \frac{x-2}{b_3}. \tag{1.2}$$

Let us suppose for the moment that this representation for $r_0(x)$ exists, and then we may define $r_1(x)$, $r_2(x)$ and $r_3(x)$ by

$$r_i(x) = b_i + \frac{(x-x_i)}{r_{i+1}(x)}, \qquad i = 0,1,2. \tag{1.3}$$

Provided $r_{i+1}(x_i) \neq 0$,

then $\quad r_i(x_i) = b_i \quad , \qquad i = 0,1,2. \qquad (1.4)$

and $r_3(x_3) = b_3$ anyway. Hence we view (1.3) as a prescription for specifying b_i,

$$b_i := r_i(x_i) \tag{1.5}$$

and a set of successive value transformations

$$r_{i+1}(x) := \frac{x-x_i}{r_i(x) - b_i}, \quad x = x_{i+1}, x_{i+2}, \ldots, x_n \tag{1.6}$$

Using (1.5), (1.6), we find the following table for determining the coefficients of (1.1) :-

x_i	4	3	2	1
f_i	$\underline{6}$	2	1	1
$r_1(x_i)$		$\underline{\frac{1}{4}}$	$\frac{2}{5}$	$\frac{3}{5}$
$r_2(x_i)$			$\underline{\frac{-20}{3}}$	$\frac{-40}{7}$
$r_3(x_i)$				$\underline{\frac{-21}{20}}$

The values underlined in this table are the values of b_i used in (1.1) .

We note that an interpolating fraction of the form

$$r(x) = b_0 + \frac{x-1}{b_1} {}_{+} \frac{x-2}{b_2} {}_{+} \frac{x-3}{b_3} \tag{1.7}$$

apparently can not be found in this way, because $b_1 = \infty$. The difficulty encountered is caused by an inept choice of the ordering of the interpolation points and values, because the original order used in example 1 is entirely satisfactory.

Example 2 Consider the function values $\{f_i \,, i = 0,1,2\}$, taken at the interpolation points $\{x_i \,, i = 0,1,2\}$, which are given by the table:-

i	0	1	2
x_i	0	1	2
f_i	0	0	1

These values are interpolated by the Newton polynomial

$$\pi_2(x) = 0 + 0(x-x_0) + \tfrac{1}{2}(x-x_0)(x-x_1)$$

$$= \tfrac{1}{2}\, x(x-1).$$

If a [1/1] type rational interpolant of the form

$$r(x) = \frac{ax + b}{cx + d} \tag{1.8}$$

fits these data, then the ratios a:b:c:d satisfy the linear equations

$$\begin{aligned} ax_0 + b &= 0 \\ ax_1 + b &= 0 \\ ax_2 + b - cx_2 - d &= 0 \end{aligned} \tag{1.9}$$

The coefficient matrix has rank 3, yet inspection of the solution shows that no [1/1] type rational interpolant fits the data. We have encountered an impossible rational interpolation problem. Furthermore, the impossibility of solving the problem can not be directly indicated by the nature of the homogeneous equations, as we shall see in §2.

Suppose that we use the method of successive value transformations on this data in the form

i	0	1	2
x_i	2	1	0
f_i	1	0	0

.

We find that

$$r(x) = 1 + \frac{x-2}{1} \underset{+}{} \frac{x-1}{-1} \; . \tag{1.10}$$

It is quite clear that the method of successive value transformations fails because $r_1(x_0) = 0$, in the language of (1.3), and the fraction $(x-2)/r_1(x)$ (which is a subfraction of (1.10)) is not zero as $x \to 2$. We learn that the method of successive value transformations must be supplemented by a consistency test. It is interesting to note, [Meinguet, 1970] that Cauchy apparently overlooked the necessity of consistency tests in his pioneering paper in 1821. We are fortunate that his paper was accepted for publication.

The morals that we draw from the example of this section turn out to be quite generally applicable to all iterative methods of rational interpolation. An algorithm for constructing a continued fraction representation of the rational interpolant must contain a facility for reordering the interpolation points when necessary, and the algorithm must have a consistency test prior to termination.

§2 Interpolation with Thiele fractions

Function values, $\{f_i, i = 0,1,\ldots,n\}$ at respective interpolation points $\{x_i, i = 0,1,\ldots,n\}$ are given. The rational interpolation problem consists of finding an interpolating rational fraction, $r^{[\ell/m]}(x)$, of type $[\ell/m]$, for which

$$r^{[\ell/m]}(x_i) = f_i, \qquad i = 0,1,\ldots n \tag{2.1}$$

and $\ell + m = n$.

$r^{[\ell/m]}(x)$ is defined to be a rational fraction of type $[\ell/m]$ if

$$r^{[\ell/m]}(x) = p^{[\ell/m]}(x) \,/\, q^{[\ell/m]}(x), \tag{2.2}$$

where $p^{[\ell/m]}(x)$, $q^{[\ell/m]}(x)$ are polynomials and

$$\partial\{p^{[\ell/m]}\} \le \ell \text{ and } \partial\{q^{[\ell/m]}\} \le m. \tag{2.3}$$

A closely allied, but by no means equivalent problem is the modified rational interpolation problem. Given the values $\{f_i, i = 0,1,\ldots,n\}$, respective interpolation points $\{x_i, i = 0,1,\ldots,n\}$, and integers ℓ, m for which $\ell + m = n$, the modified rational interpolation problem is the problem of finding polynomials $p^{[\ell/m]}(x)$ and $q^{[\ell/m]}(x)$ such that (2.3) is satisfied and

$$p^{[\ell/m]}(x_i) = q^{[\ell/m]}(x_i).f_i \quad \text{for } i = 0,1,\ldots,n. \tag{2.4}$$

The definitions used here follow Meinguet, [1970], and Warner, [1974]. The solution of the modified rational interpolation problem is uniquely associated with the solution of a set of n + 1 homogeneous linear integral equations in the n + 2 unknown coefficients of the polynomials $p^{[\ell/m]}(x)$ and $q^{[\ell/m]}(x)$, and shown in example 2 of §1. Consequently, the modified rational interpolation problem always has at least one solution. The rational interpolation problem proper may or may not have a solution, as was shown by the examples of §1. Concerning the feasibility of solution of the rational interpolation problem, we have the following familiar results:-

Theorem 2.1 The solution of the rational interpolation problem is unique, up to possible common factors of $p^{[\ell/m]}(x)$ and $q^{[\ell/m]}(x)$ in (2.2), (2.3).

Proof Suppose that $p_\ell(x)/q_m(x)$ and $p^*_\ell(x)/q^*_m(x)$ are two different solutions of (2.2) satisfying (2.1) and (2.3). Then we find that

$$p_\ell(x)q^*_m(x) - p^*_\ell(x)q_m(x) \equiv 0, \tag{2.5}$$

because the lefthand side of (2.5) is a polynomial of order at most $\ell + m$ which vanishes on $\ell + m + 1$ distinct points.

Theorem 2.2 [Werner - Schaback, 1972]. The following algorithm shows that the $[\ell/m]$ rational interpolation problem is easily reduced to an $[\ell'/m']$ superdiagonal rational interpolation problem with $m' \leq \ell'$.

Reduction Algorithm If $\ell \geq m$, there is nothing to do, so we suppose that $\ell < m$. If none of the data values are zero, we define

$$\alpha = 0 \qquad \text{and} \qquad \pi_0(x) = 1 . \tag{2.6}$$

Otherwise, we assume that the data points have been arranged so that

$$f_i = 0, \qquad i = 0,1,\ldots,\alpha-1, \tag{2.7a}$$

$$f_i \neq 0, \qquad i = \alpha,\alpha+1,\ldots,\ell+m. \tag{2.7b}$$

In this case,

$$\alpha \geq 1 \quad \text{and} \quad \pi_\alpha(x) = \prod_{j=0}^{\alpha-1} (x-x_j). \tag{2.8}$$

Since $m > \ell \geq \ell-\alpha$, we may employ an algorithm for construction of the rational interpolant $\rho^{[m/\ell-\alpha]}(x)$, satisfying

$$\rho^{[m/\ell-\alpha]}(x_i) = f_i^{-1} \cdot \pi_\alpha(x_i) \quad , \quad i = \alpha,\alpha+1,\ldots,\ell+m. \tag{2.9}$$

If the rational interpolation problem expressed by (2.9) is impossible, then the originating rational interpolation problem is impossible, (A). Otherwise, $\rho^{[m/\ell-\alpha]}(x)$ satisfies (2.9) and we consider

$$r^{[\ell/m]}(x) = \pi_\alpha(x) \ / \ \rho^{[m/\ell-\alpha]}(x) \tag{2.10}$$

As a solution of the $[\ell/m]$ rational interpolation problem expressed by (2.7), (or 2.6) for $\alpha = 0$). If

$$\rho^{[m/\ell-\alpha]}(x_i) \neq 0, \qquad i = 0,1,\ldots,\alpha, \tag{2.11}$$

then (2.10) is a solution of the rational interpolation problem under discussion, (B); otherwise the originating rational interpolation problem is impossible, (C).

<u>Proof</u> Only statements (A), (B) and (C) require proof. The case of $\alpha = 0$ is trivial, and we suppose that $\alpha \geq 1$. Suppose that $p^{[\ell/m]}(x) \ / \ q^{[\ell/m]}(x)$ is a solution of the originating rational interpolation problem, according to (2.2) and (2.3). Without loss of generality, we assume that $p^{[\ell/m]}(x)$ and $q^{[\ell/m]}(x)$ have no common factors. From (2.7a), we see that

$$p^{[\ell/m]}(x_i) = 0, \qquad i = 0,1,\ldots,\alpha$$

and that

$$p'(x) = p^{[\ell/m]}(x) \ / \ \pi_\alpha(x)$$

is a polynomial of order $\ell-\alpha$.

<u>Proof of (A)</u> $q^{[\ell/m]}(x) \ / \ p'(x)$ is a solution of the rational interpolation problem posed by (2.9), and (A) is proved by contradiction.

<u>Proof of (B)</u> Obvious from (2.9)-(2.11).

Proof of (C) For some j in the range $0 \le j < \alpha$, let

$$\rho^{[m/\ell-\alpha]}(x_j) = 0. \tag{2.12}$$

If the rational interpolation problem is soluble, then $p^{[\ell/m]}(x)$ and $q^{[\ell/m]}(x)$ have no common factor according to the hypothesis, and $q^{[\ell/m]}(x_j) \neq 0$. Therefore the solution $q^{[\ell/m]}(x) \,/\, p'(x)$ of the rational interpolation problem posed by (2.9) is also non-zero at $x = x_j$. This contradicts the uniqueness of the solution of the rational interpolation problem (2.9) established by theorem 2.1.

An immediate consequence of theorem 2.2 is that, in principle, we may restrict our attention to the case of rational interpolation with $\ell \ge m$. What is not so obvious is what should be done in practice if any of the data values are anomalously small.

Given the values $\{f_i\ ,\ i = 0,1,..,n\}$ at the respective points $\{x_i\ ,\ i = 0,1,..,n\}$, the basic rational interpolation problem consists of finding a rational fraction of type $[m+1/m]$ in the case when n is odd, or of type $[m/m]$, in the case when n is even, where $m = \text{int}(n/2)$. The rational interpolation problem with $\ell > m$ expressed by (2.1)-(2.3) can be reduced to a basic rational interpolation problem using an additive Newton interpolating polynomial of order $k = \ell-m-1$ according to the formula

$$r^{[\ell/m]}(x) = \pi_k(x) + r^{[m/m]}(x) . \prod_{j=0}^{k} (x-x_j), \tag{2.13}$$

where

$$\pi_k(x_i) = f_i \qquad , \qquad i = 0,1,...,k \tag{2.14}$$

and

$$r^{[m/m]}(x_i) = [f_i - \pi_k(x_i)] . \prod_{j=0}^{k} (x_i-x_j)^{-1},\ i = k+1,k+2,..,\ell+m. \tag{2.15}$$

No questions of existence are begged by the analysis of (2.13)-(2.15). We conclude from theorem 2.2 and the foregoing remarks that a full solution of the basic rational interpolation problem is tantamount to a full solution of the rational interpolation problem. Consequently, there is no real loss of generality in the following discussion of Thiele interpolants.

Thiele rational interpolants provide an answer to the problem of interpolating data $\{f_i\ ,\ i = 0,1,...,n\}$ at respective points $\{x_i\ ,\ i = 0,1...,n\}$ with an interpolating rational fraction of type

[m+1/m] if n is odd, or of type [m/m] if n is even, where $m \le \mathrm{int}(\frac{n}{2})$.
A Thiele fraction takes the form

$$r(x) = b_0 + \frac{x-x_0}{b_1} + \frac{x-x_1}{b_2} + \ldots + \frac{x-x_{t-1}}{b_t} \tag{2.16}$$

with $b_i \neq 0$, $i = 1,2,\ldots,t$, or else

$$r(x) = f_0 + \frac{a_1(x-x_0)}{1} + \frac{a_2(x-x_1)}{1} + \ldots + \frac{a_t(x-x_{t-1})}{1} \tag{2.17}$$

with $a_i \neq 0$, $i = 1,2,\ldots,t$. The forms (2.16) and (2.17) are equivalent in every sense.

Example 3 A Thiele interpolant for exp(x) is

$$r(x) = 1 + \frac{x}{0.0149095} + \frac{x-6}{-8.81730} + \frac{x-1}{-0.0913453} + \frac{x-5}{23.7996} + \frac{x-2}{0.215982} + \frac{x-4}{-36.0678} . \tag{2.18}$$

r(x) is a [3/3] type interpolant to exp(x) on $x = 0,1,2,3,4,5,6$.

Reliability We define reliable algorithms to be ones which, in principle, provide solutions for soluble problems, and which recognise and reject insoluble problems. The phrase "in principle" means that exact arithmetic is assumed to be used. It is important to use reliable algorithms, [Graves-Morris and Hopkins, 1978], for rational interpolation, because some of the best known algorithms, such as Stoer's second algorithm, [Stoer, 1961], may require an (unspecified) reordering of the data points. The following theorem shows that the Thiele representations (2.16), (2.17), allowing for a possible reordering of the interpolation points, provide reliable representations for the solution of the rational interpolation problem.

Theorem 2.3 [Thacher-Tukey, 1960]. Let the rational fraction $N^{(0)}(x) / D^{(0)}(x)$ take the values $\{f_i,\ i = 0,1,\ldots,n\}$ at the respective interpolation points $\{x_i,\ i = 0,1,\ldots,n\}$, where $N^{(0)}(x)$ and $D^{(0)}(x)$ are polynomials with no common factors and satisfy

$$\partial\{N^{(0)}\} \le \mathrm{int}(\tfrac{n+1}{2}) \quad \text{and} \quad \partial\{D^{(0)}\} \le \mathrm{int}(\tfrac{n}{2}). \tag{2.19}$$

Then, a Thiele fraction r(x) of the form (2.16) exists such that

$$r(x) = N^{(0)}(x) \;/\; D^{(0)}(x), \tag{2.20}$$

$$r(x_i) = f_i, \qquad i = 0,1,\ldots,n \tag{2.21}$$

and

$$t \le n. \tag{2.22}$$

Conversely, a Thiele fraction r(x) of the form (2.16) with $t \le n$ which fits the data can be expressed as a ratio of polynomials, $r(x) = N^{(0)}(x) / D^{(0)}(x)$, in which $N^{(0)}(x)$, $D^{(0)}(x)$ satisfy (2.19).

The proof of the Thacher-Tukey theorem can be based on the following algorithm. This algorithm is a modification of the Thacher -Tukey algorithm, by Graves-Morris and Hopkins, [1978].

Modified Thacher-Tukey Algorithm

Specification An integer n and values $\{f_i^{(0)},\ i = 0,1,\ldots,n\}$ taken at distinct respective interpolation points $\{x_i\ ,\ i = 0,1,\ldots,n\} \equiv X_0$ are supplied. A value $f_j^{(0)}$ is uniquely associated with each point x_j, but the n+1 pairs, $(f_j^{(0)}, x_j)$, may be sequentially reordered when necessary.

Initialisation Choose x_0 from X_0 arbitrarily. If $f_i^{(0)} = f_0^{(0)}$ for all i in $0 \le i \le n$, then the algorithm finishes with $r(x) = f_0^{(0)}$. Otherwise, define

$$b_0 = f_0^{(0)}, \qquad X_1 = X_0 \backslash x_0 \qquad \text{and} \qquad j = 1\,. \tag{2.23}$$

Iteration

If $f_i^{(j-1)} = b_{j-1}$ for all i in $j \le i \le n$, then the iteration is completed. Let $t := j-1$ and go to termination.

If either $f_i^{(j-1)} = b_{j-1}$ or $f_i^{(j-1)} = \infty$ for all i in $j \le i \le n$, then the rational interpolation problem is impossible and the algorithm terminates with this report, (IF).

Otherwise, it is possible to select $x_j \in X_j$ such that $f_j^{(j-1)} \neq b_{j-1}$

and $f_j^{(j-1)} \neq \infty$. With such a value of j, define

$$b_j = (x_j - x_{j-1}) \;/\; (f_j^{(j-1)} - b_{j-1}). \tag{2.24}$$

If j = n, terminate with t = n. Otherwise, define

$$f_i^{(j)} = (x_i - x_{j-1}) \;/\; (f_i^{(j-1)} - b_{j-1}) \text{ for } i = j+1,\, j+2, \ldots, n. \tag{2.25}$$

Note that ∞ is an allowed value for $f_i^{(j)}$ in this expression.

Define $X_{j+1} = X_j \backslash x_j$, $j := j+1$ and iterate.

Termination For j = 1,2,...,t-1, evaluate

$$R^{(j)}(x_{j-1}) = b_j + \frac{x-x_j}{b_{j+1}} + \frac{x-x_{j+1}}{b_{j+2}} + \ldots + \frac{x-x_{t-1}}{b_t}. \tag{2.26}$$

If $R^{(j)}(x_{j-1}) = 0$ for any j in $1 \le j \le t-1$, then the rational interpolation problem is impossible, (TF). Otherwise this exit from the algorithm provides a solution of the rational interpolation problem in the representation (2.16).

Proof of theorem 2.3 Suppose that an interpolating rational fraction $N^{(0)}(x)/D^{(0)}(x)$ exists and satisfies (2.19) - (2.21). The iterative phase of the algorithm is structured either to construct a Thiele fraction r(x) of the form (2.16) or else to decide that the rational interpolation problem posed is impossible.

Part 1 Suppose that the iterative phase of the algorithm yields the representation (2.16). By hypothesis, (2.19) and (2.21) are valid. For j in the range $0 \le j < n$, make the inductive hypothesis that polynomials $N^{(j)}(x)$ and $D^{(j)}(x)$ have been defined such that

$$N^{(j)}(x_j) \;/\; D^{(j)}(x_j) = b_j , \tag{2.27}$$

$$N^{(j)}(x_i) \;/\; D^{(j)}(x_i) = f_i^{(j)}, \qquad i = j+1, j+2, \ldots, n, \tag{2.28}$$

and

$$\partial\{N^{(j)}\} \le \operatorname{int}\left[\frac{n-j+1}{2}\right] , \quad \partial\{D^{(j)}\} \le \operatorname{int}\left[\frac{n-j}{2}\right] . \tag{2.29}$$

The definition (2.24) provides a finite, non-zero value of b_j. From (2.27), we see that the equations

$$N^{(j+1)}(x) = D^{(j)}(x) \tag{2.30}$$

and

$$D^{(j+1)}(x) = [N^{(j)}(x) - b_j\, D^{(j)}(x)]/(x-x_j) \tag{2.31}$$

define polynomials $N^{(j+1)}(x)$ and $D^{(j+1)}(x)$. From (2.29) - (2.31),

$$\partial\{N^{(j+1)}\} \le \operatorname{int}\left(\frac{n-j-1}{2}\right), \quad \partial D^{(j+1)} \le \operatorname{int}\left(\frac{n-j-2}{2}\right). \tag{2.32}$$

From (2.30), (2.31), we find that

$$\frac{N^{(j+1)}(x)}{D^{(j+1)}(x)} = \frac{x-x_j}{[N^{(j)}(x)/D^{(j)}(x)]-b_j} . \tag{2.33}$$

Using (2.28) and (2.24), we deduce that

$$\frac{N^{(j+1)}(x_{j+1})}{D^{(j+1)}(x_{j+1})} = \frac{x_{j+1} - x_j}{f_{j+1}^{(j)} - b_j} = b_{j+1} \tag{2.34}$$

Using (2.28) and (2.25), we deduce from (2.33) that

$$\frac{N^{(j+1)}(x_i)}{D^{(j+1)}(x_i)} = \frac{x_i - x_j}{f_i^{(j)} - b_j} = f_i^{(j+1)} \quad \text{for } i\in[j+1,n] \tag{2.35}$$

and the inductive hypothesis is established. We deduce that a successful exit from the iterative stage of the algorithm under the conditions of the theorem occurs when

$$N^{(t)}(x_i) \;/\; D^{(t)}(x_i) = b_t \,, \text{ for all } i \text{ in } t \le i \le n. \tag{2.36}$$

From (2.33), we deduce that $N^{(0)}(x)/D^{(0)}(x)$ has the representation (2.16).

Part 2 Suppose that the iterative phase of the algorithm fails via the exit (IF). This occurs if and only if $n \ge 2$ and k exists in the range $2 \le k \le n$ such that

(i) $f_i^{(k-1)} = b_{k-1}$ or $f_i^{(k-1)} = \infty$ for all $i\in[k,n]$ and

(2.37)

(ii) $f_i^{(k-1)} = \infty$ for some $i\in[k,n]$.

In this case, we can assume that

$$N^{(k-1)}(x_i) \;/\; D^{(k-1)}(x_i) \;=\; b_{k-1}\;, \quad \text{for } i = k-1,k,\ldots,\alpha.$$

and

$$N^{(k-1)}(x_i) \;/\; D^{(k-1)}(x_i) \;=\; \infty\;, \quad \text{for } i = \alpha+1,\alpha+2,\ldots,n.$$

We deduce that the polynomial

$$\pi(x) \equiv D^{(k-1)}(x)\,[N^{(k-1)}(x) - b_{k-1}\,D^{(k-1)}(x)] \tag{2.38}$$

vanishes at $x = x_i$ for $i = k-1,k,\ldots,n$. However (2.29) shows that $\partial\{\pi\} \leq n-k+1$, and so $\pi(x) \equiv 0$. We deduce that the hypotheses of the theorem preclude a failure at (IF).

Converse It is familiar that the recurrences

$$A_i(x) = b_i A_{i-1}(x) + (x-x_{i-1})\,A_{i-2}(x) \tag{2.39a}$$

$$B_i(x) = b_i B_{i-1}(x) + (x-x_{i-1})\,B_{i-2}(x) \tag{2.39b}$$

together with the (partly artificial) initial conditions

$$A_0(x) = b_0\;,\; B_0(x) = 1\;,\; A_{-1}(x) = 1\;,\; B_{-1}(x) = 0 \tag{2.40}$$

yield a representation of the right-hand side of (2.16) as $A_t(x)/B_t(x)$, where $A_t(x)$ and $B_t(x)$ satisfy (2.19).

Example 4 We consider an example which shows how it happens that the value ∞ may be used consistently in (2.25). The top two rows of the following table represent data; the other rows are computed using the algorithm.

x	0	1	2	3
$f^{(0)}(x)$	1	1	3	13
$f^{(1)}(x)$	*	∞	1	$\frac{1}{4}$
$f^{(2)}(x)$		0	*	$-\frac{4}{3}$
$f^{(3)}(x)$		$-\frac{3}{2}$		*

We find that

$$r(x) = 1 + \frac{x}{1} + \frac{x-2}{-4/3} + \frac{x-3}{-3/2} .$$

It is easily verified that r(x) passes the termination tests.

Unlike theorem 2.3, theorem 2.4 following makes no initial hypothesis about the existence of a rational interpolant to the data. Theorems 2.4 and 2.5 make quite precise the statement that the modified Thacher-Tukey algorithm is reliable.

Theorem 2.4 If the algorithm runs successfully and defines a Thiele fraction r(x) of the form (2.16), which passes the termination test, then

$$r(x_i) = f_i, \qquad \text{for } i = 0,1,\ldots,n.$$

Conversely, if the algorithm fails, either by failure in the iteration stage (IF) or termination stage (TF), then no Thiele fraction of the form (2.16) nor any rational fraction of the form $N(x)/D(x)$ with $\partial\{N\} \le \text{int}(\frac{n+1}{2})$, $\partial\{D\} \le \text{int}(\frac{n}{2})$ exists which fits the data.

Proof If the algorithm terminates successfully, the construction of (2.39) and (2.40) yields polynomials $A_t(x)$ abd $B_t(x)$ such that

$$r(x) = \frac{A_t(x)}{B_t(x)} = b_0 + \frac{x-x_0}{b_1} + \frac{x-x_1}{b_2} + \ldots + \frac{x-x_{t-1}}{b_t}$$

for all x. Provided the denominator check is passed, it follows that $r(x_0) = b_0$ and

$$r(x_k) = b_0 + \frac{x_k-x_0}{b_1} + \frac{x_k-x_1}{b_2} + \ldots + \frac{x_k-x_{k-1}}{b_k} \quad \text{for } 1 \le k \le t$$

and

$$r(x_k) = b_0 + \frac{x_k-x_0}{b_1} + \frac{x_k-x_1}{b_2} + \ldots + \frac{x_k-x_{t-1}}{b_t} \quad \text{for } t < k \le n.$$

Making repeated use of (2.25) with i = k and j = 1,2,...,k-1, using (2.24) with j = k and using the fact that $b_0 = f_0$, it follows that $r(x_k) = f^{(0)}(x_k)$ for each value of k in $0 \le k \le n$. The first part of the theorem is proved.

If the algorithm fails in the iteration phase (IF), then part 2 of theorem 2.3 shows that no rational fraction of the form $N^{(0)}(x)/D^{(0)}(x)$ exists satisfying (2.27)-(2.29). Using the whole of theorem 2.3, we see that failure via (IF) has the consequences stated in theorem 2.4.

If the algorithm fails in the termination phase (TF), then the fraction

$$R_k(x) = b_k + \frac{x-x_k}{b_{k+1}} + \frac{x-x_{k+1}}{b_{k+2}} + \ldots + \frac{x-x_{t-1}}{b_t}$$

has a zero at $x = x_{k-1}$, for some k in $1 \le k \le t-1$.
If an $N^{(0)}(x)/D^{(0)}(x)$ representation of r(x) exists satisfying the premises of theorem 2.3, then the reduction of (2.30), (2.31) and (2.32) is possible, and we find that

$$N^{(k)}(x) \;/\; D^{(k)}(x) = R_k(x)$$

$$\therefore \quad N^{(k)}(x_{k-1}) = 0$$

From (2.30), $D^{(k-1)}(x_{k-1}) = 0$.

From (2.31), $N^{(k-1)}(x_{k-1}) = 0$.

Therefore, $N^{(k-1)}(x)$ and $D^{(k-1)}(x)$ have a common factor of $(x-x_{k-1})$. From (2.30) and (2.31), we deduce that $N^{(0)}(x)$ and $D^{(0)}(x)$ share this common factor, contrary to hypothesis. Using theorem 2.3, the final part of theorem 2.4 is now proved.

This section is concluded with another reliability theorem which synthesises parts of theorems 2.3 and 2.4.

Theorem 2.5 If a rational fraction of the form N(x)/D(x) exists for which $N(x_i)/D(x_i) = f_i$, $i = 0,1,\ldots,n$ and $\partial\{N\} \le \text{int}(\frac{n+1}{2})$, $\partial\{D\} \le \text{int}(\frac{n}{2})$, then the modified Thacher-Tukey algorithm yields a Thiele fraction r(x) of the form (2.16) for which $r(x_i) = f_i$, $i = 0,1,\ldots n$, and $r(x) = N(x)/D(x)$.

Proof The hypotheses of theorem 2.5 are the same as those of theorem 2.3. Therefore, the iterative stage of the algorithm yields a Thiele fraction of the form (2.16), which interpolates properly according to (2.21). The converse of theorem 2.4 implies that the algorithm must necessarily terminate successfully in this case.

§3 Thiele-Werner Interpolation

Before starting our description and analysis of Thiele-Werner interpolants, it is instructive to review what is understood about the block structure of rational interpolation tables. There is no single clear picture which emerges as in the case of Padé approximation, but the following theorem has clear parallels with Padé's block structure theorem [1892], and helps us to understand why Werner's method is so attractive.

Theorem 3.1 [Wuytack, 1974]. Suppose that the $[\ell/m]$ modified rational interpolation problem with values $\{f_i,\ i = 0,1,\ldots,\ell+m\}$ taken at $\{x_i\ ,\ i = 0,1,\ldots,\ell+m\}$ has the solution $\{p_\ell(x), q_m(x)\}$. Define $p'(x)$, $q'(x)$ to be the polynomials with no common factors for which

$$p'(x)\ /\ q'(x) = p_\ell(x)\ /\ q_m(x). \tag{3.1}$$

Define $$\ell' = \partial\{p'\}\ ,\ m' = \partial\{q'\}. \tag{3.2}$$

Then the interpolation points may be ordered so that

$$f_i \cdot q'(x_i) = p'(x_i)\ ,\quad i = 0,1,\ldots,\ell'+m'+k, \tag{3.3}$$

and $$f_i \cdot q'(x_i) \neq p'(x_i)\ ,\quad i = \ell'+m'+k+1,\ldots,\ell+m \tag{3.4}$$

with $k \geq 0$. (Eq. (3.4)is inapplicable if $\ell'+m'+k = \ell+m$).

Let $\{p_L(x),\ q_M(x)\}$ be the solution of an $[L/M]$ modified rational interpolation problem, for which L,M satisfy

$$\ell' \leq L \leq \ell'+k \quad \text{and} \quad m' \leq M \leq m'+k, \tag{3.5}$$

and (3.3) remains valid.

Define $$r_{L,M}(x) = p_L(x)\ /\ q_M(x)\ . \tag{3.6}$$

Then $$r_{L,M}(x) = p'(x)\ /\ q'(x)\ . \tag{3.7}$$

The inequalities

$$\ell \le \ell'+k \quad \text{and} \quad m \le m'+k \tag{3.8}$$

are valid and delimit the degeneracy block.

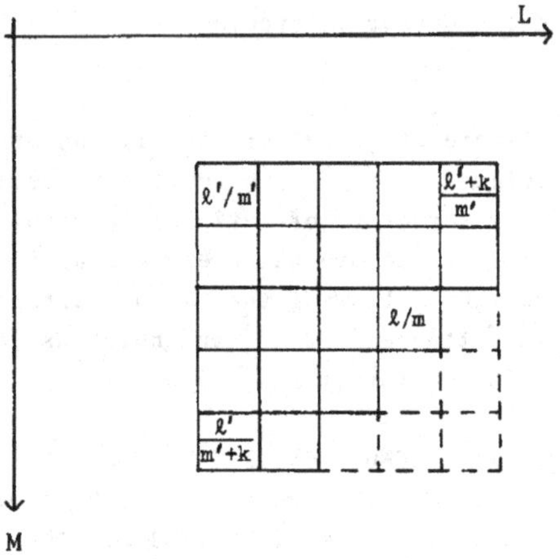

Fig.1. A block in the rational interpolation table.

Remarks For the proof, we refer to Wuytack's original paper. To understand how this theorem identifies the block structure of the rational interpolation table, we start from position (ℓ,m), and (3.1),(3.2) identify (ℓ',m') which is the top left corner of the block in fig.1. The integer k defined by (3.3),(3.4) identifies the top right and bottom left corners of the block. If $\ell+m \le \ell'+m'+k$, then the solution of the modified rational interpolation problem is also the solution of the rational interpolation problem, according to (3.3). If $\ell+m > \ell'+m'+k$, (3.4) shows that the rational interpolation problem is recognised as impossible. The inequalities (3.8) identify the $[\ell/m]$ interpolant requested as belonging to the square block shown in fig.1, subject to a proper reordering of the points according to (3.3) and (3.4). (If $\ell+m < \ell'+m'+2k$, no statement is made or implied by (3.3)-(3.8) about hypothetical interpolants interpolating at x_i, with $\ell+m < i \le \ell'+m'+2k$). Wuytack's theorem gives substance to the statement that an impossible rational interpolation problem occurs if and only if a lower order interpolant actually interpolates on more points than its order indicates. Some of these ideas originate from theorems by Maehly and Witzgall, [1960].

An impossible rational interpolation problem is characterised by a non-trivial set of interpolation points satisfying (3.4). Such points are called unattainable points. As is clear from their identification, they are uniquely associated with the particular rational interpolation problem, rather than with any particular method of solution. This was observed first, in a different context, by Meinguet, [1970]. Consequently some authors prefer to say that certain rational interpolation problems have unattainable points rather than to say that the problem is impossible. We note that blocks in the rational interpolation table which occur when the points are not ordered according to (3.3) and (3.4) are not necessarily square, [Galucci and Jones, 1976]. Interesting open problems concern whether the set of $[\ell/m]$ rational interpolation problems with $\ell,m = 0,1,\ldots$ has solutions such that the blocks in the infinite rational interpolation table are square.

Thiele Werner Interpolation

The requirement that $f_i^{(j)}$ is allowed to be ∞ in (2.25) in the modified Thacher-Tukey algorithm is unappealing from some viewpoints. We are not surprised that there is one good reason for avoiding the use of ∞ as an allowed value, which is stated in §4. The efficient reliable method of rational interpolation which avoids using ∞ as a value was proposed by Werner [1979,1980]. He considers generalised Thiele interpolants of the form

$$R^{(0)}(x) = p_0(x) + \frac{w_0(x)}{p_1(x)} + \frac{w_1(x)}{p_2(x)} + \ldots + \frac{w_{t-1}(x)}{p_t(x)} . \qquad (3.9)$$

In this formula, the polynomial

$$w_s(x) = \prod_{i=h_s}^{k_s} (x-x_i) \qquad s = 0,1,\ldots,t-1 \qquad (3.10)$$

vanishes at the (possibly reordered) interpolation points $x_{h_s}, x_{h_s+1}, \ldots, x_{k_s}$; $p_s(x)$ is a Newton interpolating polynomial which interpolates $f^{(s)}(x)$ on $x_{h_s}, x_{h_s+1}, \ldots, x_{k_s}$.

The data for each stage of the iterative construction of the fraction are defined by

$$f^{(s+1)}(x_i) = \frac{w_s(x_i)}{f^{(s)}(x_i) - p_s(x_i)} \qquad \begin{array}{l} s = 0,1,\ldots,t-1, \\ i = h_{s+1}, h_{s+1}+1, \ldots, n. \end{array} \qquad (3.11)$$

Any Thiele fraction, such as (2.18) in example 3, is an example of a fraction of the form (3.9). For greater generality, we consider two examples which exhibit developments of Thiele fractions :-

Example 5.

$$R^{(0)}(x) = 1 + \frac{x}{p_1(x)} + \frac{(x-6)(x-5)}{p_2(x)} + \frac{(x-1)(x-2)}{p_3(x)} \qquad (3.12)$$

is the [3/3] interpolant to exp(x) on $x = 0,1,2,3,4,5,6$, and

$$p_1(x) = 0.0149095 - 0.0190088\ (x-6)$$
$$p_2(x) = 42.3708 + 11.6612\ (x-1)$$
$$p_3(x) = 0.405366 - 0.0213441\ (x-4).$$

Note that formulas (2.18) and (3.12) are different representations of the [3/3] interpolant to exp(x) on the same points $x = 0,1,2,3,4,5,6$. Consequently expressions (3.12) and (2.18) are equal by theorem 2.1 (within numerical error).

Example 6.

$$R^{(0)}(x) = 1 + \frac{(x-1)(x+1)}{1.6} + \frac{(x-0.6)(x+0.6)}{0.8} \qquad (3.13)$$

interpolates $|x|$ on $x = -1, -0.6, -0.2, 0.2, 0.6, 1.0$.

In example 5, the first stage of the interpolation process involves interpolation at $x = 0$, and the data are transmitted to the second stage in the form of

$$f^{(1)}(x) = \frac{x}{\exp(x) - 1}, \quad x = 1,2,\ldots,6, \qquad (3.14)$$

and a [3/2] type interpolant for $f^{(1)}(x)$ is required. This may be achieved (schematically) by

$$[3/2]_{f^{(1)}} = p_1(x) + \frac{(x-6)(x-5)}{[2/1]_{f^{(2)}}} \qquad (3.15)$$

as in example 5, or else by

$$[3/2]_{f^{(1)}} = b_1 + \frac{x-6}{[2/2]_{f^{(2)}}}. \qquad (3.16)$$

as in example 3. We note that $p_1(x)$ in (3.15) interpolates $f^{(1)}(x)$ at two points, whereas b_1 in (3.16) interpolates $f^{(1)}(x)$ at one point. Consequently, we refer to the method suggested by (3.15) as the fast method; it normally generates fractions of the form (3.12) and we denote it by $T = 0$. The method suggested by (3.16) normally generates Thiele fractions; it is denoted by $T = 1$ and called the slow method. The nature of the representation of the Thiele-Werner fraction is therefore specified by the integer T in the algorithm. In fact, more general cases intermediate between T = 0 and T = 1 can be considered by allowing a vector parameter $\underline{T}$ with components $T_s = 0,1$ where $s = 0,1,\ldots,t-1$ which controls the representation of the interpolant at each stage s, but we are not concerned with this level of generality.

Werner's algorithm is a reliable algorithm for the generation of $[\ell/m]$ type rational interpolants with $\ell \geq m$. Our attitude is that the case $\ell < m$ is fully covered by theorem 2.2. We detail the algorithm next:-

Specification On entry, parameters T, ℓ_0 and m_0 are supplied; T denotes the nature of the representation of the $[\ell_0/m_0]$ type interpolant requested. The inequality $\ell_0 \geq m_0$ is assumed. Values $\{f_i^{(0)}, i = 0,1\ldots,n\}$ at respective points $\{x_i, i = 0,1,\ldots,n\} \equiv X_0$ are supplied, where $n = \ell_0 + m_0$.

Initialisation
$$r_0 = \max(\ell_0 - m_0 - T, 0), \tag{3.17}$$

$$h_0 = 0\ ,\ s = 0. \tag{3.18}$$

Iteration Define $p_s(x)$ to be the Newton polynomial of order r_s to the data

$$\{f_i^{(s)}\ ,\ i = h_s\ ,\ h_s+1,\ldots,h_s+r_s\} \text{ at points}$$

$$\{x_i\ ,\ i = h_s\ ,\ h_s+1,\ldots,h_s+r_s\}\ .$$

Reorder the points of X_s, if necessary, and define k_s to be the largest integer for which

$$p_s(x_i) = f_i^{(s)}\ , \text{ for } i \text{ in the range } h_s \leq i \leq k_s\ . \tag{3.19}$$

If $k_s = n$, let $t := s$ and go to the termination stage. This is a successful exit from the iterative phase of the algorithm. Otherwise,

$$p_s(x_i) \neq f_i^{(s)}\ , \text{ for } i \text{ in the range } k_s < i \leq n\ . \tag{3.20}$$

If $k_s - h_s \geq \ell_s$, exit from the algorithm. This is a failure exit from the iterative stage of the algorithm and is called (IF). It occurs when $h_s + \ell_s \leq k_s < n$ and the rational interpolation problem posed is impossible.

Define $$w_s(x) = \prod_{i=h_s}^{k_s} (x-x_i) \tag{3.21}$$

and $$h_{s+1} = k_s+1 \ . \tag{3.22}$$

The data for the next stage of the algorithm are defined by

$$f_i^{(s+1)} = \frac{w_s(x_i)}{f_i^{(s)} - p_s(x_i)} \ , \quad i = h_{s+1} \ , h_{s+1}+1,\ldots,n \tag{3.23}$$

The order and type of the next reduced interpolant are specified by

$$m_{s+1} = \ell_s - k_s + h_s - 1, \tag{3.24}$$

$$\ell_{s+1} = m_s \ , \tag{3.25}$$

$$r_{s+1} = \max\,(\ell_{s+1} - m_{s+1} - T \ , 0) \tag{3.26}$$

The next interpolation set is

$$X_{s+1} = X_s \setminus \{x_i \ , i = h_s \ , h_s+1,\ldots,k_s\} \ .$$

Resume iteration with $s := s+1$.

<u>Termination</u> Making use of quantities already computed, we define

$$R^{(s)}(x) = p_s(x) + \frac{w_s(x)}{p_{s+1}(x)} + \frac{w_{s+1}(x)}{p_{s+2}(x)} + \ldots + \frac{w_{t-1}(x)}{p_t(x)} \tag{3.27}$$

for $s = 0,1,\ldots,t-1$. The termination check consists of verifying that

$$R^{(s)}(x_i) \neq 0 \quad \text{for} \quad i = h_{s-1} \ , h_{s-1}+1,\ldots,k_{s-1} \tag{3.28}$$

for $s = 1,2,\ldots,t-1$. If this check is not passed, the specified interpolation problem is impossible, and we have a failure at termination, (TF). Otherwise, the interpolant

$$R^{(0)}(x) = p_0(x) + \frac{w_0(x)}{p_1(x)} + \frac{w_1(x)}{p_2(x)} + \ldots + \frac{w_{t-1}(x)}{p_t(x)} \qquad (3.29)$$

is defined by the algorithm, it interpolates the data specified and is of type $[\ell_0/m_0]$.

Werner's algorithm is a reliable algorithm for finding a rational interpolant, as the following theorems show. It is clear from the structure of the algorithm that computation ends either with a successful exit via (3.29) or with an iteration failure via (IF) or with a termination failure via (TF). The next theorem justifies the iteration failure test.

Theorem 3.2 [Werner, 1970a, b]. A rational interpolant of type $[\ell/m]$ is required for data $\{f_i\ ,\ i = 0,1,\ldots,n\}$ at points $\{x_i\ ,\ i = 0,1,\ldots,n\}$, where $\ell \geq m$ and $n = \ell+m$. Define $\pi(x)$ to be the Newton polynomial of order $r = \ell-m$ which interpolates the values $\{f_i\ ,\ i = 0,1,\ldots,r\}$ at the respective points $\{x_i\ ,\ i = 0,1,\ldots,r\}$. Suppose that $\pi(x)$ actually interpolates the data on precisely $\nu+1$ points, so that

$$\pi(x_i) = f_i\ , \qquad i = 0,1,\ldots,\nu \qquad (3.30)$$

and

$$\pi(x_i) \neq f_i\ , \qquad i = \nu+1,\nu+2,\ldots,n. \qquad (3.31)$$

Then, the rational interpolation problem posed in this theorem is impossible if $n > \nu \geq \ell$.

Proof Suppose that the problem posed has a solution of the form $p(x)/q(x)$, so that

$$p(x_i)/q(x_i) = f_i\ , \qquad i = 0,1,\ldots,n, \qquad (3.32)$$

where $p(x)$, $q(x)$ are polynomials with $\partial\{p\} \leq \ell$ and $\partial\{q\} \leq m$. We suppose, without loss of generality, that $p(x)$ and $q(x)$ have no common factors. Then, by (3.30) and (3.32),

$$\tilde{\pi}(x) = p(x) - \pi(x)\, q(x) \qquad (3.33)$$

is a polynomial which vanishes on $\ell+1$ points and $\partial\{\tilde{\pi}\} \leq \ell$. Therefore

$\tilde{\pi}(x) = 0$ and

$$\pi(x) = p(x)/q(x). \tag{3.34}$$

However, by (3.31) and (3.32),

$$\pi(x_n) \neq p(x_n)/q(x_n)$$

and a contradiction has been found. Hence the rational interpolation problem posed is impossible.

Corollary A solution of the *modified* rational interpolation problem posed by theorem 3.2 is

$$q^*(x) = \prod_{i=\nu+1}^{n} (x-x_i) \tag{3.35}$$

$$p^*(x) = \pi(x)\ q^*(x) \tag{3.36}$$

Proof Note that

$$p^*(x_i) - f_i\ q^*(x_i) = 0\ , \qquad i = 0,1,\ldots,n,$$

$$\partial\{q^*\} = n-\nu \le m$$

$$\partial\{p^*\} = \partial\{q^*\} + \partial\{\pi\} \le \ell.$$

q.e.d.

Next, we proceed with statements of the theorems which establish the reliability property of Thiele-Werner interpolants.

Theorem 3.3 Let the rational fraction $N^{(0)}(x)/D^{(0)}(x)$ take the values $\{f_i^{(0)}, i = 0,1,\ldots,n\}$ at the respective points $\{x_i, i = 0,1,\ldots,n\}$, where $N^{(0)}(x)$ and $D^{(0)}(x)$ are polynomials satisfying

$$\partial\{N^{(0)}\} \le \ell_0\ ,\ \partial\{D^{(0)}\} \le m_0\ ,\ \ell_0 \ge m_0\ , \tag{3.37}$$

and $n = \ell_0+m_0$. Then a Thiele-Werner fraction $R^{(0)}(x)$ of the form

$$R^{(0)}(x) = p_0(x) + \frac{w_0(x)}{p_1(x)} + \frac{w_1(x)}{p_2(x)} + \ldots + \frac{w_{t-1}(x)}{p_t(x)} \tag{3.38}$$

exists such that

$$R^{(0)}(x) = N^{(0)}(x)/D^{(0)}(x), \tag{3.39}$$

$$R^{(0)}(x_i) = f_i^{(0)} \quad , \qquad i = 0,1,\ldots,n. \tag{3.40}$$

and

$$\sum_{i=0}^{t-1} \partial\{w_i\} \le n. \tag{3.41}$$

Proof , part (i). Suppose that an interpolating rational fraction $N^{(0)}(x)/D^{(0)}(x)$ exists satisfying (3.37). Run the algorithm, and suppose that the iterative phase provides the representation (3.38). By this hypothesis,

$$N^{(0)}(x_i)/D^{(0)}(x_i) = f_i^{(0)} \quad , \qquad i = 0,1,\ldots,n, \tag{3.42}$$

$$\partial\{N^{(0)}\} \le \ell_0 \ , \quad \partial\{D^{(0)}\} \le m_0. \tag{3.43}$$

For s in the range $0 \le s \le t-1$, make the inductive hypothesis that integers ℓ_s , m_s , h_s and k_s, numbers $f_i(s)$ and a polynomial $p_s(x)$ have been defined by the algorithm for use at stage s. Suppose also that polynomials $N^{(s)}(x)$ and $D^{(s)}(x)$ exist such that

$$N^{(s)}(x_i)/D^{(s)}(x_i) = p_s(x_i) \ , \qquad i = h_s,h_s+1,\ldots,k_s \tag{3.44}$$

$$N^{(s)}(x_i)/D^{(s)}(x_i) = f^{(s)}(x_i) \ , \quad i = h_s,h_s+1,\ldots n, \tag{3.45}$$

$$\partial\{N^{(s)}\} \le \ell_s \ , \ \partial\{D^{(s)}\} \le m_s \ \text{ and } \ \ell_s \ge m_s \ . \tag{3.46}$$

By (3.21), (3.44), we see that we define polynomials by

$$N^{(s+1)}(x) = D^{(s)}(x) \tag{3.47}$$

and $$D^{(s+1)}(x) = [N^{(s)}(x) - p_s(x)D^{(s)}(x)]/w_s(x). \tag{3.48}$$

By(3.25), (3.46) and (3.47),

$$\partial\{N^{(s+1)}\} \le \ell_{s+1}.$$

By (3.24), (3.48) and because $\partial\{p_s\} \le \ell_s - m_s$, we find that

$$\partial\{D^{(s+1)}\} \le m_{s+1}.$$

By (3.19), (3.24), (3.25) and (3.26),

$$r_s \geq \ell_s - m_s - T \quad , \quad k_s \geq h_s + r_s$$

and $\quad \ell_{s+1} - m_{s+1} \geq k_s - h_s + 1 - r_s - T \geq 1 - T \geq 0.$

Therefore (3.46) is a consistent inductive hypothesis. From (3.47), (3.48), we find that

$$\frac{N^{(s+1)}(x)}{D^{(s+1)}(x)} = \frac{w_s(x)}{N^{(s)}(x)/D^{(s)}(x) - p_s(x)} \tag{3.49}$$

From (3.23), (3.45) and (3.49), we find that

$$\frac{N^{(s+1)}(x_i)}{D^{(s+1)}(x_i)} = f^{(s+1)}(x_i) \quad \text{for } i = h_{s+1}, h_{s+1}+1,\ldots,n. \tag{3.50}$$

Eq. (3.20) in the algorithm ensures that zero divisors are not encountered in (3.23). Hence the numbers $f^{(s+1)}(x_i)$ in (3.50) are finite, and can be interpolated by $p_{s+1}(x)$ on $\{x_i \ , \ i = h_{s+1},h_{s+1}+1,\ldots,h_{s+1}+r_{s+1}\}$. The inductive hypothesis expressed by (3.44)-(3.46) is now established. We deduce that a successful exit from the algorithm occurs when

$$N^{(t)}(x_i)/D^{(t)}(x_i) = p_t(x_i) \quad , \quad i = h_t,h_t+1,\ldots,n.$$

From repeated use of (3.49), it follows that $N^{(0)}(x)/D^{(0)}(x)$ has the Thiele-Werner representation (3.38).

Part (ii) Suppose that an interpolating rational fraction $N^{(0)}(x)/D^{(0)}(x)$ exists satisfying (3.37). Run the algorithm, and suppose that there is a failure at the jth iteration, because $k_j-h_j \geq \ell_j$, following (3.20). According to (3.45) and (3.46), $N^{(j)}(x)/D^{(j)}(x)$ defined in the proof of part (i) interpolates the data $\{f_i^{(j)} \ , \ i = h_j,h_j+1,\ldots,n\}$ on points $\{x_i, \ i = h_j,h_j+1,\ldots,n\}$. Theorem 3.2 precludes the possibility of a failure in the iterative stage under current hypotheses.

We have established the existence of $R^{(0)}(x)$, represented by (3.38), as well as (3.39) and (3.40). By (3.21) and (3.22), we see that $\partial\{w_s\} = h_{s+1} - h_s$, and so (3.41) follows immediately.

A converse? It seems hard to formulate a sharp converse to theorem 3.3 along the lines of the converse of theorem 2.3.

Theorem 3.4 If Werner's algorithm runs successfully and defines a fraction $R^{(0)}(x)$ of the form (3.38) which passes the termination test, then

$$R^{(0)}(x_i) = f_i^{(0)} , \qquad i = 0,1,\dots,n.$$

If the algorithm fails to run successfully, either by failure in the iteration or termination stages, then no fraction of the form $N^{(0)}(x)/D^{(0)}(x)$ satisfying (3.37) exists which fits the data.

Theorem 3.5 If a rational fraction $N^{(0)}(x)/D^{(0)}(x)$ exists satisfying the premises of theorem 3.3, then Werner's algorithm runs and yields the representation (3.38), satisfying (3.39)-(3.41).

The proofs of theorems 3.4 and 3.5 are very similar to those of theorems 2.4 and 2.5 and do not need to be elucidated further. Theorems 3.3-3.5 establish that Werner's algorithm is a reliable algorithm for rational interpolation. Furthermore, it is clear that the flexibility allowed and required by (3.19), (3.20) enables the algorithm to circumnavigate the blocks of the rational interpolation table defined by Wuytack's theorem - see (3.3) and (3.4).

If we view Werner's algorithm as a set of successive rational transformations, as in (3.49), we see that the algorithm may be used to solve the Cauchy-Jacobi-Hermite rational interpolation problem, in which confluence occurs. This problem has been fully solved along these lines by Arndt, [1980], exploiting divided difference representations which have well-defined confluent limits.

§4 Stability of Rational Interpolation

Backward error analysis Numerical floating point calculations normally involve rounding error, and computed solutions of numerical problems are usually only approximations to the true solution. In this context, a backward error analysis bounds the discrepancy between the starting values of the given problem and other starting values which lead to the computed results of the given problem in a hypothetical exact calculation. Very roughly, a backward error analysis tells one about some problem which has been solved precisely. Obviously, we wish to design our algorithms so that the problem actually solved is close to the problem posed. The backward error analysis is an invaluable principle which enables the path of an algorithm to be optimised so that we can guarantee that computed results correspond to initial starting values which differ from the actual starting values by only a small amount. Notice that a backward error analysis normally gives no information about the accuracy of the results computed by an algorithm, nor even about the solubility of the given problem.

Example 7 Calculate π+e using five bit floating point mantissa precision.

Let tildes denote numerical representations of numbers.

$$\tilde{\pi} = +.11001_2\ (+10) \tag{4.1}$$

$$\tilde{e} = +.10110_2\ (+10) \tag{4.2}$$

$$\tilde{\pi} + \tilde{e} = +.10111_2\ (+11) \tag{4.3}$$

$$\tilde{\pi} - \tilde{e} = +.11000_2\ (-01). \tag{4.4}$$

Note that rounding is used in deriving (4.1)-(4.3) but not for (4.4).

* * *

In general, let f_1, f_2 be real or complex numbers, and use tildes for numerical representations. Corresponding to example 7, we find that

$$|f_1-\tilde{f}_1| \le \varepsilon\ |f_1| \quad , \quad |f_2-\tilde{f}_2| \le \varepsilon\ |f_2| \tag{4.5}$$

and
$$\widetilde{\tilde{f}_1+\tilde{f}_2} = f_1(1+2\varepsilon_1) + f_2(1+2\varepsilon_2) \tag{4.6}$$

where $|\varepsilon_1| \le \varepsilon$, $|\varepsilon_2| \le \varepsilon$ and ε is the machine precision, following Wilkinson, [1963]. In the example above, $\varepsilon = 2^{-5} = .00001$ and rounding has been done accurately. It is dangerous to assume that computers round accurately in practice, and it is common to take ε to be the smallest positive number such that $1+\varepsilon$ is computationally distinguishable from 1. Either way, (4.6) shows that the computed value of $\tilde{f}_1 + \tilde{f}_2$ corresponds to the addition of two numbers which differ by $2\varepsilon f_1$ and $2\varepsilon f_2$ at most from f_1 and f_2 respectively. This is a typical conclusion of a backward error analysis.

<u>Linear interpolation</u> We need to consider a linear fit to data (x_0,f_0) and (x_1,f_1), expressed by the polynomial

$$\tilde{p}_0(x) = \tilde{f}_0 + (x-x_0)\,\tilde{f}_{01}. \qquad (4.7)$$

We regard $\tilde{p}_0(x)$ as being defined by its parameters $\tilde{f}_0$ and $\tilde{f}_{01}$, and it is computed from $\tilde{f}_0$ and $\tilde{f}_1$. From (4.5),

$$|\tilde{f}_0 - f_0| \le \varepsilon\, |f_0| \qquad (4.8)$$

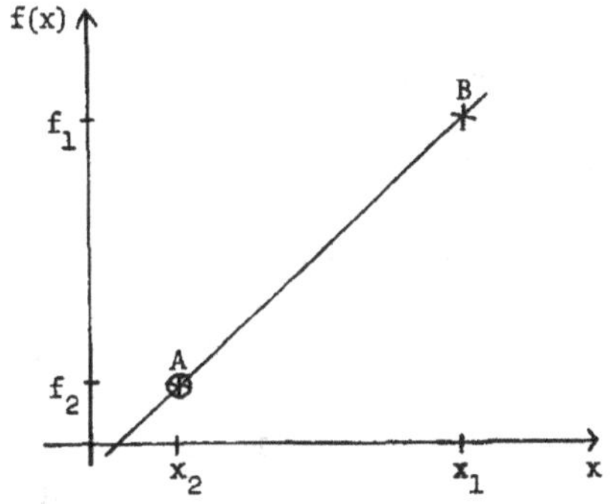

Fig.2 Linear interpolation

To simplify the exposition, we will assume throughout §4 that $x_i = \tilde{x}_i$ for $i = 0,1,\ldots,n$. We use a first order error analysis: terms in ε^2 are neglected. The precise form of (4.7) requires the value

$$f_{01} = (f_0 - f_1)/(x_0 - x_1), \qquad (4.9)$$

and so we define

$$\tilde{f}_{01} := \overbrace{(\tilde{f}_0 - \tilde{f}_1)}/\overbrace{(x_0 - x_1)}. \qquad (4.10)$$

Hence $$\tilde{f}_{01} = \frac{\tilde{f}_0 - \tilde{f}_1}{x_0 - x_1}\,(1+3\varepsilon_{01}) \qquad (4.11)$$

with $|\varepsilon_{01}| \le \varepsilon$. At the point x_1, we find the value of the computed interpolant to be

$$F_1 \equiv \tilde{p}_0(x_1) = \tilde{f}_0 + (x_1 - x_0)\tilde{f}_{01} \qquad (4.12)$$

and $F_1 - \tilde{f}_1 = 3\varepsilon_{01}(\tilde{f}_1 - \tilde{f}_0)$. (4.13)

Therefore $\dfrac{|F_1-\tilde{f}_1|}{|\tilde{f}_1|} \le 3\varepsilon \left|\dfrac{\tilde{f}_0}{\tilde{f}_1} - 1\right| \le 3\varepsilon \left(\left|\dfrac{\tilde{f}_0}{\tilde{f}_1}\right| + 1\right)$. (4.14)

Hence we see the necessity of choosing

$$|\tilde{f}_0| \le |\tilde{f}_1| \quad (4.15)$$

to avoid error build up in (4.14). We must assume that the points are either ordered or reordered so that (4.15) is valid. We then obtain the bounds

$$|F_1-\tilde{f}_1| \le 6\varepsilon\ |f_1| \quad (4.16)$$

and $|F_1-f_1| \le 7\varepsilon\ |f_1|$ (4.17)

for the accuracy of the computed interpolant at the second point of its data base. We can understand (4.15) and (4.16) in geometrical terms using fig 2. $\tilde{p}_0(x)$ given by (4.7) must be "pinned down" either at A (as shown) or at B. If $\tilde{p}_0(x)$ is "pinned down" at B, and (4.15) remains valid, small errors in the slope of AB will lead to small errors in the position of A, but more significant relative errors in the ordinate of A.

<u>Quadratic interpolation</u> We consider a quadratic Newton interpolating polynomial,

$$\tilde{p}_0(x) = \tilde{f}_0 + \tilde{f}_{01}\ (x-x_0) + \tilde{f}_{012}(x-x_0)(x-x_1) \quad (4.18)$$

analogous to (4.7). Again we use the computational equation

$$\tilde{f}_{012} := \frac{\tilde{f}_{01}-\tilde{f}_{02}}{x_1-x_2} \quad (4.19)$$

as well as

$$\tilde{f}_{01} := \frac{\tilde{f}_0-\tilde{f}_1}{x_0-x_1} \quad ; \quad \tilde{f}_{02} := \frac{\tilde{f}_0-\tilde{f}_2}{x_0-x_2} . \quad (4.20)$$

Hence we find the precise results

$$\tilde{f}_{012} = \frac{\tilde{f}_{01}-\tilde{f}_{02}}{x_1-x_2}\ (1+3\varepsilon_{012}), \quad (4.21)$$

$$\tilde{f}_{01} = \frac{\tilde{f}_0-\tilde{f}_1}{x_0-x_1}(1+3\varepsilon_{01}) \quad , \quad \tilde{f}_{02} = \frac{\tilde{f}_0-\tilde{f}_2}{x_0-x_2}(1+3\varepsilon_{02}) \qquad (4.22)$$

with $|\varepsilon_{012}| \le \varepsilon$, $|\varepsilon_{01}| \le \varepsilon$ and $|\varepsilon_{02}| \le \varepsilon$. (4.23)

The error analysis for $\tilde{p}_0(x)$ at $x = x_0$ and $x = x_1$ is exactly the same as for the linear case, because the quadratic term of (4.18) vanishes exactly. However, on $x = x_2$,

$$F_2 \equiv \tilde{p}_0(x_2) = \tilde{f}_0 + \tilde{f}_{01}(x_2-x_0) + \tilde{f}_{012}(x_2-x_0)(x_2-x_1).$$

Using (4.21)-(4.23), we find that

$$F_2 = \tilde{f}_2(1+3\varepsilon_{02}+3\varepsilon_{012}) - \tilde{f}_0(3\varepsilon_{02}+3\varepsilon_{012}) - \tilde{f}_{01}(x_2-x_0)3\varepsilon_{012} \qquad (4.24)$$

We obtain the easy bound

$$|F_2-\tilde{f}_2| \le 6\varepsilon|\tilde{f}_2| + 6\varepsilon|\tilde{f}_0| + 3\varepsilon|\tilde{f}_{01}|.|x_2-x_0| \ . \qquad (4.25)$$

The aim of the backward error analysis is to find a strategy to make (4.25) as small as possible. To this end, we arrange the interpolation points so that $|\tilde{f}_0| < |\tilde{f}_2|$. Thus x_0 is chosen so that $|\tilde{f}_0| < |\tilde{f}_1|,|\tilde{f}_2|$. To minimise (4.25), we also choose x_1 so that $|\tilde{f}_{01}| \le |\tilde{f}_{02}|$. In other words, the points are ordered so that the divided difference used is smallest in modulus. It then follows that

$$|F_2-f_2| \le 19\varepsilon|f_2|. \qquad (4.26)$$

<u>Polynomial Interpolation</u> The previous results generalise in a very obvious way to polynomial interpolation. We define $F_k = \tilde{p}_0(x_k)$. We find that the error analysis indicates that at every stage of the construction of the Newton interpolating polynomial, the points must be ordered such that the divided difference which is smallest in modulus is used. With this simple strategy, we find that

$$|F_k-f_k| \le |f_k|(6\times2^k-5)\varepsilon. \qquad (4.27)$$

<u>Thiele Interpolation</u> Let us suppose that either the Werner algorithm with T = 1 or the modified Thacher Tukey algorithm has been used to construct a Thiele type interpolant, and that the

computed numerical result is

$$R^{(0)}(x) = \tilde{f}_0 + \frac{x-x_0}{\tilde{f}_1^{(1)}} + \frac{x-x_1}{\tilde{f}_2^{(2)}} + \ldots + \frac{x-x_{n-1}}{\tilde{f}_n^{(n)}} \tag{4.28}$$

We will see that it is arguable that the most stable numerical construction is based on an ordering of the interpolation points at each stage j of the interpolation algorithm, such that

$$|\tilde{f}_j^{(j)}| = \min_{i:j\le i\le n} |\tilde{f}_i^{(j)}|, \tag{4.29}$$

for all j in the range $0 \le j \le n$; the quantities $f_i^{(j)}$ are defined iteratively by (3.23) or (2.25). Let us consider the final stage of the construction of (4.28). The final computational equation is

$$\tilde{f}_n^{(n)} := \frac{x_n-x_{n-1}}{\tilde{f}_n^{(n-1)}-\tilde{f}_{n-1}^{(n-1)}} . \tag{4.30}$$

As before, we deduce that

$$\tilde{f}_n^{(n-1)} - \tilde{f}_{n-1}^{(n-1)} = \frac{x_n-x_{n-1}}{\tilde{f}_n^{(n)}} (1+3\varepsilon_n^{(n)}) \tag{4.31}$$

with $|\varepsilon_n^{(n)}| \le \varepsilon$. Corresponding to (4.31), we have an equality pertaining to exact values of the calculated interpolant, denoted by $F_i^{(j)}$, which is

$$F_n^{(n-1)} - \tilde{f}_{n-1}^{(n-1)} = \frac{x_n-x_{n-1}}{\tilde{f}_n^{(n)}} . \tag{4.32}$$

From (4.31), (4.32),

$$F_n^{(n-1)} - \tilde{f}_n^{(n-1)} = (\tilde{f}_{n-1}^{(n-1)} - F_n^{(n-1)})3\varepsilon_n. \tag{4.33}$$

Because the interpolation points are ordered such that

$$|\tilde{f}_{n-1}^{(n-1)}| = \min\{|\tilde{f}_{n-1}^{(n-1)}| , |\tilde{f}_n^{(n-1)}|\} , \tag{4.34}$$

we find that

$$|F_n^{(n-1)} - \tilde{f}_n^{(n-1)}| \le |f_n^{(n-1)}| \times 6\varepsilon . \tag{4.35}$$

Provided the termination test is passed,

$$F_{n-1}^{(n-1)} = \tilde{f}_{n-1}^{(n-1)}, \tag{4.36}$$

and we have in (4.35) and (4.36) the error bounds relevant to the $(n-1)^{th}$ stage of the computation. With these preliminaries computed, we can now prove the main theorem:-

Theorem 4.1 Using current notation, the value of a Thiele interpolant (4.28) constructed according to the principle of (4.31) at an interpolation point x_i is bounded by

$$|F_i-\tilde{f}_i| \le 6\varepsilon(2^i-1)|\tilde{f}_i|, \qquad i = 0,1,\dots,n. \tag{4.37}$$

Proof The result (4.37) is a special case of the more general result that

$$|F_i^{(j)}-\tilde{f}_i^{(j)}| \le 6\varepsilon(2^{i-j}-1)|f_i^{(j)}|, \; i = j,j+1,\dots,n. \tag{4.38}$$

This result bounds the errors at stage j, and is proved by induction. It is established by (4.35), (4.36) for the case of $j = n-1$. Suppose that (4.38) is true for $j = k$, and we must deduce that it holds for $j = k-1$. The differences at stage k are computed from

$$\tilde{f}_i^{(k)} := \frac{x_i-x_{k-1}}{\tilde{f}_i^{(k-1)}-\tilde{f}_{k-1}^{(k-1)}}, \qquad i = k,k+1,\dots,n. \tag{4.39}$$

Therefore,

$$\tilde{f}_i^{(k-1)} - \tilde{f}_{k-1}^{(k-1)} = \frac{x_i-x_{k-1}}{\tilde{f}_i^{(k)}} (1+3\varepsilon_i^{(k)}), \tag{4.40}$$

where $|\varepsilon_i^{(k)}| \le \varepsilon$ for $i = k,k+1,\dots,n$. The corresponding result for the computed fraction (4.28) using a tail-to-head evaluation is

$$F_i^{(k-1)} - \tilde{f}_{k-1}^{(k-1)} = \frac{x_i-x_{k-1}}{F_i^{(k)}} \tag{4.41}$$

Substitute (4.41) into (4.40) and subtract $F_i^{(k-1)}-\tilde{f}_{k-1}^{(k-1)}$ from each side to obtain

$$\tilde{f}_i^{(k-1)} - F_i^{(k-1)} = [F_i^{(k-1)} - \tilde{f}_{k-1}^{(k-1)}]\left\{\frac{F_i^{(k)}}{\tilde{f}_i^{(k)}} (1+3\varepsilon_i^{(k)})-1\right\}. \tag{4.42}$$

Using the first order error analysis and the hypothesis (4.38), valid for $j = k$, we find that

$$|F_i^{(k-1)} - \tilde{f}_i^{(k-1)}| \le |F_i^{(k-1)} - \tilde{f}_{k-1}^{(k-1)}|(6\times 2^{i-k} -3)\varepsilon. \tag{4.43}$$

As a consequence of the principle (4.29), we deduce that

$$|F_i^{(k-1)} - \tilde{f}_i^{(k-1)}| \le |\tilde{f}_i^{(k-1)}|(6\times 2^{i-k+1}-6)\varepsilon, \tag{4.44}$$

valid for $i = k,k+1,\ldots,n$. Provided the termination test is passed, $F_{k-1}^{(k-1)} = \tilde{f}_{k-1}^{(k-1)}$, and we see that (4.38) is proved. The result (4.37) follows as the special case for $j = 0$.

We emphasise that there is no a priori guarantee in theorem 4.1 that either the modified Thacher-Tukey algorithm of §2 or the Werner algorithm of §3 yields a Thiele interpolant in any particular case. If the implication of (4.29) in the context of the modified Thacher-Tukey algorithm is that

$$f_j^{(j)} = 0, \tag{4.45}$$

as may well occur, then theorem 4.1 does not apply. Likewise, the presence of degeneracy may prevent Werner's algorithm applied with $T = 1$ from generating the pure Thiele fraction (4.28). The error bound (4.37) is only proved for the case in which the numerical algorithm terminates with $t = n$.

Error analysis for Thiele-Werner interpolation Largely because of the similarity of the results (4.27) and (4.37), we find that the previous analysis can be extended to the case of Thiele-Werner interpolation, and a result of the form (4.37) is valid quite generally. At every juncture of each stage of the construction of the Thiele-Werner fraction, when a divided difference coefficient of $p_s(x)$ in (3.9) is to be calculated, the interpolation points are so ordered that each coefficient in each polynomial $p_s(x)$ in the Thiele-Werner fraction has the least possible modulus. The details of the proof are murky, [Graves-Morris, 1980], and we omit them. The advantage of the Werner algorithm (over the modified Thacher-Tukey algorithm) which allows the error bound (4.37) to follow in all cases for which an interpolant is constructed is that (4.45) cannot occur.

Prospective users of the algorithms described in §2, §3 and §4 are warned to expect large floating point numbers to occur in the representation of nearly degenerate but non-degenerate rational interpolation problems. Although the presence of large numbers is normally a signal that a routine is badly in error, this is not necessarily the case here. If $|f_n^{(n)}|$ is anomalously large in (4.28), we regard $R^{(0)}(x)$ in (4.28) as a near miss at a Thiele fraction with $t = n-1$; such a fraction would interpolate identically to the original fraction at the points $\{x_i, i = 0,1,\dots,n-1\}$ but interpolation would be only approximate at $x = x_n$. However, no parameters should normally exceed $O(\varepsilon^{-1})$, which is always representable by floating point numbers on modern computers.

§5 Practical Considerations

If the function values for rational interpolation are integers or simple fractions, then exact arithmetic can be used. Degeneracy tests are decisive in these situations, and no error analysis is needed. Methods such as the modified Kronecker algorithm and the modified Viskovatov algorithm, such as were reviewed by Bultheel, [1979], and Graves-Morris, [1979], are efficient and reliable. The Kronecker algorithm uses an antidiagonal sequence in the rational interpolation table. With the Euclidean modification for reliability, it becomes an attractive method for computing a particular $[\ell/m]$ interpolant. Viskovatov's algorithm uses a diagonal staircase sequence in the rational interpolation table. With Bultheel's modification for reliability, it generates sequences whose empirical convergence may suggest an appropriate stopping criterion when the number of data points needed for given accuracy is not known a priori. The recent availability of ALTRAN makes implementation of such methods attractive, [Geddes, 1979].

The modified Thacher-Tukey algorithm for real valued rational interpolation will probably be available as a NAG library subroutine at Mark 9. Selection of an algorithm for a subroutine library involves many criteria, which sometimes conflict. In this case, the decision about which algorithm is "the best" for general purpose use is especially difficult, not least because of the merits of Werner's algorithm.

Termination stage degeneracy tests in the modified Thacher-Tukey and Werner algorithm should be implemented by a running error analysis, [Peters and Wilkinson, 1971]. In this philosophy, if it can happen that the parameters of the computed interpolant may correspond to data values of an insoluble rational interpolation problem within numerical error, the actual interpolation problem is declared to be insoluble within the available numerical precision.

It is important to realise that rounding error can convert insoluble rational interpolation problems into soluble problems and vice-versa. Baker, [1975], defines a defect of a rational interpolant to be a pole with a nearby zero, which is tantamount to a pole with a small residue. If the computed interpolant has a defect near an interpolation point, we regard the interpolant as being close to a degenerate interpolant in a suitable parameter space. Likewise, near misses in degeneracy tests indicate defects. These remarks are mainly justified by consideration of simple examples. Such examples show that small variations in the parameter space may allow defects in a nearly degenerate problem to coalesce into unattainable points in an impossible problem. The importance of the detection and avoidance of defects is a significant problem nowadays, and the existence of a stable rational interpolation algorithm must surely help in this respect.

§6 Conclusions and acknowledgement

We have seen that Thiele fractions can be reliably used as representations for rational interpolants when provision is made for reordering the interpolation points. We have seen how further reordering of the points improves the stability of the representation, and how the use of Thiele-Werner interpolants combines all these merits with unconditional stability.

I am grateful to T.R. Hopkins and D.J. Winstanley for helpful discussions and expert coding.

References

Arndt, H., (1980), "Ein verallgemeinerter Kettenbruch - Algorithmus zur rationalen Hermite - Interpolation", Num. Math. 36, 99-107.

Baker, G.A., Jr., (1975), "Essentials of Padé approximants", Academic Press, N.Y..

Bultheel, A., (1979), "Recursive algorithms for the Padé table : two approaches" in "Padé approximation and its applications", ed. L. Wuytack, Springer-Verlag, 211-230.

Gallucci, M.A. and Jones, W.B., (1976), "Rational approximations corresponding to Newton series (Newton-Padé approximants)", J. Approx. Theory 17, 366-392.

Geddes, K.O., (1979) "Symbolic computation of Padé approximants", A.C.M. Trans. Math. Software 5, 218-233.

Graves-Morris, P.R., (1980), "Practical, reliable, rational interpolation", J. Inst. Maths. Applics., 25, 267-286.

Graves-Morris, P.R. and Hopkins, T.R., (1978), "Reliable rational interpolation", Kent preprint; Num. Math., to be published.

Maehly, H. and Witzgall, Ch., (1960), "Tschebyscheff-Approximationen in kleinen Intervallen II", Num. Math. 2, 293-307.

Meinguet, J., (1970), "On the solubility of the Cauchy interpolation problem" in "Approximation theory", ed. A. Talbot, Academic Press, London, 137-163.

Padé, H., (1892), "Sur la représentation approchée d'une fonction par des fractions rationelles", Ann. Ecole Norm. 9 suppl., 1-93.

Peters, G. and Wilkinson, J.H., (1971) "Practical problems arising in the solution of polynomial equations", J. Inst. Math. Applics., 8, 16-35.

Stoer, J., (1961), "Uber zwei Algorithmen zur Interpolation mit rationalen Funktionen", Num. Math. 3, 285-304.

Thacher, H.C. and Tukey, J.W., (1960) "Rational interpolation made easy by a recursive algorithm", unpublished.

Warner, D.D., (1974), "Hermite interpolation with rational functions", Univ. of California thesis.

Werner, H., (1963), "Rationale Tschebyscheff-Approximation, Eigenwerttheorie und Differenzenrechnung", Archive for Rational Mechanics and Analysis 13, 330-347.

Werner, H., (1979), "A reliable method for rational interpolation" in "Padé approximation and its applications", ed. L. Wuytack, Springer-Verlag, 257-277.

Werner, H., (1980). "Ein Algorithmus zur rationalen Interpolation" in "Numerische Methoden der Approximationtheorie 5", eds. L. Collatz, G. Meinardus and H. Werner", Birkhäuser, 319-337.

Werner. H. and Schaback, R., (1972), "Praktische Mathematik II", Springer-Verlag, p. 71.

Wilkinson, J.H., (1963), "Rounding errors in algebraic processes", Notes in applied science 32, H.M.S.O. London, chap. 1.

Wuytack, L., (1974), "On some aspects of the rational interpolation problem", SIAM J. Num. Anal. 11, 52-60.

NON-LINEAR SPLINES, SOME APPLICATIONS TO SINGULAR PROBLEMS

Helmut Werner

Institute for Applied Mathematics
University of Bonn
D-5300 Bonn 1, Wegelerstraße 6

Introduction

In approximating the solution of ordinary differential equations rational functions have proven to give results superior to linear combinations of functions (e.g. polynomials) in regions where the approximated solution has a pole-like singularity. Obviously if there are more general types of singularities one would like to use approximating functions with a singularity of a corresponding type. To increase the accuracy of approximation on a fixed interval one must introduce more and more parameters into the approximating function. This may be done either by taking more complicated functions, e.g. increasing the degrees of polynomials and rationals, or by retaining simple functions but splitting the interval into more and more subintervals on each of which the approximation is determined separately. If, in addition, we ask for smooth transitions between these approximations of adjacent intervals we arrive at the concept of non-linear splines. These appear as a natural generalization of rational function approximation. I may therefore be allowed to present these results on non-linear splines instead of the talk on the reliability and stability of rational interpolation I had announced, since this topic is covered by another paper. Hopefully my paper will show new directions of applications and stimulate new research.

Here we will present some examples of such non-linear spline functions (slightly different from the splines we used in the papers [5,6]) in their application to initial value problems of non linear differential equations. We investigate the asymptotic behavior of the attained accuracy, in particular in the neighborhood of the singularity.

The splines are generated by functions of the form

$$u(z,x,u_o,u_o',u_o'',u_o''') = u_o + u_o' \cdot z + \frac{u_o'' \cdot b^2}{\alpha(\alpha-1)} \left[(1+\frac{z}{b})^{\alpha} - \frac{\alpha z}{b} - 1\right] ,$$

z the local coordinate about x,
with u_o, u_o', u_o'' and b the spline parameters while α is fixed and characteristic for the spline class. The parameter b may easily be expressed by $u_o''', \ldots, u_o$.
Estimates for the location of the singularity are obtained and the error of this procedure is appraised. Our results also furnish an explanation for the monotonicity of the estimated locations of the singularity in dependence on the parameter α of the spline family thus giving an asymptotic inclusion of the singularity.

If α is chosen as the exponent that is characteristic for the singularities produced by the differential equation we get optimal estimation. Furthermore we describe an extrapolation procedure for this case. Its quality is supported by numerical examples.

Definition of Problem and Numerical Method

We set out to solve the initial value problem

$$y' = f(x,y)$$
$$y(x_o) = y_o$$

where $f(x,y)$ is given in a domain G of $\mathbb{R}^2$ and $(x_o,y_o)\in G$.

We ask for a solution $y(x)$ to be determined in some interval $I := [x_o,x^+]$, with $x^+ > x_o$ given.

For the sake of completeness we briefly review the spline methods available for this problem but instead of stating them with a very general set up we concentrate on one simple case that parallels the case handled by Loscalzo-Talbot [3] for cubic splines. A general survey may be found in [4,6].

To define the spline functions we first cover the interval I by subintervals of length h (this stepsize could even vary as we go along). Let knots be given by

$$x_j = x_o + j \cdot h, \quad j = 0,\dots,N,$$

with $x_N \geq x^+$.

Then we define the spline $u(x,h)$ to be of class $C^2(I)$ so that its restriction to any subinterval $I_j := [x_{j-1},x_j]$ should equal a member of a class of functions that depends upon 4 parameters. (One could call them spline parameters.) For theoretical considerations it is convenient to assume that the parameterization is via the value of u and its derivatives at the left-hand endpoint of the interval.

$$u(x,h) = \tilde{u}(z;x_j,u_j,u_j',u_j'',u_j'''), \quad z = x - x_j\in[0,h].$$

The function $\tilde{u}$ should be at least of class $C^4(I_{j+1})$, and also be smooth with respect to the parameters, but in our following examples we even assume it to be holomorphic. Actually, by proceding from I_j to I_{j+1} it is seen that three parameters are fixed by the required order of differentiability of u across the knots.

The algorithm for the approximate solution of the initial value problem starts out by deriving the value of u and its first two derivatives from the given initial value and the differential equation:

$$u_o = y_o ,$$
$$u_o' = f(x_o, y_o) ,$$
$$u_o'' = D_x f(x, y(x))|_{x=x_o} := f_x(x_o, y(x_o)) + f_y(x_o, y(x_o)) \cdot u_o' .$$

With this information we start the recursive definition of u. In I_j (for $j = 1,\dots,N$) we assume u_{j-1}, u_{j-1}', u_{j-1}'' (where $u_j := u(x_j)$ etc) to be already known and determine u_{j-1}''' such that the differential equation is satisfied at least at x_j, i.e. we solve the equation

$$(*) \quad u'(x_j,h) = f(x_j, u(x_j,h))$$

for u_{j-1}'''. Of course this might not always be possible, but we assume for our given problem that a solution exists. The new piece of u(x,h) defined in I_j will then determine u_j, u_j', u_j'', the data needed to carry on the recursion until x^+ is reached.

There may be an abnormal end to the algorithm if there is no solution to the equation (*) or if the solution becomes singular. (We elaborate on this case below.)

If there exists a solution u(x,h) for every sufficiently small h in the above algorithm and if f(x,y), the right-hand side of the differential equation, is sufficiently smooth along (x,y(x)) for $x_o \le x \le x^+$ then it can be seen that

$$\| D^i(u(x,h) - y(x)) \| = O(h^{4-i}), \quad i = 0,1,2,3,$$

holds uniformly in I (compare [4]). This regular case may be extended to higher order equations as was done by H. Arndt [1].

Application to Polynomial Differential Equations, Selection of Spline Families

To exhibit examples we apply the method to differential equations with

$$f(x,y) = p_m(x) \cdot y^m + p_{m-1}(x)y^{m-1} + \ldots + p_o(x), \quad m>1$$

and $p_j(x)$ polynomials. These differential equations have been investigated in the complex domain and they are known to have solutions with algebraic singularities. According to the theory of Painlevé a solution having a singularity at x* may be expanded with respect to

$$t = x - x^*$$

into a series

$$y(t) = c \cdot t^\mu \cdot (1+c_1 t^\gamma + c_2 t^{\gamma+|\mu|} + \ldots)$$

where

$$\mu = \frac{1}{1-m}$$

and

$$c^{m-1} = \frac{\mu}{p_m(x^*)} \text{ if } p_m(x^*) \neq 0.$$

This may be found heuristically by inserting the formal series y(t) into the differential equation. In the same way one establishes that γ is a multiple of $|\mu|$. For the proofs the interested reader is referred to books on differential equations in the complex domain, e.g. E. Hille [2].

The form of the above expansion suggests a family of splines that contains terms of the form $(z + \text{const})^\alpha$, where α may be chosen in accordance with the right-hand side of the differential equation or may be treated as one of the spline parameters. In either case the spline may become singular in the interval I.

Since we have dealt with the second case in other papers we will elaborate here mainly on the first case. In accordance with the said normalization we construct the splines from pieces of the form

$$\tilde{u}(z;x,u,u',u'',u''') = u+u'\cdot z+\frac{u''\cdot b^2}{\alpha(\alpha-1)}\cdot [(1+\tfrac{z}{b})^{\alpha} - \frac{\alpha\cdot z}{b} -1] \quad \text{for } \alpha \neq 0,1,2,$$

$$= u+u'\cdot z+u''b^2[\tfrac{z}{b} - \ln(1+\tfrac{z}{b})] \quad \text{for } \alpha = 0,$$

$$= u+u'\cdot z+u''b^2[\tfrac{z}{b} - (1+\tfrac{z}{b})\cdot \ln(1+\tfrac{z}{b})] \quad \text{for } \alpha = 1,$$

$$= u+u'\cdot z+u''\frac{z^2}{2} + u''' \cdot \frac{z^3}{3!} \quad \text{for } \alpha = 2.$$

Here z is the local coordinate at x.
It is easy to see that u, u', and u" are the corresponding derivatives of $\tilde{u}$ at $z = 0$. Furthermore we find it convenient for our further purposes to introduce b instead of u''' , which may be expressed by b and u". For $\alpha=2$ we take any function containing four parameters, e.g. a cubic polynomial that exhibits the dependence on the given data $u,\ldots,u'''$.
The geometric meaning of b is clear - it determines the location $\tilde{x}$ of the singularity of $\tilde{u}$, namely

$$\tilde{x} - x_j = \tilde{z} = - b,$$

if the function is determined in $[x_j,x_{j+1}]$.

These splines now prove useful in estimating the locations where the solution of the initial value problems becomes singular. Instead of x^*, the singularity of y, we use $\tilde{x}_j = x_j - b_j$, the place where the restriction of $u(x,h)$ to I_{j+1} would become singular.

In performing the recursion we stop if an $\tilde{x}_j$ is found that would lie in the interval I_{j+2} that is next to I_{j+1}.

Determination of b_j

From the above functions (for $\alpha \neq 1,2$) it is easily seen that

$$u'(x,h) = u_j' + \frac{u_j'' b_j}{\alpha-1}\left[\left(1+\frac{z}{b_j}\right)^{\alpha-1} - 1\right], \qquad \text{for } x \in I_{j+1}.$$

The parameter b_j is chosen such that the equation

$$u'(x_{j+1},h) = u_j' + \frac{u_j'' \cdot b_j}{\alpha-1}\left[\left(1+\frac{h}{b_j}\right)^{\alpha-1} - 1\right] = f(x_{j+1}, u_{j+1})$$

holds. It transcribes to

$$\frac{1}{\alpha-1} \cdot \frac{\left(1+\frac{h}{b_j}\right)^{\alpha-1} - 1}{h/_{b_j}} = \frac{f(x_{j+1},u_{j+1}) - u_j'}{h \cdot u_j''}$$

and this equation is actually solved, if possible, in each subinterval I_j.

We want to exploit this equation to get an appraisal of b_j in comparison to $x_j - x^*$, the distance of x_j from the singularity of y.

To illustrate the quality of the results obtained we treat an example in which one knows the precise location of the singularity, hence the error can be seen explicitly.

Example

$$y' = 1 + y^2 + y^4$$
$$y(0) = 1 \;.$$

An elementary but lengthy calculation shows

$$x^* = \frac{\pi\sqrt{3}}{12} - \frac{\ln 3}{4} = 0.178\ 796\ 769 \ldots \quad .$$

With $t = x - x^*$ the solution y has the expansion

$$y(t) = -(3t)^{-1/3} + \frac{1}{5}(3t)^{1/3} + \frac{3}{25} \cdot (3t) + \ldots$$

that is, $\mu = -1/3$ and $\gamma = 2/3$.

If spline approximations are calculated with h = 0.015 625 up to

$x_j = 0.15625$ and then $\tilde{x}_j = x_j - b_j$ is used to estimate x^* we get the following results for different values of α.

α	-1	-1/2	-1/3	-1/5	-1/10
$\tilde{x}_j$	.18234	.17970	.17896	.17815	.17763

It is not too surprising that $\alpha = \mu$ gives the best approximation. What is remarkable, however, is the monotonicity of $\tilde{x}_j$ in dependence on the exponent α.

We set out first to give an explanation of this phenomenon.

It was stated before that 4th order convergence of $u(x,h)$ to $y(x)$ will take place in any compact interval in which $y(x)$ is regular. In fact it can even be seen that $u'(x,h)$ converges to $y'(x)$ with 4th order at every knot x_j, if convergence takes place at all.

This suggests that we replace $u(x,h)$ by $y(x)$ in the previous equation under the assumption that the systematic error between b_j and $x_j - x^*$ is mainly controlled by another effect. After this substitution b_j appears only on the left-hand side of the preceding equation.

Let

$$V(x,h) := \frac{f(x+h,y(x+h)) - y'(x)}{h\cdot y''(x)} \quad \text{for fixed } x,$$

then

$$V(x,h) = \frac{y'(x+h) - y'(x)}{h\cdot y''(x)}$$

$$= 1 + \frac{h}{2}\,\frac{y'''}{y''} + \frac{h^2}{6}\cdot\frac{y^{IV}}{y''} + \ \dots \ .$$

This series converges for $|h| < |x - x^*|$, assuming that there is no other singularity of y close by.

With $v := \frac{h}{b}$ we find an expansion of the left-hand side of the above equation from

$$G(v) := \frac{1}{\alpha-1} \cdot \frac{(1+v)^{\alpha-1} - 1}{v}$$

$$= 1 + \frac{\alpha-2}{2} \cdot v + \frac{\alpha-2}{2} \cdot \frac{\alpha-3}{3} \cdot v^2 + \ldots \quad .$$

The equation

$$G(\tfrac{h}{b}) = V(x,h)$$

gives

$$\frac{\alpha-2}{2} \cdot \frac{1}{b} + \frac{\alpha-2}{2} \cdot \frac{\alpha-3}{3} \cdot \frac{h}{b^2} + \ldots = \frac{1}{2} \cdot \frac{y'''}{y''} + \frac{h}{6} \cdot \frac{y^{IV}}{y''} + \ldots \quad .$$

We summarize these findings:

for fixed x as $h \to 0$ we get

$$b = (\alpha-2) \cdot \frac{y''}{y'''} \qquad (\alpha \neq 2,1)$$

and the convergence is linear.

If $t = x - x^*$ denotes the distance of x from the singularity of y then

$$b = \frac{2-\alpha}{2-\mu} \cdot t(1+c^* \cdot t^\gamma + \ldots) \ .$$

The last relation is found by differentiation of the Painlevé-expansion of y to get

$$\frac{y''(x)}{y'''(x)} = \frac{c \cdot \mu(\mu-1) \cdot t^{\mu-2} + c_1(\mu+\gamma)(\mu+\gamma-1) \cdot t^{\mu+\gamma-2} + \ldots}{c \cdot \mu(\mu-1)(\mu-2)t^{\mu-3} + c_1(\mu+\gamma)(\mu+\gamma-1)(\mu+\gamma-2)t^{\mu+\gamma-3} + \ldots} \quad .$$

Furthermore we may compare

$$x^* = x - t \quad \text{and} \quad \tilde{x} = x - b.$$

By the above formula

$$\tilde{x} = x - \frac{2-\alpha}{2-\mu} \cdot t + O(t^{1+\gamma}) \cong x - t + \frac{\alpha-\mu}{2-\mu} \cdot t$$

$$\cong x^* + \frac{\alpha-\mu}{2-\mu} \cdot t \quad .$$

For small negative t, i.e. if we integrate the initial value problem with $h > 0$ and approach a singularity then we have (asymptotically) the <u>monotonicity relation:</u>

If $\mu < 2$ and $\alpha < \mu$ then $\tilde{x} > x^*$,
$\alpha > \mu$ then $\tilde{x} < x^*$.

This explains the numbers seen in the previous example.

On the other hand one may use the obtained information to subject $\tilde{x}$ to a correction

$$x^* \cong \tilde{x} - \frac{\alpha-\mu}{2-\mu} \cdot t \cong \tilde{x} - \frac{\alpha-\mu}{2-\alpha}\, b =: \hat{x} \quad .$$

In the above example, where $x^* = 0.178796 \ldots$ this correction yields the following table of results:

	α	-1	-1/2	-1/3	-1/5	-1/10
$h = .015625$	$\tilde{x}_j$	.18234	.17970	.17896	.17815	.17763
$x_j = .156250$	$\hat{x}_j$	.17654	.17814	———	.17948	.18001
$h = .0078125$	$\tilde{x}_j$	.18180	.17956	.178825	.17824	.17780
$x_j = .1640625$	$\hat{x}_j$	.17786	.17853	———	.17910	.17933

It is seen that the estimation of x^* by $\hat{x}$ is again monotonic, which is understandable again from the above expansions. It should, however, be kept in mind that we neglected the errors of the numerical integration, therefore caution in using this procedure is advisable.

The correction does not work if $\alpha = \mu$.
In the next section we propose an extrapolation scheme that is applicable in this case.

The Extrapolation Technique in Case $\alpha = \mu$

If $\alpha = \mu$ the expansion of b is simplified to be

$$b = t + c_1^* \cdot t^{1+\gamma} + \ldots .$$

This is the value b corresponding to a place $x = x^* + t$ for $h = 0$. We do not know, however, and do not wish to compute c_1^* by cumbersome calculations that would have to be repeated for every equation.

The expansion of b may be inverted to give

$$t = b + d_1(-b)^{1+\gamma} + \ldots$$

(remember $b < C$).

To get more information we may consider b_j associated with x_j for different values of j. Though t_j is unknown, we can infer that

$$t_j - t_i = (j-i)h.$$

This makes it feasible to eliminate the higher order terms with the unknown coefficients $d_1, \ldots$. If we take two terms, for example,

$$t_j = b_j + d_1(-b_j)^{1+\gamma}$$

$$t_{j+1} = b_{j+1} + d_1(-b_{j+1})^{1+\gamma}$$

then

$$h = b_{j+1} - b_j + d_1[(-b_{j+1})^{1+\gamma} - (-b_j)^{1+\gamma}] ,$$

so that d_1 is found and may be used to calculate

$$\tilde{\tilde{x}}_{j+1} = \tilde{x}_{j+1} + d_1(-b_{j+1})^{1+\gamma} .$$

It is clear how to generalize this method to more terms.

One may systematize the elimination procedure by rewriting

$$x^* \cong x_j + t_j = \tilde{x}_j + d_1(-b_j)^{1+\gamma} + d_2(-b_j)^{1+\gamma+|\mu|} + \dots$$

into the form of a system of homogeneous linear equations for the "unknowns" $(1, d_1, d_2, \dots)$, i.e.

$$0 = 1 \cdot (\tilde{x}_j - \tilde{x}^*) + d_1(-b_j)^{1+\gamma} + d_2(-b_j)^{1+\gamma+|\mu|} + \dots$$

$$0 = 1 \cdot (\tilde{x}_{j+1} - \tilde{x}^*) + d_1(-b_{j+1})^{1+\gamma} + d_2(-b_{j+1})^{1+\gamma+|\mu|} + \dots$$

$$\dots .$$

Since there is a nontrivial solution of this system its determinant must vanish.

$$\det \begin{pmatrix} \tilde{x}_j - \tilde{x}^* & (-b_j)^{1+\gamma} & (-b_j)^{1+\gamma+|\mu|} & \dots \\ \vdots & \vdots & \vdots & \\ \tilde{x}_{j+n} - \tilde{x}^* & (-b_{j+n})^{1+\gamma} & (-b_{j+n})^{1+\gamma+|\mu|} & \dots \end{pmatrix} = 0$$

Expanding with respect to the first column we see that

$$\tilde{x}^* \cdot \sum_{i=0}^{n} A_{i1} = \sum_{i=0}^{n} \tilde{x}_{j+i} \cdot A_{i1} ,$$

where the A_{i1} are the obvious cofactors. It is beyond the scope of this talk to give a detailed analysis and comparison of the different sources of error or to provide exhaustive proofs.

We conclude by giving the result of the extrapolation in case $\alpha = \mu = -\frac{1}{3}$, and $n = 1$ for the previously given example.

From

x_j	$\tilde{x}_j$	$\lvert b_j \rvert$
0.1640625	0.178871	0.014808
0.171875	0.178825	0.006950

we find $\tilde{x}^* = 0.178807 \dots$ as compared to $x^* = 0.178796 \dots$.

The same formal extrapolation technique was already used in [6] for the case of splines in which the exponent α was one of the spline parameters,

$$\tilde{u}(z;x,u,u',b,\alpha') = u + u'\frac{b}{\alpha}[(1+\frac{z}{b})^{\alpha}-1]$$

for $b \neq 0$, $\alpha \neq 0,1$. There the numerical results for the above treated initial value problem and this class of splines are given.

REFERENCES

[1] Arndt, H., Lösung von gewöhnlichen Differentialgleichungen mit nichtlinearen Splines, Num. Math. 33, 323-333 (1979).

[2] Hille, E., Ordinary Differential Equations in the Complex Domain, J. Wiley & Sons, New York - London - Sydney - Toronto (1976).

[3] Loscalzo, F. R. and Talbot, T. D., Spline Function Approximat-ions for Solutions of Ordinary Differential Equations, SIAM J. Numer. Anal. 4, 433-445 (1967).

[4] Werner, H., An Introduction to Non Linear Splines, in: Polynomial and Spline Approximation, ed. by B. N. Sahney, Reidel Publ. Co. Dordrecht 1979, (this article contains many references).

[5] Werner, H., Extrapolationsmethoden zur Bestimmung der beweglichen Singularitäten von Lösungen gewöhnlicher Differentialgleichungen, in: Numerische Mathematik, ed. by R. Ansorge - K. Glashoff - B. Werner, ISNM 49, Birkhäuser Verlag, Basel 1979.

[6] Werner, H., The Development of Non Linear Splines and Their Applications, in: Approximation Theory III, ed. by W. Cheney, Academic Press, New York - London - Toronto - Sydney - San Francisco 1980, 125 - 150.

ON THE CONDITIONING OF THE PADE APPROXIMATION PROBLEM

LUC WUYTACK
Department of Mathematics
University of Antwerp
Universiteitsplein 1
B-2610 WILRIJK
(Belgium)

ABSTRACT

Several aspects of the conditioning of the Padé approximation problem are considered. The first is concerned with the operator that associates a power series f with its Padé approximant of a certain order. It is shown that this operator satisfies a local Lipschitz condition, in case the Padé approximant is normal.

The second aspect is the conditioning of the Padé approximant itself. It is indicated how this rational function r should be represented such that changes in its coefficients will effect changes on r as less as possible. A "condition number" for this problem is introduced.

The third aspect is the problem of the representation of the Padé approximant, such that the determination of its coefficients be a well-conditioned problem. It is known that the choice of powers of x as base functions can result in an ill-conditioned problem for the determination of the coefficients. The possibility of using other base functions is analysed.

I. INTRODUCTION.

Let f be a given series or $f(x)=c_0+c_1.x+c_2.x^2+\dots$ with $c_0\neq 0$. Let $R_{m,n}$ be the class of ordinary rational functions $r=\frac{p}{q}$, where p and q are polynomials of degree at most m and n respectively, such that $\frac{p}{q}$ is irreducible.

The Padé approximation problem consists in finding an element $r=\frac{p}{q}$ in $R_{m,n}$ such that

$$f(x).q(x)-p(x)=O(x^{m+n+1+j}), \tag{1}$$

where j is an integer, which is as large as possible.

It is known [11] that there exists a unique solution to this problem. It is called the Padé approximant for f of order (m,n) and it will be denoted by $r_{m,n}=\frac{p_{m,n}}{q_{m,n}}$. We assume that the representation of $r_{m,n}$ is normalized such that $q_{m,n}(0)=1$.

In case of a normal Padé approximant we have j=0 in (1) and, moreover, $p_{m,n}$ and $q_{m,n}$ have degree m and n exactly.

II. LIPSCHITZ CONTINUITY OF THE PADE OPERATOR.

Let T be the operator which associates $r_{m,n}$ to f, for fixed values of m and n, or $Tf=r_{m,n}$. In [3] it was conjectured that T satisfies a local Lipschitz condition. This conjecture will be proved here in the real case. An extension to the complex case seems also to be possible.

Let $r_{m,n}=\frac{p}{q}$ then the normalization $q(0)=1$ implies the existence of a finite interval [a,b] about the origin, such that $q(x)>0$ for all x in [a,b]. Let $c=[c_0, c_1, c_2, \dots, c_{n+m}]$ and

$$\|c\|=\max_{1\le i\le n+m} |c_i|.$$

The norm $\|g\|$ of a continuous function g on [a,b] is defined as follows (denoting both norms by $\|.\|$, causes no ambiguities),

$$\|g\|=\max_{x\in[a,b]} |g(x)|.$$

__Theorem.__ Let $r_{m,n}=\frac{p}{q}$ be normal and $q>0$ in [a,b]. There exist constants K and d (only depending on $c_0, c_1, \dots, c_{n+m}$ and [a,b], such that for every f', with $f'(x)=\sum_{i=0}^{\infty} c_i' x^i$ and $\|c-c'\|\le d$,

$$\|Tf-Tf'\|\le K.\|c-c'\|.$$

__Proof.__ Let D' be the determinant of the matrix M' defined as follows

$$M' = \begin{bmatrix} c'_m & c'_{m-1} & \cdots & c'_{m-n+1} \\ c'_{m+1} & c'_m & \cdots & c'_{m-n+2} \\ \cdot & \cdot & & \cdot \\ \cdot & \cdot & & \cdot \\ \cdot & \cdot & & \cdot \\ c'_{m+n-1} & c'_{m+n-2} & \cdots & c'_m \end{bmatrix}$$

Since Tf is normal there exists a $d_0>0$ such that Tf' is also normal for every f' with $\|c-c'\|\le d_0$ (see [12]). The normality of Tf' implies that $D'\neq 0$. If $Tf' = \frac{p'}{q'}$, then p' and q' can be represented as follows [12, p. 430] :

$$p'(x) = \frac{1}{D'}\cdot \begin{vmatrix} f'_m(x) & x.f'_{m-1}(x) \quad \cdots \quad x^n.f'_{m-n}(x) \\ \hline c'_{m+1} & \\ c'_{m+2} & \\ \cdot & M' \\ \cdot & \\ \cdot & \\ c'_{m+n} & \end{vmatrix}$$

and

$$q'(x) = \frac{1}{D'}\cdot \begin{vmatrix} 1 & x \quad x^2 \quad \cdots \quad x^n \\ \hline c'_{m+1} & \\ c'_{m+2} & \\ \cdot & M' \\ \cdot & \\ \cdot & \\ c'_{m+n} & \end{vmatrix}$$

where $f'_k(x)=\sum_{i=0}^{k} c'_i.x^i$ for every $k\ge 0$ and $f'_k(x)=0$ if $k<0$. Let the coefficient of x^i in p' and q' be denoted by $A_i(c')$ and $B_i(c')$ respectively, then these are differentiable functions in c', hence there exists a constant L such that

$$|A_i(c)-A_i(c')|\le L.\|c-c'\| \text{ and}$$
$$|B_i(c)-B_i(c')|\le L.\|c-c'\|,$$

for every c' with $\|c-c'\|\le d_0$. Consider now

$$|p(x)-p'(x)|\le \sum_{i=0}^{m} |A_i(c)-A_i(c')|\cdot|x^i| \text{ and}$$

$$|q(x)-q'(x)|\leq\sum_{i=0}^{n}|B_i(c')-B_i(c')|.|x^i|,$$

for every x in [a,b]. The above relations and the finiteness of [a,b] imply the existence of a constant M such that

$$\|p-p'\|\leq M.\|c-c'\| \text{ and} \tag{2}$$

$$\|q-q'\|\leq M.\|c-c'\|, \tag{3}$$

for every c' with $\|c-c'\|\leq d_0$. Since q>0 in [a,b] and q' is a continuous function of c' with $q'_c=q$, we get the existence of constants $d\leq d_0$ and E such that $q'_{c'}\geq E$ in [a,b] for every c' with $\|c-c'\|\leq d$. This implies the existence of a constant $N=\frac{1}{E}$ such that

$$\|\frac{1}{q'}\|\leq N \tag{4}$$

for every c' with $\|c-c'\|\leq d$. Since

$$\frac{p}{q}-\frac{p'}{q'}=\frac{(p-p').q+(q'-q).p}{q.q'}$$

we get, from (2), (3) and (4)

$$\|Tf-Tf'\|\leq\|\frac{1}{q}\|.(\|q\|+\|p\|).M.N.\|c-c'\|.$$

Hence there exists a constant K such that

$$\|Tf-Tf'\|\leq K.\|c-c'\| \tag{5}$$

for every c' with $\|c-c'\|\leq d$.

The above proof uses similar arguments as the proof of a related result for rational interpolation (see [10], p. 297). In the case of Chebyshev approximation it was even possible to prove that the Chebyshev operator (defined in the same way as the Padé operator) is continuous if and only if f is normal or belongs to $R_{m,n}$ (see [14]).

The above theorem implies the continuity of the Padé operator in case $r_{m,n}$ is normal. If also the inverse is true seems to be an open problem. Other aspects of the continuity of Padé approximants and related approximants are treated e.g. in [5] and [6].

Remark that the above result does not hold if $r_{m,n}$ is not normal, which means that normality is essential for the local Lipschitz continuity of the Padé operator T. This remark will be illustrated in the following example, which was suggested by Al. Magnus.

Example. Let $f(x)=1+c.x+x^2$ and m=n=1. Then

$$r_{1,1}(x)=\frac{1+(c-\frac{1}{c}).x}{1-\frac{1}{c}.x} \quad \text{if } c\neq 0,$$

and $r_{1,1}(x)=1$ if $c=0$.

Let $r(x)=1$ and $r_c(x)=\frac{c+(c^2-1).x}{c-x}$, then $\lim_{c\to 0} r_c(x)=r(x)$ for every x. However there exists no interval $[a,b]$ such that $\|r_c-r\|$ remains bounded if $c\to 0$.

The practical meaning of the theorem is as follows. Assume that the normal Padé approximant r for f of order (m,n) has been computed and that the computed result is r'. Assume also that r' can be considered as the exact Padé approximant of order (m,n) for $f'(x)=\sum_{i=0}^{\infty} c'_i.x^i$ with $\|c-c'\|\leq d$, where d depends on the precision used during the computation. The above theorem then says that r' will be "close" to r if d is small. In case r is not normal, then this might not be true.

III. THE CONDITIONING OF RATIONAL FUNCTIONS.

1. Statement of the problem.

Since every Padé approximant is a rational function, the problem of the conditioning of rational functions can be of interest. This problem can be stated as follows. Consider a rational function

$$r=\frac{p}{q}=\frac{\sum_{i=0}^{m} a_i.p_i}{\sum_{i=0}^{n} b_i.q_i}$$

where p_i and q_i are polynomials of degree at most i. If we now make "small" changes δa_i, δb_i in the coefficients $a_i.b_i$, what is then the effect on r, or what is the difference between r and r' where

$$r'=\frac{p'}{q'}=\frac{\sum_{i=0}^{m}(a_i+\delta a_i).p_i}{\sum_{i=0}^{n}(b_i+\delta b_i).q_i}$$

Connected with this conditioning problem is the following practical problem. Since every rational function can be represented in different ways, depending on the choice of p_i, q_i, it is important to know what choice is "best possible". This means that it would be interesting to know for what choice of base functions p_i, q_i the effect on r of changes in the coefficients is "as small as possible".

Several kinds of approaches can be followed to attack this problem. First there is an experimental approach as e.g. given by J.R. Rice in [13]. One can compute the difference r-r' for certain choices of p_i, q_i, δa_i, δb_i. Based on these computational results one can then make some statements about the conditioning of the rational functions under consideration.

Another type of approach is more analytic and based on a first order perturbation analysis. This approach is followed e.g. by W. Kammler and R.J. McGlinn in [9] for the problem of the conditioning of nonlinear parametric forms.

The approach that will be followed here is a generalization of a technique used for polynomials by C. De Boor in [4], p. 17-19. The application of this technique to rational functions will result in the definition of a "condition number" for rational functions, similar to the "condition number" for polynomials. If this number is small then the effect of a relative change in the coefficients will also be small.

2. A condition number for rational functions.

Let $r=\frac{p}{q}$, with $p=\sum_{i=0}^{m} a_i.p_i$ and $q=\sum_{i=0}^{n} b_i.q_i$, where p_i, q_i are polynomials of degree at most i and r be continuous in the closed interval I. The norms used in this section are maximum-norms or

$$\|a\|=\max_{0\le i\le k}|a_i| \quad \text{for every } a=[a_0, a_1, \ldots, a_k]^T \text{in } R^{k+1},$$

$$\|b\|=\max_{x\in I}|g(x)| \quad \text{for every } g \text{ in } C(I).$$

Let W be a subspace of C(I) generated by the functions $w_0, w_1, \ldots, w_k$ then we define m_W, M_W, n_W and N_W as follows :

$$m_W=\min_c \frac{\|\sum_{i=0}^{k} c_i.w_i\|}{\|c\|} \qquad M_W=\max_c \frac{\|\sum_{i=0}^{k} c_i.w_i\|}{\|c\|}$$

$$n_W=\min_c\left(\frac{\min_{x\in I}|\sum_{i=0}^{k} c_i.w_i(x)|}{\|c\|}\right) \qquad N_W=\max_c\left(\frac{\min_{x\in I}|\sum_{i=0}^{k} c_i.w_i(x)|}{\|c\|}\right)$$

Let P = span $\{p_0, p_1, \ldots, p_m\}$, Q = span $\{q_0, q_1, \ldots, q_n\}$ then these definitions imply, for every a in R^{m+1},

$$m_P.\|a\|\le\|\sum_{i=0}^{m} a_i.p_i\|\le M_P.\|a\|. \tag{6}$$

We also have, for every b in R^{n+1},

$$m_Q.\|b\|\le\|\sum_{i=0}^{n} b_i.q_i\|\le M_Q.\|b\|,$$

$$n_Q\cdot\|b\|\le\min_{x\in I}\left|\sum_{i=0}^{n} b_i\cdot q_i(x)\right|\le N_Q\cdot\|b\|. \tag{7}$$

Since $r=\frac{p}{q}$ we have

$$\frac{\|p\|}{\|q\|}\le\|r\|\le\|p\|\cdot\frac{1}{\min_{x\in I}|q(x)|}\ . \tag{8}$$

Combining (6), (7) and (8) we get

$$\frac{m_P}{M_Q}\cdot\frac{\|a\|}{\|b\|}\le\|r\|\le\frac{M_P}{n_Q}\cdot\frac{\|a\|}{\|b\|}. \tag{9}$$

Consider now $r'=\frac{p'}{q'}=\frac{\sum_{i=0}^{m} a'_i\cdot p_i}{\sum_{i=0}^{n} b'_i\cdot q_i}$, where $a'_i=a_i+\delta a_i$ and $b'_i=b_i+\delta b_i$,

then $$r'-r=\frac{\delta p-r.\delta q}{q'}, \tag{10}$$

with $\delta p=\sum_{i=0}^{m}\delta a_i.p_i$ and $\delta q=\sum_{i=0}^{n}\delta b_i.q_i$. We will now derive an upper and lower bound for $\|r'-r\|$ in two different situations. Let $V=P+rQ=\text{span}\,\{p_0, p_1, \dots, p_m, r.q_0, r.q_1, \dots, r.q_n\}$.

2.a. <u>first order analysis.</u>

Neglecting higher order perturbation terms in (10) we get

$$r'-r=\frac{\delta p-r.\delta q}{q}$$

Using the same arguments as for deriving (9), with P replaced by V, we get

$$\frac{m_V}{M_Q}\cdot\frac{\|\delta c\|}{\|b\|}\le\|r'-r\|\le\frac{M_V}{n_Q}\cdot\frac{\|\delta c\|}{\|b\|} \tag{11}$$

where $\delta c=[\delta a_0, \delta a_1, \dots, \delta a_m, \delta b_0, \delta b_1, \dots, \delta b_n]^T$. Combining (9) and (11) we get $\frac{m_V}{M_P}\cdot\frac{n_Q}{M_Q}\cdot\frac{\|\delta c\|}{\|a\|}\le\frac{\|r'-r\|}{\|r\|}\le\frac{M_V}{m_P}\cdot\frac{M_Q}{n_Q}\cdot\frac{\|\delta c\|}{\|a\|}$

Since $P\subset V$ we have $M_P\le M_V$ and $m_V\le m_P$ and this implies

$$\frac{m_V}{M_V}\cdot\frac{n_Q}{M_Q}\cdot\frac{\|\delta c\|}{\|a\|}\le\frac{\|r'-r\|}{\|r\|}\le\frac{M_V}{m_V}\cdot\frac{M_Q}{n_Q}\cdot\frac{\|\delta c\|}{\|a\|} \tag{12}$$

These inequalities imply that $\frac{M_V}{m_V}\cdot\frac{M_Q}{n_Q}$ can be regarded as a (first order) condition number for the rational function r. The larger this condition number, the greater the effect of a "small" relative change in the coefficients of r can be.

2.b. Exact analysis.

Using (10) and the same technique as for deriving (11) we get

$$\frac{m_V}{M_Q}\cdot\frac{\|\delta c\|}{\|b'\|}\le\|r'-r\|\le\frac{M_V}{n_Q}\cdot\frac{\|\delta c\|}{\|b'\|}\ .$$

Combining this result with (9) and using $M_P \le M_V$ and $m_V \le m_P$ we get

$$\frac{m_V}{M_V}\cdot\frac{n_Q}{M_Q}\cdot\frac{\|\delta c\|}{\|a\|}\cdot\frac{\|b\|}{\|b'\|}\le\frac{\|r'-r\|}{\|r\|}\le\frac{M_V}{m_V}\cdot\frac{M_Q}{n_Q}\cdot\frac{\|\delta c\|}{\|a\|}\cdot\frac{\|b\|}{\|b'\|} \qquad (13)$$

This result reduces to (12) if $\frac{\|b\|}{\|b'\|}$ is replaced by 1. We remark that also in this case the condition number can be defined as

$$\frac{M_V}{m_V}\cdot\frac{M_Q}{n_Q}$$

2.c. Choice of base functions.

As a result of the above analysis it is favourable to take the base functions p_i, q_i in such a way that the condition number is as small as possible. This means that p_i, q_i must be taken such that M_Q, M_V are as small as possible and n_Q, m_V as large as possible.

If we assume $q>o$ in I we could take n_Q to be a lower bound for q on I. The requirement that n_Q should be as large as possible indicates that one should stay away from zero as far as possible.

In order to make M_Q and $\frac{M_V}{m_V}$ as small as possible it seems to be of interest not to use $p_i(x)=q_i(x)=x^i$ as base functions, since it is known from the polynomial case (see [4], p. 18 and [9], p. 843) that this choice results in large condition numbers. It is an open question for what choices of base functions we will get the smallest condition number.

Remark also that in the above analysis we did not use the fact that p_i, q_i are assumed to be polynomials. The same results hold if $\{p_i\},\{q_i\}$ are any set of base functions, different from polynomials.

IV. CONDITIONING OF THE SYSTEM FOR DETERMINING THE COEFFICIENTS.

4.1. Statement of the problem.

Let $r_{m,n}=\frac{p}{q}$, with $p=\sum_{i=o}^{m} a_i.x^i$ and $q=\sum_{i=o}^{n} b_i.x^i$, be normal then it follows from (1) that the coefficients a_i, b_i have to satisfy the following linear system of equations :

$$\left.\begin{array}{ll} c_0.b_0 & = a_0 \\ c_1.b_0+c_0.b_1 & = a_1 \\ \text{-----------------} & \\ c_m.b_0+c_{m-1}.b_1+\ldots+c_{m-n}.b_n & = a_m \end{array}\right\} \qquad (14.a)$$

$$\left.\begin{array}{ll} c_{m+1}.b_0+c_m.b_1+\ldots+c_{m-n+1}.b_n & = o \\ \text{-----------------} & \\ c_{m+n}.b_0+c_{m+n-1}.b_1+\ldots+c_m.b_n & = o \end{array}\right\} \qquad (14.b)$$

where c_i=o if i<o. The coefficients $b_0, b_1, \ldots, b_n$, with b_0=1, can be found by solving (14.b), which is a linear system with a Toeplitz matrix. The coefficients $a_0, a_1, \ldots, a_m$ can then be found from (14.a).

In some recent papers A. Bultheel [1], [2] showed that most of the algorithms for computing Padé approximants can be seen as algorithms for factorizing the Toeplitz or Hankel matrix, connected with the system (14.b).

As a consequence of the normality of $r_{m,n}$ the system (14.b) has a unique solution if the coefficients are changed "slightly" (see [3]). The system will therefore be well-conditioned if the condition number of its matrix of coefficients is not too large ([8], p. 38). There are however examples for which the condition number of the matrix of system (14.b) is very large ([7], p. 242).

The problem that will be considered here is the effect of another choice of base functions in p and q, such that the determination of its coefficients will be a well-conditioned problem.

4.2. Derivation of the system of equations.

Let us define the Toeplitz matrix M as follows

$$M = \begin{bmatrix} c_{m+1} & c_m & \cdots & c_{m-n+1} \\ c_{m+2} & c_{m+1} & \cdots & c_{m-n+2} \\ \vdots & \vdots & & \vdots \\ c_{m+n} & c_{m+n-1} & \cdots & c_m \end{bmatrix} \quad \text{and } b = \begin{bmatrix} b_0 \\ b_1 \\ \vdots \\ b_n \end{bmatrix}$$

then (14.b) can be written in the form

$$M.b=o \qquad (15)$$

Consider now $p=\sum_{i=0}^{m} a_i'.p_i$, $q=\sum_{i=0}^{n} b_i'.q_i$ where p_i, q_i are polynomials of degree i. This implies the existence of coefficients $c_{i,j}$

such that

$$\begin{bmatrix} q_0 \\ q_1 \\ \vdots \\ q_n \end{bmatrix} = \begin{bmatrix} c_{0,0} & o & \cdots & o \\ c_{1,0} & c_{1,1} & \cdots & o \\ \vdots & \vdots & & \vdots \\ c_{n,0} & c_{n,1} & \cdots & c_{n,n} \end{bmatrix} . \begin{bmatrix} 1 \\ x \\ \\ x^n \end{bmatrix} \quad \text{or } \bar{q}=C.\bar{x}$$

We will now try to find the conditions for $b_0', b_1', \ldots, b_n'$ such that (1) holds.

Since $q=\sum_{i=0}^{n} b_i.x^i=[b]^T.\bar{x}$ we get $q=[b]^T.C^{-1}.\bar{q}$. We also have

$q=\sum_{i=0}^{n} b_i'.q_i$ or $q=[b']^T.\bar{q}$. This implies

$$[b]^T.C^{-1}=[b']^T \text{ or } [b]^T=[b']^T.C \text{ or } b=C^T.b'.$$

Consequently (15) implies $M.C^T.b'=o$. Using the normalization $b_0'=1$ we get the following linear system of n equations in the n unknowns $b_1', b_2', \ldots, b_n'$:

$$M^*.C^T.b^*=d^*, \tag{16}$$

where $b^*=[b_1', b_2', \ldots, b_n']^T$, d^* is the opposite of the first column of the matrix $M.C^T$ and M^* is equal to M without its first column.

4.3. Choice of base functions.

As a result of the preceeding analysis we get that the problem of finding $b_1', b_2', \ldots, b_n'$ will be well-conditioned if $M^*.C^T$ has a small condition number. How $q_0, q_1, \ldots, q_n$ must be chosen in practice such that this condition holds is an open question. Here we will try to find the choice of base functions such that (16) is a triangular or diagonal system.

(a) choice of base functions such that (16) has triangular form

Let us define M' as follows

$$M' = \begin{bmatrix} c_m & c_{m-1} & \cdots & c_{m-n} \\ \hline & M & & \end{bmatrix}$$

then M' is a square matrix of Toeplitz form. In order to find the matrix C such that $M^*.C^T$ is lower triangular we first try to find C such that

$$M'.C^T=L, \tag{17}$$

where L is a lower triangular matrix. Since by omitting the first row and column in $M'.C^T$ we get $M^*.C^T$, this matrix will then also be lower triangular. The condition (17) implies $M'=L.(C^T)^{-1}$ or M' is the product of a lower and an upper triangular matrix. Using Bultheel's result [1] we get that the elements $c_{i,j}$ must be the coefficients of the denominaters in a row in the Padé table for f. This implies that in order to make $M^{\star}.C^T$ of lower triangular form we must take q_i to be the denominators of Padé approximants in a row in the Padé table. This however is exactly the problem that must be solved, which means that in order to make $M^{\star}C^T$ of lower triangular form we have to know the solution of (15).

(b) choice of base functions such that (16) has diagonal form

In this case C must be chosen such that $M'.C^T=D$ where D is a diagonal matrix. This implies that $M'=D.(C^T)^{-1}$ or M' must have upper triangular form. This condition means $c_{m-i}=0$ for $i=1, 2, \ldots, n$. The result is that (16) will have diagonal form only in a very special situation. Remark that it is also possible to take C a full matrix or q_i to be a polynomial of degree at most n. The condition $M'.C^T=D$ then implies that the computation of C is related to the computation of the inverse of a Toeplitz matrix. This also requires that the solution of (15) must be known before (16) can be put into diagonal form.

4.4. Remarks.

Certain classes of algorithms which compute Padé approximants in the form of a continued fraction are also recursive in nature and related to the factorization of a Toeplitz or Hankel matrix ([1]). The condition of this problem is therefore related to the above considered problem.

It is an open question whether the choice of base functions which make (16) well-conditioned also represent the Padé approximant in such a form that its condition number (see section III) is small.

REFERENCES.

1 BULTHEEL A. : Fast algorithms for the factorization of Hankel and Toeplitz matrices and the Padé approximation problem. Report TW 42, Applied Mathematics and Programming Division, University of Leuven, 1978.

2 BULTHEEL A. : Recursive algorithms for the matrix Padé problem. Mathematics of computation 35 (1980), pp. 875-892.

3 BULTHEEL A., WUYTACK L. : Stability of numerical methods for computing Padé approximants. In the Proceedings of the Austin Conference on Approximation Theory (Cheney E.W., ed.), Academic Press, 1980.

4 DE BOOR C. : A practical Guide to Splines. Springer Verlag, Berlin, 1978.

5 FOSTER L.V. : The convergence and continuity of rational functions closely related to Padé approximants. Journal of Approximation Theory 28 (1980), pp. 120-131.

6 GALLUCCI M.A. and JONES W.B. : Rational approximations corresponding to Newton series (Newton-Padé approximants). Journal of Approximation Theory 17 (1976), pp. 366-392.

7 GRAVES-MORRIS P. : The numerical calculation of Padé approximants. In "Padé approximation and its Applications" (ed. L. Wuytack, Springer Verlag, Berlin, 1979), pp. 231-245.

8 ISAACSON E. and KELLER H.B. : Analysis of numerical methods. J. Wiley, New York, 1966.

9 KAMMLER D.W. and McGLINN R.J. : Local conditioning of parametric forms used to approximate continuous functions. American Mathematical Monthly 86 (1979), pp. 841-845.

10 MAEHLY H. und WITZGALL Ch. : Tschebyscheff-Approximationen in kleinen Intervallen II. Numerische Mathematik 2 (1960), pp. 293-307.

11 PADE H. : Sur la représentation approchée d'une fonction par des fractions rationnelles. Annales Scientifiques de l'Ecole Normale Supérieure de Paris 9 (1892), pp. 1-93.

12 PERRON O. : Die Lehre von den Kettenbrüchen. Teubner, Stuttgart, 1929.

13 RICE J.R. : On the conditioning of polynomial and rational forms. Numerische Mathematik 7 (1965), pp. 426-435.

14 WERNER H. : On the rational Tschebyscheff operator. Mathematisches Zeitschrift 86 (1964), pp. 317-326.

PADE-APPROXIMATIONS IN NUMBER THEORY

F. Beukers
Mathematisch Instituut
Rijksuniversiteit Leiden
Wassenaarseweg 80
2333 AL Leiden
The Netherlands

INTRODUCTION. In 1873 Hermite [H] was the first to construct explicit simultaneous Padé-approximations to the system of functions $1, e^z, e^{2z}, \ldots, e^{nz}$ and discovered the transcendence of e. Later Lindemann [L] in 1882 extended Hermite's work to show that π is transcendental, thus providing the negative answer to the ancient problem of squaring the circle. An elegant exposition of these methods can be found in Siegel [Si1], Chapter I. The work of Hermite-Lindemann was largely extended by the work of Siegel and later Shidlovski on the algebraic independence of values of so-called E-functions. In this extension however, the authors use non-explicitly constructed rational approximations and we shall not proceed along these lines. We would like to refer interested readers to [Ba2], Chapter 11.

Very recently there has been an upsurge of interest in the use of Padé-approximations in irrationality and transcendence proofs, stimulated by Apéry's remarkable irrationality proof for $\zeta(3) = 1^{-3} + 2^{-3} + 3^{-3} + \ldots$. Literature on this proof can be found in R. Apéry [A], A.J. van der Poorten [P], E. Reyssat [R] and F. Beukers [Be1]. In attempts to generalize Apéry's method the role played by Padé-approximations in irrationality theory began to be re-appreciated. We first define what we mean by Padé-approximation.

Let $f_1, f_2, \ldots, f_k$ be a system of functions analytic around $z = 0$ and suppose $f_1(0) \neq 0$. We distinguish two kinds of Padé-approximations (see K. Mahler [M], H. Jager [J]) type I : polynomials $P_1(z),\dots,P_k(z)$ of degree $n_1,\dots,n_k$ such that

$$P_1(z)f_1(z) + \dots + P_k(z)f_k(z) = O(z^{N+k-1})$$

where $N = \Sigma n_i$

type II: polynomials $P_1(z),\dots,P_k(z)$ of degree $N-n_1,\dots,N-n_k$ with $\Sigma n_i = N$, such that

$$P_i(z)f_j(z) - P_j(z)f_i(z) = O(z^{N+1}) \qquad i,j=1,\dots,k .$$

Notice that if $k = 2$, then both types coincide, and we have in fact the classical Padé-table of the function $f_2(z)/f_1(z)$.

In the following we shall use the abbreviation P.A. for Padé-approximation.

G.V. Chudnovsky [C1] [C2] has constructed a very wide class of explicit type I and II-approximations of systems of generalised hypergeometric functions, which can be applied to obtain irrationality-results. To quote a few of them,

1) $\Gamma(\frac{1}{4})^4/\pi^2$ is irrational, [C2]

2) $\mathrm{dil}(a^{-1}) \notin \mathbb{Q}$ for $a \in \mathbb{Z}$, $|a| \geq 14$, [C1] [C2]

where

$$\mathrm{dil}(z) = \frac{z}{1^2} + \frac{z^2}{2^2} + \frac{z^3}{3^2} + \dots$$

3) $|\pi - \frac{p}{q}| > q^{-19.89}$ for all $\frac{p}{q} \in \mathbb{Q}$, $q > q_0$. [C1]

Also, the importance of Chudnovsky's work on the theoretical side of approximation theory should be stressed [C3] [C4]. One of the main features is the close connection that exists between P.A.'s to systems of functions satisfying a linear differential equation and the monodromy group of this differential equation.

In Section 1 of this note we give an impression of the applications of Padé-fractions in irrationality theory by showing, $e^a \notin \mathbb{Q}$ for $a \in \mathbb{Q}$, $a \neq 0$.

In Section 2 we will review some of the results that have been obtained by application of P.A.'s of $(1-z)^{1/n}$ to some diophantine equations.

Despite the interest in P.A.'s that Apéry's irrationality proof for $\zeta(3)$ has aroused, it was hitherto unclear how to formulate Apéry's proof naturaly in terms of P.A.'s. In Section 3 we indicate how this might be achieved, although we must extend our définition of Padé-approximation a little.

SECTION 1.

THEOREM 1. *Let* $a \in \mathbb{Q}$, $a \neq 0$. *Then* e^a *is irrational.*

PROOF. Notice that it is sufficient to prove this theorem for $a \in \mathbb{N}$. The proof for $a \in \mathbb{Q}$ then follows easily by noticing that $(e^a)^{\mathrm{den}(a)} \notin \mathbb{Q}$ (where den(a) = denominator of a) and so we certainly have $e^a \notin \mathbb{Q}$. The $[n,n]$ P.A. of e^z can be found as follows. Consider

$$(1) \qquad I_n(z) = z^{n+1} \int_0^1 e^{zt} P_n(t)dt$$

where $P_n(t)$ is the Legendre polynomial defined by $P_n(t) = \frac{1}{n!} (\frac{d}{dt})^n t^n(1-t)^n$. Notice that degree $P_n(t) = n$ and $P_n(t) \in \mathbb{Z}[t]$. By repeated partial integration we obtain,

$$(2) \qquad I_n(z) = (-1)^n \frac{z^{2n+1}}{n!} \int_0^1 e^{zt} t^n(1-t)^n dt .$$

On the other hand, it is straightforward to see that

$$z^{n+1} \int_0^1 t^m e^{zt} dt = z^{n-m} \int_0^z e^t t^m dt$$

$$= Q_n(z)e^z + (-1)^{m+1} m! \, z^{n-m}$$

where $Q_n(z) \in \mathbb{Z}[z]$ has degree n. Therefore, term by term integration of (1) yields

$$(3) \qquad I_n(z) = A_n(z) + B_n(z)e^z ,$$

where $A_n(z), B_n(z) \in \mathbb{Z}[z]$ have degree $\leq n$.

We now substitute $z = a$. From (1) it is easy to see that $I_n(a) \neq 0$. Suppose $e^a = p/q \in \mathbb{Q}$, then (3) yields

$$\frac{1}{q} \leq |A_n(a)+B_n(a)\frac{p}{q}| = |I_n(a)| ,$$

which, for sufficiently large n, is in contradiction with the upper bound we obtain from (2),

$$|I_n(a)| < \frac{a^{2n+1}}{n!} e^a.$$

Hence e^a is irrational.

With a similar method it is also possible to show the irrationality of π^2 and the zeros of Bessel-functions of integer order, see [Be2]. We can also show the irrationality of $\log 2$ and $\pi/\sqrt{3}$ by using the Padé-table for $\log(1-z)$ and then substituting $z = -1$ and $z = e^{\pi i/3}$ respectively. Moreover, by refining the arguments in this case we can show theorems of the following type,

THEOREM 2. *For every* $\varepsilon > 0$ *there exist explicitly calculable numbers* $q_0(\varepsilon), q_1(\varepsilon)$ *such that*

$$|\log 2 - \frac{p}{q}| > |q|^{-4.660137\ldots-\varepsilon} \quad \textit{for} \quad |q| > q_0(\varepsilon)$$

$$|\frac{\pi}{\sqrt{3}} - \frac{p}{q}| > |q|^{-8.30998\ldots-\varepsilon} \quad \textit{for} \quad |q| > q_1(\varepsilon) .$$

For a clear derivation of these irrationality measures, see [A-R]. They were found independently by G.V. Chudnovsky [C1], and several others.

SECTION 2. In 1964 A. Baker [Ba] used P.A.'s to $(1-z)^{1/3}$ in order to prove

THEOREM 3. *For any* $\frac{p}{q} \in \mathbb{Q}$ *we have*

$$\left|\frac{p}{q} - \sqrt[3]{2}\right| > \frac{10^{-6}}{q^{2.955}} .$$

Multiplication of this inequality with $q(p^2+pq\sqrt[3]{2}+q^2\sqrt[3]{4})$ yields

$$|p^3 - 2q^3| > 10^{-6}\, q^{0.045} \quad \text{for any } p,q \in \mathbb{N} .$$

This implies that the diophantine equation $x^3 - 2y^3 = k$ (k given integer) has only finitely many solutions. Moreover, if x,y is a solution then $|y| < 10^{138}\, |k|^{23}$. Now let a,b,c,n be given integers with $n \geq 3$. In general we can use P.A.'s to $(1-z)^{1/n}$ in order to study the diophantine equation

$$(4) \qquad ax^n - by^n = c$$

in the unknown integers x,y . It is only possible however to give upper bounds for the number of solutions of (4) and not for the size of the solutions. In 1937 C.L. Siegel [Si2] was the first to study equation (4) in this way. By elaborating Siegel's methods one can show that if $n \geq 5$ and $c = 1$, equation (4) has at most 2 solutions (with $x,y \geq 0$ if n is even). See [D]. Very recently J. Evertse showed that if c is a prime-power then there are at most $2n + 6$ solutions (private communication).

In 1977 the author, using P.A.'s to $\sqrt{1-z}$, obtained,

THEOREM 4. *For any* $x, r \in \mathbb{N}$ *we have*

$$\left|\frac{x}{2^r} - \sqrt{2}\right| > \frac{2^{-43.9}}{2^{1.8r}} .$$

Multiplication of this inequality with $2^{2r}(\sqrt{2}+x2^{-r})$ yields

$$|x^2 - 2^{2r+1}| > 2^{0.2r}\, 2^{-43.4}$$

from which we easily derive

COROLLARY. *Let* $D \in \mathbb{Z}$ *and let* $x, n \in \mathbb{N}$ *be a solution of the diophantine equation* $x^2 + D = 2^n$. *Then* $n < 435 + 10 \log|D|/\log 2$.

As a consequence we see that for given $D \in \mathbb{Z}$ the diophantine equation $x^2 + D = 2^n$ can be solved in finitely many steps. Moreover, after some technical considerations it is possible to show that $x^2 + D = 2^n$ has at most four solutions, unless $D = 7$ in which case the solutions read $(x,n) = (1,3),(3,4),(5,5),(11,7),(181,15)$. All this can be found in [Be3].

SECTION 3. The by now traditional way to prove the irrationality of $\zeta(3)$ can be sketched as follows. Define

$$a_n = \sum_{k=0}^{n} \binom{n}{k}^2 \binom{n+k}{k}^2 .$$

Then there exist numbers $b_n \in [1,\ldots,n]^{-3}\mathbb{Z}$ (here $[1,\ldots,n]$ denotes the lcm.) with

$$(5) \qquad 0 < |a_n - b_n\zeta(3)| < 3(\sqrt{2}-1)^{4n} .$$

If $\zeta(3)$ were rational, say p/q then $|a_n - b_n\zeta(3)| \geq q^{-1}[1,\ldots,n]^{-3}$, contradicting the upper bound in (5) for n sufficiently large. For full details, see [R] or [Be1].

We will now show how the numbers a_n and b_n can be derived from Padé-type approximations. Define

$$L_k(z) = \frac{z}{1^k} + \frac{z^2}{2^k} + \frac{z^3}{3^k} + \dots .$$

Notice that $L_2(1) = \zeta(2)$ and $L_3(1) = \zeta(3)$. We look for polynomials $A_n(z), B_n(z), C_n(z), D_n(z)$ of degree n such that

$$
\text{(6)} \quad
\begin{aligned}
A_n(z)L_2(z) + B_n(z)L_1(z) + C_n(z) &= O(z^{2n+1}) \\
2A_n(z)L_3(z) + B_n(z)L_2(z) + D_n(z) &= O(z^{2n+1})
\end{aligned}
$$

and $B_n(1) = 0$. The four polynomials have $4(n+1)$ coefficients and the system (6) together with $B_n(1) = 0$ gives $2(2n+1) + 1 = 4n + 3$ linear conditions, so that the polynomials A_n, B_n, C_n, D_n really exist. Write

$$A_n(z) = \sum_{r=0}^{n} \alpha_r z^r \quad \text{and} \quad B_n(z) = \sum_{r=0}^{n} \beta_r z^r .$$

Since degree $C_n, D_n \le n$, the Taylor coefficient of z^m $(n+1 \le m \le 2n)$ in $A_nL_2 + B_nL_1$, $2A_nL_3 + B_nL_2$ respectively, must be zero, i.e.

$$
\text{(7)} \quad
\begin{aligned}
\sum_{r=0}^{n} \frac{\alpha_r}{(m-r)^2} + \frac{\beta_r}{m-r} &= 0 \\
\sum_{r=0}^{n} \frac{2\alpha_r}{(m-r)^3} + \frac{\beta_r}{(m-r)^2} &= 0
\end{aligned}
\qquad m=n+1,\dots,2n .
$$

Furthermore, $B_n(1) = 0$ implies $\Sigma\beta_r = 0$. This system of linear equations for α_r and β_r is easy to solve. Consider the rational function

$$R_n(t) = \sum_{r=0}^{n} \frac{\alpha_r}{(t-r)^2} + \frac{\beta_r}{t-r} = \frac{Q_n(t)}{t^2(t-1)^2\dots(t-n)^2} .$$

The conditions (7) now imply that $R_n(t)$ and its derivative are zeró for $t = n+1, n+2, \ldots, 2n$. This implies that $Q_n(t)$ is a multiple of $(t-n-1)^2(t-n-2)^2\ldots(t-2n)^2$. If we put $Q_n(t)$ equal to this product then degree $Q_n(t) = 2n$, whereas the denominator of $R_n(t)$ has degree $2n+2$. This automatically implies $\Sigma\beta_r = 0$. Therefore, the coefficients α_r, β_r can be obtained from the partial fraction expansion of

$$\frac{(t-n-1)^2(t-n-2)^2\ldots(t-2n)^2}{t^2(t-1)^2\ldots(t-n)^2}.$$

In particular it is easy to see that

$$\alpha_r = \binom{n}{r}^2\binom{2n-r}{n}^2.$$

Substitute $z = 1$ in (6) and use $B_n(1) = 0$. Then the second line yields

(8) $$2A_n(1)\zeta(3) + D_n(1) = \text{remainder}$$

where

$$A_n(1) = \sum_{r=0}^{n}\binom{n}{r}^2\binom{2n-r}{n}^2 = \sum_{k=0}^{n}\binom{n}{k}^2\binom{n+k}{n}^2 = a_n.$$

Thus we have recovered the number a_n from the approximations (6). It is now a matter of straightforward computation to show that the approximation (8) is actually the same as (5).

REFERENCES

[A-R] K. Alladi, M. Robinson, On certain irrational values of the logarithm, Lecture Notes in Math. 751, 1-9.

[A] R. Apéry, Irrationalité de $\zeta(2)$ et $\zeta(3)$. "Journées arithmétiques de Luminy", Astérisque n° 61, 1979, 11-13.

[Ba1] A. Baker, Rational approximations to $\sqrt[3]{2}$ and other algebraic numbers, Quart. J. Math. Oxford, 15(1964), 375-383.

[Ba2] A. Baker, Transcendental Number Theory (Cambridge, 1975).

[Be1] F. Beukers, A note on the irrationality of $\zeta(2)$ and $\zeta(3)$, Bull. London Math. Soc., 11(1979), 268-272.

[Be2] F. Beukers, Legendre polynomials in irrationality proofs, Bull. Australian Math. Soc. (to appear).

[Be3] F. Beukers, The generalised Ramanujan-Nagell equation, Thesis, University of Leiden (1979), also to appear in Acta Arithmetica.

[C1] G.V. Chudnovsky, C.R. Acad. Sc. Paris, 288(1979), 607-609, 965-967, 1001-1003.

[C2] G.V. Chudnovsky, Padé-approximations to the generalized hypergeometric functions I, J. Math. pures et appl. 58(1979), 445-476.

[C3] G.V. Chudnovsky, Rational and Padé-approximations to solutions of linear differential equations and the monodromy theory, Lecture Notes in Physics 126, 136-169.

[C4] G.V. Chudnovsky, Padé-approximation and the Riemann monodromy problem, Proceedings of the NATO Advanced Study Institute, held at Cargèse, Corsica, France, June 24-July 7, 1979.

[D] Y. Domar, On the diophantine equation $|Ax^n-By^n| = 1$, $n \geq 5$, Math. Scand. 2(1954), 29-32.

[H] Ch. Hermite, Sur la fonction exponentielle, Oeuvres III, 150-181.

[J] H. Jager, A multidimensional generalization of the Padé table, Thesis, University of Amsterdam (1964).

[L] F. Lindemann, Ueber die Zahl π, Math. Ann. 20(1882), 213-225.

[M] K. Mahler, Application of some formulae by Hermite to the approximation of exponentials and logarithms, Math. Ann. 168(1976), 200-227.

[P] A.J. van der Poorten, A proof that Euler missed ... Apéry's proof of the irrationality of $\zeta(3)$, Math. Intelligencer, 1(1978), 195-203.

[R] E. Reyssat, Irrationalité de $\zeta(3)$ selon Apéry, Sém. Delange-Pisot-Poitou, 20e année, 1978/79, n° 6.

[Si1] C.L. Siegel, Transcendental Numbers (Princeton 1949).

[Si2] C.L. Siegel, Die Gleichung $ax^n - by^n = c$, Math. Ann. 114(1937), 57-68.

ERROR ANALYSIS OF INCOMING AND OUTGOING SCHEMES FOR THE TRIGONOMETRIC MOMENT PROBLEM

A. Bultheel
K.U.LEUVEN
Afd. Toegepaste Wiskunde
en Programmatie
Celestijnenlaan 200 A
B-3030 Heverlee
BELGIUM

Abstract

The solution of the trigonometric moment problem involves the computation of a (0/n) Laurent-Padé approximant for a positive real function on the complex unit circle. The incoming scheme is equivalent with the recursion for Szegö's orthogonal polynomials, while the outgoing scheme is equivalent to the Schur recursion for contractions of the unit disc. The numerical stability of both algorithms is proved under certain conditions via a backward error analysis.

1. Introduction

The trigonometric moment problem (TMP) is one of the classical moment problems [1-3]. It consists of asking whether a prescribed sequence ... $t_{-2}, t_{-1}, t_0 = 1, t_1, t_2, \ldots$ with $t_k + t_{-k}$ real can be represented as

$$t_k = \frac{1}{2\pi} \int_{-\pi}^{\pi} e^{-jk\theta}\, d\mu(\theta) \qquad , k = 0, \pm 1, \pm 2, \ldots \qquad (j = \sqrt{-1})$$

with $d\mu \geq 0$, a positive measure of the unit circle.
A constructive way to study the problem is to define the linear functional $\overline{L(z^{-k}) = t_k}$ and the polynomials orthogonal with respect to the inner product $(f,g) = L(f(z)\overline{g(1/\bar{z})})$. If $d\mu$ exists, we may also represent this product as

$$(f,g) = \frac{1}{2\pi} \int_{-\pi}^{\pi} f(e^{j\theta})\overline{g(e^{j\theta})}\, d\mu(\theta) \quad .$$

The corresponding orthonormal system $\phi_0, \phi_1, \ldots$ is the set of Szegö polynomials [4]. Note that ϕ_n depends only upon t_k, $|k| \leq n$, and is related to them in a Padé sense. Namely : if $w(z)$ is the weight function, representing the absolutely continuous part of $d\mu$, i.e. $d\mu = w(e^{j\theta})d\theta$, then $w_n(z) = [\phi_n(z)\overline{\phi_n(1/\bar{z})}]^{-1}$ is a [0/n] Laurent-Padé approximant for $w(z)$ [5]. For the TMP, one continues to study the convergence of $w_n(z)$ (or $\phi_n(z)$) to establish the existence and/or uniqueness of $d\mu$.
We shall not dig deeper into the analysis than strictly necessary. For our purpose we consider the problem as an approximation problem. We suppose $w(e^{j\theta})$ exists and

we study numerical aspects of algorithms constructing the successive approximants w_n (i.e. ϕ_n) of w for n = 0,1,2,... . As we mentioned, this is a special case of a Laurent-Padé approximation problem [8], but it is also related to Padé-type approximation [6], 2 point Padé approximation, T-fractions [7] least squares approximation [2] etc. . One of the classical applications of the TMP is in the theory of linear prediction of stationary stochastic processes [2,9]. In practical situations, the algorithms we are going to consider compute the parameters of a digital filter which generates the linear predictor. (i.e. ϕ_n) [10,11]. It is from this point of view that we shall look at the algorithms. We place ourselves in the simplest situation and formulate the algorithms in that case. Suppose that $d\mu = w d\theta$ exists and that $t_0 = 1$ and $t_k = t_{-k}$ is real for all k, then $\phi_0 = 1$ and the Szegö polynomials have real coefficients. Denote by $q_n(z) = \phi_n(z).r_n^{1/2}$ the monic version of the orthonormal Szegö polynomial $\phi_n(z)$. I.e.

$$q_n(z) = Q_{0n} + Q_{1n}\, z + \dots + Q_{nn}\, z^n = \mathcal{Q}_n^T[1\ z \dots z^n]^T$$

with $Q_{nn} = 1$ and $\mathcal{Q}_n^T = [Q_{0n} \dots Q_{nn}]$. We define the reciprocal of a real polynomial of degree n as $\hat{q}_n(z) = z^n q(1/z)$ and for the corresponding coefficient vector $\hat{\mathcal{Q}}_n^T = [Q_{nn} \dots Q_{0n}]$.

The first algorithm which we call *incoming algorithm* is the recursion scheme for the Szegö polynomials $q_n(z)$:

$$\mathcal{Q}_0 = r_0 = 1$$

for i = 0,1,2,...,n-1

$$p_i = [t_1 \dots t_{i+1}]\mathcal{Q}_i \quad ; \quad \rho_{i+1} = p_i\, r_i^{-1}$$

$$\mathcal{Q}_{i+1}^T = [0\ \ \mathcal{Q}_i^T] - [\hat{\mathcal{Q}}_i^T\ \ 0]\rho_{i+1}$$

$$r_{i+1} = r_i(1-\rho_{i+1}^2)\ .$$

This may also be written as

$$\begin{bmatrix} q_{i+1}(z) \\ \hat{q}_{i+1}(z) \end{bmatrix} = \begin{bmatrix} 1 & -\rho_{i+1} \\ -\rho_{i+1} & 1 \end{bmatrix} \begin{bmatrix} z & 0 \\ 0 & 1 \end{bmatrix} \begin{bmatrix} q_i(z) \\ \hat{q}_i(z) \end{bmatrix} = \prod_{j=1}^{i+1} \begin{bmatrix} z & -\rho_j \\ -z\rho_j & 1 \end{bmatrix} \begin{bmatrix} 1 \\ 1 \end{bmatrix} .$$

Given the ρ_i parameters, (i.e. *reflection coefficients*), the latter recursion can be seen as a method to evaluate $q_{i+1}(z)$. It can be implemented efficiently as a ladder filter [10,11] such that it becomes a scheme competitive with the classical Horner scheme to evaluate $q_{i+1}(z)$. There are some other advantages such as low sensitivity etc. so that we shall use the incoming scheme *not* as an algorithm to evaluate the coefficients $\mathcal{Q}_i$ but as an algorithm to compute the set of reflection coefficients ρ_i.

Another algorithm which computes the reflection coefficients is the algorithm of Schur [12], also given as one of the applications of the TMP in [2]. We call it the *outgoing algorithm.* If we set $a_0(z) = \sum_0^\infty t_k z^k$ and $b_0(z) = \sum_1^\infty t_k z^k$ while recursively

$$\begin{bmatrix} a_{i+1}(z) \\ b_{i+1}(z) \end{bmatrix} = \begin{bmatrix} 1 & -\rho_{i+1} \\ -\rho_{i+1} & 1 \end{bmatrix} \begin{bmatrix} z & 0 \\ 0 & 1 \end{bmatrix} \begin{bmatrix} a_i(z) \\ b_i(z) \end{bmatrix},$$

then $a_i(z) = r_i z^i + O(z^{i+1})$ and $b_i(z) = p_i z^{i+1} + O(z^{i+2})$.
Of course, in practical situations one cannot use an infinite sum. To formulate the finite version we need cutting operators to delete the first or last element of a vector. They are

$$\Delta = \begin{bmatrix} 0 & 1 & & \\ \vdots & & \ddots & \\ 0 & & & 1 \end{bmatrix} \quad \text{and} \quad \nabla = \begin{bmatrix} 1 & & & 0 \\ & \ddots & & \vdots \\ & & 1 & 0 \end{bmatrix}.$$

e_0^T is a unit vector $e_0^T = [1\ 0\ 0\ \ldots]$. The algorithm is :

$$r_0 = 1 \qquad\qquad p_0 = t_1$$

$$A_0^T = [t_0 \ldots t_n] \qquad\qquad B_0^T = [t_1 \ldots t_{n+1}]$$

for $i = 0,1,\ldots,n-1$

$$\rho_{i+1} = p_i\, r_i^{-1}$$

$$A_{i+1} = \nabla(A_i - B_i\, \rho_{i+1}) \qquad\qquad B_{i+1} = \Delta(B_i - A_i\, \rho_{i+1})$$

$$r_{i+1} = e_0^T A_{i+1} \qquad\qquad p_{i+1} = e_0^T B_{i+1}$$

We call the first algorithm "incoming" because in every iteration, new information (t_{i+1}) comes in and this is accumulated in Q_i, a vector of increasing length. The second algorithm starts with all the data at the beginning (A_0, B_0) and extracts in every step some information (ρ_i) so that something of the data can be left (cutting operations). The A_i, B_i are outgoing vectors (of decreasing length).

The outgoing algorithm uses two vectors that need updating, but no evaluation of an inner product to find p_i. The incoming algorithm needs only one vector Q_i but n inner products of increasing length to evaluate all p_i. These inner products involve the summation of about n(n+1)/2 products. In the outgoing algorithm we have in the first step n inner products ahead. One product/sum operation is executed for each of these products. In the second step only (n-1) inner products remain and again one product/sum operation is executed for each of them, etc. . This results in the updating of a decreasing length vector. Although the computations involved are not exactly the same after the first step, for the incoming and outgoing algorithms, the main difference is something like reordering the computations.

A similar duality exists between the Gauss and Crout methods for triangular factorization of a matrix or the Gram-Schmidt and modified Gram-Schmidt method for orthogonalisation.

There exists a matrix interpretation for the incoming and outgoing algorithm. If T_n is the symmetric Toeplitz matrix based on the parameters $\{t_0,t_1,\ldots,t_n\}$, U_n a unit upper triangular with columns $\mathcal{Q}_i$ and L_n a lower triangular with the $\mathcal{A}_i$ as columns (thus with r_i as i-th diagonal element) we have

$$T_n U_n = L_n \quad \text{and} \quad L_n = U_n^{-T} D_n \quad \text{with } D_n = \operatorname{diag}(r_0,\ldots,r_n).$$

For

$$\underline{T}_n = \begin{bmatrix} t_1 & \cdots & t_{n+1} \\ t_0 & \ddots & \vdots \\ \vdots & \ddots & \\ t_{n-1} \cdots t_0 & & t_1 \end{bmatrix}, \quad \underline{L}_n = \begin{bmatrix} \star & & 0 \\ \vdots & \ddots & \\ 1 \cdots & [\mathcal{Q}_i] & \cdots 1 \end{bmatrix} \quad \text{and } \underline{U}_n = \begin{bmatrix} \ddots & [\hat{\mathcal{B}}_i] & \cdots \\ & \ddots & \vdots \\ 0 & & \ddots \end{bmatrix}$$

we have

$$\underline{T}_n \underline{L}_n = \underline{U}_n .$$

The algorithms given above are special forms of $O(n^2)$ factorization algorithms for general Toeplitz matrices (see [13] and the references therein). These algorithms are known to be unstable in general situations [14,15]. However the special structure of the problem does not only give symmetric forms of the algorithms but the existence of $w(e^{j\theta})$ also seems to have important implications for the numerical stability. Indeed this existence implies $T_n > 0$ for $n = 0,1,\ldots$ or equivalently all $|\rho_i| < 1$, and thus r_i is decreasing but nonnegative. The latter is also equivalent with all the zeros of $q_n(z)$ being inside the closed unit disc.

The previous algorithms thus compute the Cholesky factors of T_n or T_n^{-1}. The Cholesky algorithm is known to be stable, without pivoting [16]. In practical applications incoming and outgoing algorithms have been used frequently and no stability problems have been reported. However, a theoretical justification by an error analysis in the sense of Wilkinson [16] has not been given, except for [17]. In this paper we try to answer the question: If we want to find the set of parameters $\{\rho_1,\rho_2,\ldots,\rho_n\}$, given $\{t_0 = 1,t_1,\ldots,t_n\}$ is there a numerical preference for one of both, the incoming or outgoing algorithm. This paper is related to the one of Graves-Morris and of L. Wuytack in these proceedings.

In the next section we introduce the notions of stability and conditioning and a criticism on the error analysis performed in [17]. Section 3 and 4 give a backward error analysis for the incoming and outgoing scheme.

2. Conditioning, Stability, Error Analysis

In the next sections we shall do a backward error analysis in the sense of

Wilkinson [16] and conclude from it that the incoming and outgoing algorithms are stable [18], provided the condition number remains bounded. We recall briefly these definitions and introduce some notations.

For our problem, the input data are vectors $t_n = \{t_0 = 1, t_1, \dots, t_n\}$ such that the symmetric Toeplitz matrices $T_0 = 1, T_1, \dots, T_n$ that are constructed from these parameters are all positive semi-definite. The result is a vector of reflection coefficients $p_n = \{\rho_1, \rho_2, \dots, \rho_n\}$ with $|\rho_j| \leq 1$. The *condition number* is a measure for the sensitivity of p_n w.r.t. small perturbations of t_n. Note that our condition number is not the "spectral sensitivity" of the reflection coefficients [11, p. 234] which is the sensitivity of $w_n(e^{j\theta})$ w.r.t. changes in p_n. Neither is it the classical condition number of T_n, because this measures the sensitivity of Q_n or A_n w.r.t. changes (possibly non Toeplitz) in T_n. The latter condition number is $\kappa(T_i) = \max_j \lambda_j(T_i)/\min_j \lambda_j(T_i)$, where $\lambda_j(T_i), j = 0,1,\dots,i$, are the eigenvalues of T_i. It becomes infinite if T_i is singular. Although in that situation, both algorithms break down, a solution theoretically exists. I.e. if T_i is singular, then $r_i = 0$, $A_i = B_i = 0$, $|\rho_i| = 1$ and $\rho_j = 0$ for $j > i$. The approximation is exact : $w_i = w$. Thus for T_i close to a singular matrix, the solution is not even close to non existence, which usually is the same as ill conditioning. To make the algorithms reliable we should introduce a test to check whether $r_i = 0$ and eventually stop the algorithm. In practical programs we shall replace it by the test $r_i < r^i$ for some $r < 1$ (see further). We do not propose a condition number for our problem and leave its definition as an open problem, but we remember from the foregoing that it is certainly not $\kappa(T_n)$. $\kappa(T_n)$ is only an upperbound for the condition of $T_n \to Q_n$.

Using the definitions of de Jong [18], it is sufficient to prove forward or backward stability to establish the numerical stability of the method. However, the forward error usually contains the condition number as a factor. Since we have no plausible condition number at this moment, we prefer to do a backward error analysis because this yields more information about the stability of the algorithm. For our algorithms, backward stability means the following. Suppose $p_n = f(t_n)$ is the exact relation between p_n and t_n. If the numerical implementation of f is $\bar{f}$, giving $\bar{p}_n = \bar{f}(t_n)$ and if $\bar{p}_n = f(\bar{t}_n)$ then $\bar{f}$ is called backward stable if $E_n = \bar{t}_n - t_n$ can be made as small as we like by increasing the machine precision ε, and this for all possible t_n in the set of input data. It is clear from this that we cannot allow input vectors t_n such that r_i becomes arbitrary close to 0, because then $\kappa(T_i)$ is arbitrarily large, causing uncontrolable errors in Q_i or A_i. This justifies why we shall only prove stability for a restricted set of input data. I.e. the set of t_n such that $\kappa(T_n)$ remains bounded. This means that we may suppose that $|\rho_i| \leq \rho < 1$ or $r_i = \prod_{j=1}^{i} (1-\rho_j^2) \geq r^i > 0$ (with $r = 1-\rho^2$) for all $i = 0,1,\dots,n-1$ and in the last step either $|\rho_n| = 1$ (and $r_n = 0$) or $|\rho_n| \leq \rho$. r should be "reasonably" large compared with n and ε. In this case we may replace the stop criterion $r_i = 0$ by $r_i < r^i$.

This is reasonable because r_i has the meaning of a least squares error [2] and thus we stop if the approximation is good enough. Our criterion means that w_n stays reasonably far away from w, and if it converges, it may not be slow in the last steps. The following property gives an asymptotic bound for $\kappa(T_n)$:

<u>property 2.1</u>

The spectral condition number $\kappa(T_n)$ is non decreasing with n and if $\lim_{n\to\infty} w_n \neq w$ then $\kappa(T_n)$ is bounded by ess $\sup_\theta w(e^{j\theta^n})$/ess $\inf_\theta w(e^{j\theta})$. [2, p. 64].

An error analysis for the Durbin algorithm (this is almost the same as our incoming algorithm) was done in [17]. The outcome was that the algorithm is comparable with the Cholesky method which is known to be stable [16]. However, the error analisys which was done there has some drawbacks. 1° The statement is based on an error estimate of $\mathcal{Q}_n$ (forward error). 2° The result was derived by a computation of the residual of the linear system of Yule-Walker equations. This is a kind of backward error analysis, but it does not respect the Toeplitz structure of the problem (the t_k in the right hand side have backward errors but the same t_k in the left hand side have not). 3° The condition number $\kappa(T_n)$ is not the correct one for our purposes. 4° We are interested in the computation of p_n and not of $\mathcal{Q}_n$. This justifies the error analysis we are going to describe in the next sections.

3. Error analysis for the incoming algorithm

As described in the previous section we compute $\bar{p}_n = \bar{f}(\mathcal{t}_n)$ or componentwise : $\bar{\rho}_i = \bar{f}_i(\mathcal{t}_i)$, and we are looking for $\bar{\mathcal{t}}_n = \mathcal{t}_n + E_n$ such that $\bar{p}_n = f(\bar{\mathcal{t}}_n)$. The analysis is recursive. Suppose we know $E_i = [\varepsilon_0, \varepsilon_1, \ldots, \varepsilon_i]^T$ then we try to find ε_{i+1} such that $\bar{\rho}_{i+1} = f_{i+1}(\bar{\mathcal{t}}_i, t_{i+1}+\varepsilon_{i+1})$.

Define $u_i = \nabla\mathcal{Q}_i$ and let fl denote the floating point result obtained with the rules of Wilkinson [16] and let ε be the machine precision.
We have for step i (fl_2 [16] simplifies the analysis a bit)

$$\bar{\rho}_{i+1} = fl\left(\frac{fl_2(t_{i+1} + [t_1 \ldots t_i]\bar{u}_i)}{fl(\bar{r}_{i-1}.fl(1 - fl(\bar{\rho}_i^2)))}\right)$$

$$\bar{u}_i = \nabla\bar{\mathcal{Q}}_i = fl([0\ \bar{u}_{i-1}^T]^T - fl([1\ \bar{u}_{i-1}^T]^T\ \bar{\rho}_i)).$$

The definition of ε_{i+1} is given via the equality

$$\bar{\rho}_{i+1} = \frac{t_{i+1} + \varepsilon_{i+1} + [\bar{t}_1 \ldots \bar{t}_i]u_i'}{r_{i-1}'(1 - \bar{\rho}_i^2)} .$$

The quantities with a prime represent the exact values that would have been obtained if $\bar{\mathcal{t}}_n$ was used as the input. Note $\bar{\rho}_i = \rho_i'$ by definition of $\bar{\mathcal{t}}_n$.

Using the rules of floating point computations, we can bring the first form into the second and thus find that

$$\varepsilon_{i+1} = \varepsilon^{(1)} + \varepsilon^{(2)} + \varepsilon^{(3)} .$$

$\varepsilon^{(1)}$ originates from the error in the denominator : $|\varepsilon^{(1)}| \lesssim 3i\, \bar{r}_{i-1}\, \varepsilon$

$\varepsilon^{(2)}$ comes from the inner product and division : $|\varepsilon^{(2)}| \lesssim 2\, \bar{r}_{i+1}\, \varepsilon$

$\varepsilon^{(3)}$ accumulates the errors made in previous operations :

$$\varepsilon^{(3)} = [\bar{t}_1 \ldots \bar{t}_i](\bar{u}_i - u_i') - [\varepsilon_1 \ldots \varepsilon_i]\bar{u}_i, \text{ thus with } y_j = \bar{u}_i - u_i' :$$

$$|\varepsilon^{(3)}| < \|y_i\|_1 + \|E_i\|_\infty . \|\bar{u}_i\|_1 .$$

It was shown in [17] that $\|\bar{u}_i\|_1 \leq \Pi_1^i (1+|\bar{\rho}_i|) - 1 < 2^i - 1$. Bounding the first term in $\varepsilon^{(3)}$ involves a kind of forward error analysis for the evaluation of u_i, given p_{i-1}. To find a bound for $\|y_i\|_1$ we do an error analysis on the fl-expression for $\bar{u}_i$, given above and write $\bar{u}_i$ as

$$\bar{u}_i = [0(u_{i-1}')^T]^T - \bar{\rho}_i[1(\hat{u}_{i-1}')^T]^T + y_i = u_i' + y_i .$$

We find (only first order terms)

$$\|y_i\|_1 \leq \varepsilon\|u_i'\|_1 + \|y_{i-1}\|_1(1-\bar{\rho}_i) + \bar{\rho}_i\varepsilon\|[1\ \bar{u}_{i-1}^T]^T\|_1 .$$

Thus, if $\|y_i\|_1 < \eta_i\, \varepsilon$, then $\eta_i < 2(2^i+\eta_{i-1})$ or $\eta_i < i\ 2^{i+1}$. All this results in a bound for ε_{i+1} viz.

$$|\varepsilon_{i+1}| < (i\ 2^{i+1}+3i+2)\varepsilon + \|E_i\|_\infty(2^i-1) = G(i)\varepsilon.$$

For finite n, G(i), i = 1,2,...,n-1 is bounded and thus, according to its definition, we may conclude that the incoming algorithm is stable.

Although the estimate derived for $|\varepsilon_{i+1}|$ may not be sharp, the analysis is instructive because it shows that 1° The factor 2^{i+1} mainly comes from the growth of the vector u_i. If all $\rho_j \approx -1$, then this bound is closely followed. This rapid growth is associated with the ill-conditioning of the problem $t_i \to u_i$. 2° The evaluation of the inner product in fl_2 is recommended because possibly large numbers ($\bar{u}_i$) are used to find a small number ($\bar{p}_{i+1}$). Note also that a scaling of u_i is of no use because this factor would also appear as a multiplier of t_{i+1} in the numerator of ρ_{i+1}.

4. Error analysis for the outgoing algorithm

The situation for the outgoing algorithm is inverted. Here the components of A_i and B_i will be decreasing in magnitude, together with $r_i^{1/2}$. We have the following lemma's (see [15]) :

Lemma 4.1 : Let $T = L\,L^T$ be the lower-upper and $T = U\,U^T$ the upper-lower Cholesky factorization of a Toeplitz matrix $T > 0$, then $|L_{ji}| \leq L_{ii}$, $\forall j \geq i$, i = 0,1,... and $|U_{ij}| \leq U_{jj}$, $\forall i \leq j$, j = 0,1,... .

Lemma 4.2 : With the functions $a_i(z)$ and $b_i(z)$ as defined in section 1, we have that $c_i(z) = b_i(z)/a_i(z)$ is a Schur function (i.e. analytic and contractive on the unit disc). Its Taylor series coefficients are bounded by 1 in magnitude.

The first lemma and the matrix interpretation of the outgoing scheme gives bounds for the elements in the vectors A_i. If $A_i^T = [a_{0i} \dots a_{n-i,i}]$ then $|a_{ij}| < r_i^{1/2}$. If we set $B_i^T = [b_{0i} \dots b_{n-i,i}]$ and find these coefficients from $b_i(z) = c_i(z)a_i(z)$ we may bound them by $|b_{ij}| \leq (j+1)r_i^{1/2}$.

If we write out the computations to obtain $\bar{\rho}_{i+1}$ we obtain

$$\bar{\rho}_{i+1} = fl\left(\frac{b_{i,0}-fl(\sum_{j=1}^{i})fl(\bar{\rho}_{i-j+1}\,\bar{a}_{j,i-j})}{fl(\bar{a}_{0,i-1}-fl(\bar{b}_{0,i-1}\,\bar{\rho}_i))}\right) .$$

ε_{i+1} is defined by the equality

$$\bar{\rho}_{i+1} = \frac{(b_{i,0}+\varepsilon_{i+1}) - \sum_{j=1}^{i} \bar{\rho}_{i-j+1}\, a'_{j,i-j}}{a'_{0,i-1} - b'_{0,i-1}\,\bar{\rho}_i} ,$$

where again primed quantities are exact values obtained from the input vector $\bar{t}_n$. Now ε_{i+1} has the following composition

$$\varepsilon_{i+1} = \varepsilon^{(1)} + \varepsilon^{(2)} + \varepsilon^{(3)} + \varepsilon^{(4)} .$$

$\varepsilon^{(1)}$ is the error from the division and the evaluation of the denominator :

$$|\varepsilon^{(1)}| < 3|\bar{\rho}_{i+1}|\varepsilon .$$

$\varepsilon^{(2)}$ comes from the evaluation of the numerator :

$$|\varepsilon^{(2)}| < |t_{i+1}|i\varepsilon + |\bar{\rho}_1\, a_{i,0}|(i+1)\varepsilon + |\bar{\rho}_2\,\bar{a}_{i-1,1}|i\varepsilon + \dots + |\bar{\rho}_i\,\bar{a}_{1,i-1}|2\varepsilon$$
$$< (i+5)i\varepsilon/2.$$

$$\varepsilon^{(3)} \simeq \sum_{j=1}^{i} \bar{\rho}_{i+1-j}(a'_{j,i-j}-\bar{a}_{j,i-j}) \quad \text{and} \quad \varepsilon^{(4)} \simeq (\bar{a}_{0,i-1}-a'_{0,i-1})-\bar{\rho}_i(\bar{b}_{0,i-1}-b'_{0,i-1}).$$

Bounding $\varepsilon^{(3)}$ and $\varepsilon^{(4)}$ requires again a forward analysis for the computation of A_j and B_j. From its definition :

$$\bar{A}_j = \nabla fl(\bar{A}_{j-1} - fl(\bar{B}_{j-1}\,\bar{\rho}_j)) \overset{\text{def}}{=} \nabla(A'_j + y_j) \quad \text{and}$$

$$\bar{B}_j = \Delta fl(\bar{B}_{j-1} - fl(\bar{A}_{j-1}\,\bar{\rho}_j)) \overset{\text{def}}{=} \Delta(B'_j + z_j) .$$

An error analysis gives, up to higher order terms :

$$\|y_j\|_\infty < \|y_{j-1}\|_\infty + \|z_{j-1}\|_\infty + \varepsilon\|B_{j-1}\|_\infty + \varepsilon\|A'_j\|_\infty$$
$$< \|y_{j-1}\|_\infty + \|z_{j-1}\|_\infty + \varepsilon(i+2).$$

Because of the symmetry we have the same bound for $\|z_j\|_\infty$.
Thus if $\|y_j\|_\infty$ and $\|z_j\|_\infty < \eta_j\ \varepsilon$, then $\eta_j < 2\eta_{j-1} + (i+2)$, or $\eta_j < (2^j-1)(i+2)$. This gives for $\varepsilon^{(3)}$ and $\varepsilon^{(4)}$

$$|\varepsilon^{(3)}| < \sum_{j=1}^{i} \|y_{i-j}\|_\infty < (2^i-i-1)(i+2)\varepsilon \quad \text{and}$$

$$|\varepsilon^{(4)}| < 2\ (2^{i-1}-1)(i+2)\varepsilon\ .$$

This adds up to the following bound for ε_{i+1} :

$$|\varepsilon_{i+1}| < \frac{\varepsilon}{2}\ (i+2)\{2^{i+2}-i-3\} = H(i)\varepsilon\ .$$

Again $H(i)$ can be bounded for finite i and thus the outgoing algorithm is backward stable.

As in the analysis of the incoming algorithm, the main term in the error comes from a forward error for computing (A_n, B_n) from t_n. No fl_2-trick can be done now but the bound for ε_{i+1} does not contain $\|E_i\|_\infty$.

5. Conclusion

In the previous sections we have done a backward error analysis of both the incoming and outgoing algorithm to compute the reflection coefficients from the moments of a positive definite measure. In both cases we had to introduce a kind of forward error which caused the main term in the backward error to be of the type $i\ 2^i\ \varepsilon$. However for the incoming algorithm, the bound for ε_{i+1} depended on $\|E_i\|_\infty$ so that potentially this bound can grow faster than the corresponding bound in the outgoing scheme. The forward error contains the condition of the problems associated with it, i.e. evaluation of Q_i or (A_i, B_i) from t_i. That is why we could only prove stability on input data for which $\kappa(T_n)$ remains bounded. $\kappa(T_n)$ is bounded because we required $|\rho_i| \leq \rho < 1$. We said that ρ should be "reasonable" in view of n and ε. We used this fact in the analysis because we supposed that the perturbations were such that the bounds that were valid for the exact values remain valid for the perturbed values. This is exactly what we mean by reasonable.

The fact that the upper bound for $|\varepsilon_{i+1}|$ is potentially larger in the case of the incoming algorithm than in the case of the outgoing one is no proof that the outgoing algorithm is better than the incoming one. However, because the components of Q_i are growing and simultaneously have to produce the numbers ρ_i that are decreasing we are not in a numerically favorable situation. In the outgoing algorithm, all numbers are decreasing together with $r_i^{1/2}$. Thus intuitively and partly because of the analysis, we should recommend the outgoing algorithm above the incoming one. Numerical experiments have not justified this conjecture, unless $\kappa(T_n)$ was small.

Finally, notice that we have used certain bounds in our analysis that were only true if all $T_j > 0$, $j = 0,1,\ldots$. Thus a generalization of this analysis to arbitrary Toeplitz matrices (the incoming and outgoing algorithms for the Padé problem) is more than just an asymmetric version of the foregoing. The definition of a suitable condition number for our problem and for the general Padé problem, and also the determination of situations where recursive methods are stable in a general Padé context is still an open problem.

References

[1] N.I. Akhiezer, The classical moment problem, Oliver and Boyd, Edinburgh, London, 1965.

[2] U. Grenander, G. Szegö, Toeplitz forms and their applications, University of California Press, Berkeley, 1958.

[3] H.J. Landau, The classical moment problem : Hilbertian proofs, Journ. of Funct. Anal. 38, (1980), pp. 255-272.

[4] G. Szegö, Orthogonal polynomials, A.M.S. Colloquium publ. XXIII, AMS Providence, Rhode Island, 1939.

[5] W.B. Gragg, Laurent-, Fourier- and Chebyshev-Padé tables, in E.B. Saff, R.S. Varga (eds.), Padé and rational approximation, Theory and applications, Academic Press, New York, 1977, pp. 61-72.

[6] C. Brezinski, Padé type approximation and general orthogonal polynomials, Birkhauser Verlag, Basel, 1980.

[7] W.J. Thron, Two-point Padé tables, T-fractions and sequences of Schur, in E.B. Saff, R.S. Varga (op. cit.), pp. 215-226.

[8] A. Bultheel, P. Dewilde, On the relation between Padé approximation algorithms and Levinson/Schur recursive methods, in Proc. Conf. Signal Processing, EUSIPCO-80, M. Kunt, F. de Coulon (eds.), North-Holland, 1980, pp. 517-523.

[9] N. Wiener, P. Masani, The prediction theory of multivariate stochastic processes, Acta Math. 98, (1957), pp. 111-150 and 99, (1958), pp. 93-139.

[10] V. Cappellini, A.G. Constantinides, P. Emiliani, Digital filters and their applications, Ac. Press, New York, 1978.

[11] J.D. Markel, A.H. Gray Jr., Linear prediction of speech, Springer-Verlag, Berlin, 1976.

[12] J. Schur, Ueber Potenzreihen die im Innern des Einheitskreises beschränkt sind, J.f.d.R.u.Angew.Math. 147, (1917), pp. 205-232 and 148 (1918),pp. 122-145.

[13] A. Bultheel, Recursive algorithms for the matrix Padé problem, Math. of Comp. 35 (151), 1980, pp. 875-892.

[14] L.S. de Jong, Numerical aspects of the recursive realization algorithm, SIAM J. Cont. and Opt. 16, (1978), pp. 646-659.

[15] A. Bultheel, Towards an error analysis of fast Toeplitz factorization, Rept. TW 44, K.U.Leuven, Afd. Toeg. Wisk. & Progr., May 1979.

[16] J.H. Wilkinson, The algebraic eigenvalue problem,Clarendon Press, Oxford, 1965.

[17] G. Cybenko, Error analysis of Durbin's algorithm, to appear in SIAM J. for Scient. and Stat. Comp.

[18] L.S. de Jong, Towards a formal definition of numerical stability, Numer. Math. 28, (1977), pp. 211-219.

GENERALIZED RATIONAL CORRECTORS

T.H. Clarysse

Dept. Applied Mathematics
K.U.Leuven
Celestijnenlaan 200A
B-3030 Heverlee, Belgium

1. Introduction.

In [1], [2] and [3] the approximation of a real function y(x) by a rational function of the form:

$$r(x) = \frac{\sum_{j=0}^{n} a_j . x^j}{\sum_{j=0}^{m} b_j . x^j} \tag{1}$$

was treated, using osculatory interpolation conditions. Both predictor and corrector formulas were proposed. The correctors were obtained as the roots of a quadratic equation. Hence, they can fail in case of a negative discriminant. This occurs when no interpolant of the form (1) exists, satisfying all the imposed conditions.

In this paper we try to diminish this drawback, by deriving some correctors based on a generalized rational function of the form:

$$r(x) = \frac{\sum_{j=0}^{n} a_j . k_j(x)}{\sum_{j=0}^{n} b_j . l_j(x)} \tag{2}$$

We take the same number of basis functions in numerator and denominator in order to simplify later calculations.

In section 2, the problem in general is stated and a simplified condition for the existence of a solution is derived. In section 3, a second order corrector is proposed for the case $k_j(x)=l_j(x), j=0,..,n$.

In Section 4, a fourth order corrector is derived for the case $k_j(x)=l_j(x)=t(x)^j$, $j=0,\ldots,n$. In Sections 5 and 6, numerical results are presented respectively for numerical integration and Volterra integral equations. Applications can also be found in ordinary differential equations, but will not be considered here.

2. The General Problem.

Let $J=\{x_0,x_1,\ldots,x_n\}$ be a set of n+1 distinct base points and y_i, f_i denote respectively approximations of $y(x_i)$ and its derivative value $f(x_i)=y'(x_i)$.

We impose the following osculatory interpolation conditions on the generalized rational interpolant (2):

$$\begin{cases} r(x_i) = y_i \\ r'(x_i) = f_i \end{cases} ;\ x_i \in J \tag{3}$$

We want to find an expression for $y_n=r(x_n)$, if the values y_i $(i=0,\ldots,n-1)$ and f_i $(i=0,\ldots,n)$ are given.

If we set:

$$r(x) = \frac{p(x)}{q(x)}$$

and provided that $q(x_i)\neq 0$ $(i=0,\ldots,n)$, the conditions (3) can be written as: (see [2])

$$\begin{cases} p(x_i) - y_i.q(x_i) = 0 \\ p'(x_i) - y_i.q'(x_i) - f_i.q(x_i) = 0 \end{cases} ;\ x_i \in J$$

or explicitly:

$$\begin{cases} \sum_{j=0}^{n} a_j.k_j(x_i) - y_i.\sum_{j=0}^{n} b_j.l_j(x_i) = 0 \\ \sum_{j=0}^{n} a_j.k_j'(x_i) - \sum_{j=0}^{n} b_j.[y_i.l_j'(x_i) + f_i.l_j(x_i)] = 0 \end{cases} ;x_i \in J \tag{4}$$

This homogeneous system of 2n+2 equations in the 2n+2 unknowns a_j, (j=0,1,...,n) has a solution iff the determinant of the coefficientmatrix A vanishes, or:

$$| A | = 0 \tag{5}$$

This condition reduces in general to a quadratic equation in the only unknown y_n (see [2]). Hence, depending on the choice of the basis functions, this problem can have no, one or two real solutions.
The direct calculation of this determinant for general basis sets is rather difficult. Therefore, we try to simplify this calculation using the special structure of the determinant.
Let K, K', L and L' be matrices defined as:

$$K = (k_j(x_i))_{ij} \quad ; \quad K' = (k_j'(x_i))_{ij}$$

$$L = (l_j(x_i))_{ij} \quad ; \quad L' = (l_j'(x_i))_{ij}$$

for i=0,...,n and j=0,...,n. Let Y and F be square diagonal matrices containing respectively the elements y_i and f_i (i=0,...,n).

It then follows from (4), that A has the following block structure:

$$A = \begin{vmatrix} K & -Y.L \\ K' & -(Y.L'+F.L) \end{vmatrix}$$

Due to the fact that we took the same number of basis functions in numerator and denominator, all blocks are square of order n+1.
Hence, (5) is equivalent with: (see [4])

$$|K| . |Y.L' + F.L - K'.K^{-1}.Y.L| = 0 \tag{6}$$

If the set $\{k_j\}$ is a Tchebycheffsystem (see [5],p.74), then $|K| \neq 0$ for any choice of J, and (6) reduces to:

$$|Y.L' + F.L - K'.K^{-1}.Y.L| = 0 \tag{7}$$

If moreover $\{l_j\}$ is also a Tchebycheffsystem, then we can multiply (7) with $|L^{-1}| \neq 0$, and we obtain:

$$|Y.L'.L^{-1} - K'.K^{-1}.Y + F| = 0 \tag{8}$$

Setting:

$$S_L = L'.L^{-1} \quad ; \quad S_K = K'.K^{-1}$$

we conclude:

> If $\{k_j\}$ and $\{l_j\}$ are Tchebycheffsystems, then there exists an interpolant (2) satisfying (3)
>
> iff
>
> $$|Y.S_L - S_K.Y + F| = 0 \tag{9}$$

As this last condition involves only a determinant of order n+1, it becomes easier to obtain the resulting quadratic equation in y_n from it. One may notice that (9) is a Lyapunov matrix equation, if we leave away the determinant signs (see [6], p.239 and [7]). However, it seems not possible to make any use of this fact.
In order to find explicit expressions for the coefficients of the quadratic equation, we take from now on:

$$k_j(x) \equiv l_j(x) \quad ; \; j=0,\ldots,n.$$

Hence,

$$K = L \quad \text{and} \quad S_K = S_L = S$$

We set:

$$\Delta_n^m y = y_n - y_{m-1}$$

$$T_{n+1} = Y.S - S.Y + F$$

where the index of T refers to its order.
We will also use the following column vectors, where $s_{ij} \in S$ and τ denotes the transposed:

$$\underline{d}_n = {}^{\tau}(s_{n0}, \ldots, s_{n,n-1})$$

$$\underline{p}_n = {}^{\tau}(s_{0n}, \ldots, s_{n-1,n})$$

$$\underline{g}_n = {}^{\tau}(s_{n0}.\Delta^1_{n-1}\, y, \ldots, s_{n,n-2}.\Delta^{n-1}_{n-1}\, y, 0)$$

$$\underline{u}_n = {}^{\tau}(s_{0n}.\Delta^1_{n-1}\, y, \ldots, s_{n-2,n}.\Delta^{n-1}_{n-1}\, y, 0)$$

One can then verify, that (9) can be written as:

$$|T_{n+1}| = \begin{vmatrix} T^{\star}_n & -\underline{u}_n - \underline{p}_n.\Delta^n_n\, y \\ {}^{\tau}\underline{g}_n + {}^{\tau}\underline{d}_n.\Delta^n_n\, y & f_n \end{vmatrix} = 0 \qquad (10)$$

The unknown y_n now only occurs in the element $\Delta^n_n\, y = y_n - y_{n-1}$. Applying elementary properties of determinants on the sums in the last row and column, we obtain:

$$\begin{vmatrix} T^{\star}_n & -\underline{p}_n \\ {}^{\tau}\underline{d}_n & 0 \end{vmatrix} .(\Delta^n_n y)^2 + \left[\begin{vmatrix} T^{\star}_n & -\underline{p}_n \\ {}^{\tau}\underline{g}_n & 0 \end{vmatrix} + \begin{vmatrix} T^{\star}_n & -\underline{u}_n \\ {}^{\tau}\underline{d}_n & 0 \end{vmatrix} \right] .\Delta^n_n y + \begin{vmatrix} T^{\star}_n & -\underline{u}_n \\ {}^{\tau}\underline{g}_n & f_n \end{vmatrix} = 0 \qquad (11)$$

Calculating the four resulting block determinants, we can conclude:

> If $\{k_j\}$ is a Tchebycheffsystem and $|T^{\star}_n| \neq 0$, then there exists an interpolant (2) satisfying (3)
>
> iff
>
> there exists a real solution for $\Delta^n_n y$ in:
>
> $$F.(\Delta^n_n\, y)^2 + G.\Delta^n_n\, y + H = 0 \qquad (12)$$
>
> with: $F = {}^{\tau}\underline{d}_n.T^{\star -1}_n.\underline{p}_n$
>
> $G = {}^{\tau}\underline{g}_n.(T^{\star}_n)^{-1}.\underline{p}_n + {}^{\tau}\underline{d}_n.(T^{\star}_n)^{-1}.\underline{u}_n$
>
> $H = {}^{\tau}\underline{g}_n.(T^{\star}_n)^{-1}.\underline{u}_n + f_n$

In the following sections we will apply this approach for n<2, in order to obtain generalized formulas of practical use.

3. <u>Second Order Corrector</u>.

For n=1, we have:

$$r(x) = \frac{a_0.k_0(x) + a_1.k_1(x)}{b_0.k_0(x) + b_1.k_1(x)}$$

and the coefficients of (12) become:

$$F = \frac{s_{01}.s_{10}}{f_0} \quad ; \quad G = 0 \quad ; \quad H = f_1$$

Calculation of the elements s_{ij} from $S=K'.K^{-1}$ and solving the obtained quadratic equation, gives the following corrector formula:

$$y_1 = y_0 \pm |K| . \sqrt{\frac{f_0.f_1}{z(x_0).z(x_1)}} \tag{13}$$

with: $z(x) = k_1'(x).k_0(x) - k_0'(x).k_1(x)$

$|K| = k_0(x_0).k_1(x_1) - k_0(x_1).k_1(x_0)$

From this formula we can derive the next two special cases:

a) $k_0(x)=1$; $k_1(x)=x$.

Substitution of this choice for equidistant base points $x_i = x_0 + i.h$ gives: (h stepsize)

$$y_1 = y_0 \pm h. \sqrt{f_0.f_1} \tag{14}$$

which agrees with the results proposed in [2].

b) $k_0(x)=1$; $k_1(x) = \frac{(f_1-f_0)}{2.h}.x^2 + \frac{(x_1f_0-x_0f_1)}{h}.x$

This choice has the property, that $z(x_0)=f_0$ and $z(x_1)=f_1$.

Hence, the square root disappears from the formula (13),

and taking the positive sign, we find:

$$y_1 = y_0 + \frac{h}{2}.(f_0+f_1)$$

This is the well known trapezoid rule.

Although (13) represents in general two solutions, we find that only one choice of sign gives useful results. This choice becomes obvious while determining the order of (13). For brevity we omit these calculations and state only the result, which was obtained by series expansion and linearizing the resulting expression. We assumed equidistant base points.

<u>Theorem:</u>

> If $\left| h.\frac{z'(x_0)}{z(x_0)} \right| < 1$
>
> and $\left| h.\left(\frac{f_0'}{f_0} - \frac{z'(x_0)}{z(x_0)}\right) \right| < 1$
>
> then (13) is of second order if we take:
>
> $$\text{sign} = \text{sgn}(f_0.z(x_0))$$

4. <u>Fourth Order Corrector.</u>

For n=2, we have:

$$r(x) = \frac{a_0.k_0(x)+a_1.k_1(x)+a_2.k_2(x)}{b_0.k_0(x)+b_1.k_1(x)+b_2.k_2(x)}$$

and the coefficients of (12) become:

$$\left\{\begin{array}{l} F = s_{20}s_{02}f_1+s_{21}s_{12}f_0+(s_{20}s_{01}s_{12}-s_{02}s_{10}s_{21}).\Delta_1^1\, y \\ G = \Delta_1^1\, y.[\,2s_{20}s_{02}f_1+(s_{20}s_{01}s_{12}-s_{02}s_{10}s_{21})\Delta_1^1\, y] \\ H = f_2.[\,f_0f_1+s_{10}s_{01}(\Delta_1^1 y)^2]+s_{20}s_{02}f_1(\Delta_1^1 y)^2 \end{array}\right. \qquad (15)$$

For a general basis $\{k_0,k_1,k_2\}$ the expressions of the elements s_{ij} in function of $k_j(x_i)$ and $k'_j(x_i)$ become rather complicated. Therefore, we give here only the results for the special case:

$$\begin{cases} k_0(x) = 1 \\ k_1(x) = t(x) \\ k_2(x) = t(x)^2 \end{cases} \tag{16}$$

If we denote:

$$t_i = t(x_i) \qquad \text{and} \qquad t'_i = t'(x_i)$$

the elements, we need for computing the coefficients (15), take the following form:

$$\begin{cases} \alpha = s_{10}s_{01} = -\dfrac{t'_0 . t'_1}{(t_1-t_0)^2} \\ \beta = s_{20}s_{02} = -\dfrac{t'_2 . t'_0}{(t_2-t_0)^2} \\ \gamma = s_{21}s_{12} = -\dfrac{t'_2 . t'_1}{(t_1-t_2)^2} \\ \delta = s_{20}s_{01}s_{12} = -s_{02}s_{10}s_{21} = \dfrac{t'_0 t'_1 t'_2}{(t_0-t_1)(t_1-t_2)(t_2-t_0)} \end{cases}$$

Substitution in (15) and solving the resulting equation (12) gives:

$$y_2 = y_1 - \frac{\Delta_1^1 y(\beta f_1 + \delta\Delta_1^1 y) + \operatorname{sgn}[\alpha(\Delta_1^1 y)^2 + f_0 f_1] . \sqrt{DIS}}{\beta f_1 + \gamma f_0 + 2\delta\Delta_1^1 y} \tag{17}$$

with:

$$DIS = -[\alpha(\Delta_1^1 y)^2 + f_0 f_1] . [\beta\gamma(\Delta_1^1 y)^2 + 2\delta f_2 \Delta_1^1 y + \beta f_1 f_2 + \gamma f_0 f_2]$$

The choice of the sign preceeding the square root is based on practical experience. Although theoretical proof is difficult, numerical results imply clearly that (17) is a fourth order method.
For the special case $t(x)=x$, we find back the formulas proposed in [2]. In case the denominator of (17) becomes zero, which means that $F=0$, equation (12) becomes linear and (17) has to be replaced by:

$$y_2 = y_1 - \frac{f_2[f_1 f_0 + \alpha(\Delta_1^1 y)^2] + \beta f_1(\Delta_1^1 y)^2}{2\Delta_1^1 y(\beta f_1 + \delta\Delta_1^1 y)} \tag{18}$$

5. Numerical Integration.

First we look at some applications in numerical integration. In this case, the function $y(x)$ we want to approximate is defined as:

$$y(x) = \int_a^x f(u).du$$

where $f(u)$ is given. The integral over an interval $[a,b]$ is then $I=y(b)$. The approximations are calculated in equidistant points $x_i=a+i.h$. To use (13) only the startingvalue $y_0=y(a)=0$ is required. For (17) we need also $y_1 \approx y(a+h)$. It can be obtained using a Runge-Kutta method of suitable order.

All numerical results were obtained from a PDP 11/60 digital computer of the Dept. of Applied Math. and Progr. (K.U.L.) in double precision (16 digits of accuracy).

In tables 1 upto 5, the first column contains the abscis value x_i. The other columns give the relative errors which were obtained in these points for the choice of basis functions given at the top of each column.

For the first two problems we used the second order formula (13).

Problem 1: $$y(x) = \int_{0.1}^{x} \frac{1}{(u - \frac{\pi}{2})}.du \; ; \; b=3.0 \; ; \; h=0.02$$

Table 1: Relative errors using (13)

$k_0(x)=1$ in all cases.

x_i	$k_1(x)=x$	$k_1(x)=\ln\lvert\cos(x)\rvert$	$k_1(x)=\ln\lvert\sin(x)-1\rvert$
0.5	1.1D-5	9.3D-4	1.7D-5
1.5	5.4D-4	1.5D-4	1.2D-5
3.0	---	3.3D-3	4.3D-5

In the first problem we find that for the basis {1,x} the corrector fails to give a real solution in the neighbourhood of the singularity. On the other hand we obtain good results beyond this singularity if we take basis functions of logarithmic type.

Problem 2: $$y(x) = \int_{0.1}^{x} \cos(u).du \; ; \; b=3.0 \; ; \; h=0.02$$

Table 2: Relative errors using (13)

$k_0(x)=1$ in all cases.

x_i	$k_1(x)=x$	$k_1(x)=(x-\frac{\pi}{2})^2$	$k_1(x)=\ln\lvert\cos(x)\rvert$
0.5	3.9D-5	7.4D-6	9.8D-4
1.5	1.6D-4	4.3D-6	2.7D-4
3.0	---	1.0D-5	3.7D-3

In the second problem, which involves no singularities, we see that (14) fails because $f(u)=\cos(u)$ changes sign in the point $\frac{\pi}{2}$.

Taking other basis sets, which cause an additional change of sign of $z(x)$ in this point, resolves this problem.

In the following two problems we used the fourth order corrector.

Problem 3:

$$y(x) = \int_{0.1}^{x} \frac{1}{(u-\frac{\pi}{2})}.du \quad ; \quad b=3.0 \quad ; \quad h=0.02$$

Table 3: Relative errors using (17)

x_i	t(x)=x	t(x)=ln |cos(x)|	t(x)=ln |sin(x)-1|
0.5	2.6D-10	3.8D-6	6.9D-10
1.5	1.8D-6	5.0D-7	8.9D-9
3.0	---	3.9D-4	5.1D-5

For the third problem we find, as for the first one, that a logarithmic basis set makes it possible to integrate through the singularity. We observe a considerable loss of accuracy while passing the singular point. This loss depends on the type of singularity and is not always as great as in this case.

Problem 4:

$$y(x) = \int_{0.6}^{x} \frac{\cos(u)-e^{(u-\frac{\pi}{2})}}{\sin(u)-e^{(u-\frac{\pi}{2})}}.du \quad ; \quad b=1.57 \quad ; \quad h=0.01$$

Table 4: Relative errors using (17)

x_i	t(x)=x	t(x)=ln $\|x-\frac{\pi}{2}\|$
1.00	2.3D-8	2.5D-8
1.25	1.5D-8	5.1D-8
1.50	---	1.5D-8
1.57	---	6.0D-5

In problem 4, we found that the calculations in column 2 failed at x=1.4, although there is no singularity present. A more appropriate basic set continues the calculation through this point.

From these and other testproblems we find that in general a good choice of basis sets is important in order to obtain the expected accuracy. Specially if a singularity is present this choice can be essential to avoid a failure of the method. Although the computational effort increases with the complexity of the basis sets, the fact that we have this choice remains an advantage of these generalized correctors.

6. Volterra Equations

Consider the nonlinear Volterra integral equation of the second kind:

$$f(x) = g(x) + \int_{x_0}^{x} K(x,y,f(y)).dy \quad ; \; x_0 \leq x \leq a$$

In [8] , I treated the solution of this problem, using the fourth order corrector (17) with t(x)=x. We noticed that if the corrector failed with this type of problem, we could still continue calculations because of the availability of a predictor value. However, we found that we lost accuracy in doing so. In the next problem we reconsider problem 5 mentioned in [8], for which this was the case.

Problem 5:

$$\begin{cases} f(x) = \dfrac{2}{x_0 - 1} - \dfrac{1}{x-1} - \displaystyle\int_{x_0}^{x} \dfrac{2f(y)}{y-1}.dy \\[2ex] f(x) = \dfrac{1}{x - 1} \end{cases}$$

Table 5: $x_0=0$; $h=0.01$

Relative errors using (17)

x_i	$t(x)=x$	$t(x)=\exp(x)$	$t(x)=\sin(x)$	$t(x)=\frac{1}{x-1}$
0.23	1.8D-8	6.9D-13	1.6D-10	1.6D-7
0.53	7.0D-7	4.6D-11	7.9D-10	4.1D-7
0.73	1.3D-6	1.3D-10	2.4D-9	9.3D-7
0.98	1.9D-5	3.6D-9	8.3D-8	1.3D-5

We find the best results in columns 3 and 4. These are precisely the cases where the corrector never failed. For the other results the predictor (the same for all columns) had sometimes to be used without correction, which gives a loss of accuracy.

As with numerical integration, we obtain in general only the required accuracy if our choice of basis set avoids failures of the corrector.

7. Conclusions

Since it is in many cases possible to avoid difficulties during the calculation by choosing an appropriate basis set, the proposed generalized correctors diminish the drawback of earlier proposed formulas. If no singularities are present, the accuracy will be of the indicated order. Otherwise, it will depend on the type of the singularity, but in many cases the results beyond this singularity will still be very useful.

References:

1. J.D.LAMBERT and B.SHAW: On the numerical solution of y'=f(x,y) by a class of formulae based on rational approximation. Mathematics of Computation 19 (1965), pp.456-462

2. H.C.THACHER,Jr.: Closed rational integration formulas. The computer Journal, Vol.8 (1966), pp.362-367

3. Y.L.LUKE, W.FAIR, J.WIMP: Predictor-corrector formulas based on rational interpolants. Comp.& Maths.with Appl.,Vol.1 (1975), pp.3-12

4. R.C.WEAST: CRC-Handbook of tables for mathematics.(4 th.ed) 1970

5. E.W.CHENEY: Introduction to approximation theory, 1966.

6. R.BELLMAN: Introduction to Matrix analysis, 1970.

7. F.R.GANTMACHER: The theory of matrices, Vol. I, 1960.

8. T.H.CLARYSSE: Rational predictor-corrector methods for nonlinear Volterra integral equations of the second kind. in L.N.M. 765,"Padé approximation and its applications", Proceedings Antwerp 1979, pp.278-294

SUR UNE GÉNÉRALISATION DE L'INTERPOLATION RATIONNELLE

Florent CORDELLIER

-=-

U.E.R. I.E.E.A. - Informatique
University of Lille
59655 Villeneuve d'Ascq Cédex
FRANCE

-=-

0. INTRODUCTION

Ce travail répond au double souci de présenter un formalisme qui soit assez général pour couvrir une vaste classe de problèmes d'interpolation et de proposer une algorithmique (c'est-à-dire un ensemble de méthodes numériques) qui permette de les résoudre en ne quittant pas le plan de ce formalisme général.

Rappelons tout d'abord que la restriction que constitue l'utilisation des seuls polynômes dans les problèmes d'interpolation (ou d'extrapolation) est de plus en plus fréquemment levée, que ce soit dans le cas polynomial [7,11,2], rationnel [2] ou des splines [10]. D'autre part, les affinités que présentent la recherche d'approximants de Padé et l'interpolation rationnelle ont été mis en évidence dans les récents travaux sur l'interpolation rationnelle d'Hermite [14,13,3].

Le formalisme que nous proposons ici permet de rendre compte de ce dernier problème comme de celui de l'interpolation polynomiale généralisée.

Nous commençons par introduire les notions de polynôme et de forme rationnelle généralisées (§1), ce qui nous permet de poser le (ou les) problème(s) de l'interpolation rationnelle généralisée (§2) et de vérifier comment quelques problèmes classiques sont pris en compte par ce formalisme (§3). Après avoir fourni les éléments d'une solution théorique des problèmes posés (§4), nous nous contentons de fournir quelques indications sur les aspects algorithmiques de leur solution (§5).

1. POLYNOMES ET FORMES RATIONNELLES GENERALISES

1.0 - Le contexte

Nous nous donnons :

K : un corps commutatif de caractéristique infinie et qui pratiquement sera $\mathbb{R}$ ou $\mathbb{C}$.

G : un ensemble arbitraire.

Nous notons :

$A(G,K)$: l'ensemble des applications de G dans K ; c'est un espace vectoriel sur K. De plus, on munit cet ensemble de l'opération $*$ définie par :

$$f,g \in A(G,K) \Longrightarrow h = f * g \in A(G,K)$$

$$\text{où } h(x) = f(x) * g(x),\ \forall\, x \in G.$$

Alors $A(G,K)$ est une algèbre commutative sur K.

F : une sous-algèbre de $A(G,K)$.

F' : le dual (algébrique) de F.

1.1 - Les polynômes généralisés

Soit ϕ une suite d'éléments $\phi_i \in F$, $\forall\, i \in \mathbb{N}$. Pour tout $p \in \mathbb{N}$, on note $V_p(\phi)$ le sous-espace de F engendré par les p+1 éléments ϕ_i, $i=0,\ldots,p$. La suite ϕ est dite *régulière* si, $\forall\, p \in \mathbb{N}$, la dimension de $V_p(\phi)$ est égale à p+1. Nous supposons désormais que cette condition de régularité est satisfaite.

Introduisons encore les notations suivantes :

$V_{-1}(\phi) = \{0\}$ (fonction identiquement nulle)

$V_p^*(\phi) = V_p(\phi) \setminus V_{p-1}(\phi)$, $\forall\, p \in \mathbb{N}$

$$V(\phi) = \bigcup_{i=-1}^{\infty} V_i^*(\phi)$$

Il est immédiat que :

$$V_p(\phi) = \bigcup_{i=-1}^{p} V_i^*(\phi),$$

$$p, q \in \mathbb{N} \cup \{-1\},\ p \neq q \Longrightarrow V_p^*(\phi) \cap V_q^*(\phi) = \emptyset.$$

Tout élément $u \in V(\phi)$ sera appelé *polynôme généralisé* relativement à la suite régulière ϕ. Le *degré effectif* de u est l'unique entier p tel que $u \in V_p^*(\phi)$.

1.2 - Les formes rationnelles généralisées

La notion de polynôme généralisé que nous venons d'introduire permet d'étendre sans difficulté la définition des formes rationnelles qui ont été introduites en [9].

Soient ϕ et ψ deux suites régulières d'éléments de F. Tout couple (u,v) où $u \in V_p(\phi)$ et $v \in V_q(\psi)$ est appelé *bi-polynôme généralisé* de degré (p,q) relativement aux deux suites ϕ et ψ (dans l'ordre). Son degré effectif est (p,q) si et seulement si $u \in V_p^*(\phi)$ et $v \in V_q^*(\psi)$.

Sur l'ensemble $B(\phi,\psi)$ de tous les bi-polynômes généralisés relativement à deux suites régulières fixées, on peut définir une relation d'équivalence $\sim$ par : $(u,v) \sim (u',v') \Longleftrightarrow \exists\, a \in K \setminus \{0\}$ tel que $u' = au$ et $v' = av$.

Cette relation induit sur $B^*(\phi,\psi) = B(\phi,\psi)\setminus\{(0,0)\}$ un espace quotient $Q(\phi,\psi)$ dont tout élément sera appelé *forme rationnelle généralisée* relativement aux deux suites régulières ϕ et ψ. Puisque tous les éléments d'une même classe d'équivalence ont même degré effectif, ce *degré effectif* sera aussi celui de la forme rationnelle correspondant à cette classe d'équivalence.

La manipulation des éléments d'un espace quotient n'est guère commode, et il est préférable de lui substituer celle d'éléments de l'espace initial. On s'y ramène grâce à l'artifice suivant :

Soit θ une application de $B^*(\phi,\psi)$ dans $K \setminus \{0\}$ telle que $\theta((au,av)) = a\theta((u,v))$, $\forall\, a \in K \setminus \{0\}$. Cette application θ sera appelée normalisation. Tout élément de $B^*(\phi,\psi)$ sera dit θ-*unitaire* si $\theta((u,v)) = 1$. Chaque classe d'équivalence contenant un et un seul élément θ-unitaire, il y a isomorphisme entre $Q(\phi,\psi)$ et l'ensemble des éléments θ-unitaires de $B^*(\phi,\psi)$. Cet unique élément θ-unitaire est appelé θ-*représentant* de la forme rationnelle généralisée.

Il est clair qu'on peut étendre aux polynômes généralisés les autres notions introduites en [9] (ou [4]) et, en particulier, la notion de fraction rationnelle. Toutefois, ces notions plus fines ne sont indispensables que si l'on se propose de rendre compte de la structure des tables d'interpolants, question que nous n'aborderons pas ici.

1.3 - Les fonctionnelles

Soit σ une suite régulière de fonctionnelles $\sigma_k \in F$. On notera $V_r(\sigma)$ le sous-espace de dimension r+1 engendré par les r+1 termes σ_k, k=0,...,r.

2. LES PROBLEMES

2.0 - Le problème de base P_o

Etant donnés :

$$\begin{cases} \phi,\psi \ : \text{deux suites régulières d'éléments de } F, \\ \sigma \ : \text{une suite régulière d'éléments de } F', \\ p,q \in \mathbb{N} \\ f \in F. \end{cases}$$

déterminer $(u,v) \in B^*(\phi,\psi)$ tel que :

$$\begin{cases} \text{degré } ((u,v)) \le (p,q) \\ \theta((u,v)) = 1 \text{ où } \theta \text{ est une normalisation donnée} \\ s(u-f * v) = 0, \ \forall \ s \in V_{p+q}(\sigma) \end{cases}$$

Remarques.

1. La recherche d'un élément θ-unitaire de $B^*(\phi,\psi)$ équivaut à celle de la forme rationnelle, le choix de θ n'intervenant que dans la représentation de la solution.

2. Pour p et q fixés, la connaissance des p+1 premiers termes de ϕ, des q+1 premiers termes de ψ et des p+q+1 premiers termes de σ suffit. L'introduction des 3 suites ϕ, ψ et σ n'est utile que dans la mesure où on se propose d'étudier la table des approximations associées aux 3 suites.

2.1 - Les autres problèmes

Existence et unicité : le problème P_1

Les entiers p et q, les suites ϕ, ψ et σ étant fixées, préciser la condition que doit vérifier la fonction $f \in F$ pour que le problème P_o ait une solution unique.

Expression de l'interpolant : le problème P_2

Proposer une représentation de la solution unique du problème P_o lorsque la condition du problème P_1 est satisfaite.

Valeurs de l'interpolant : le problème P_3

Outre les données du problème P_o, on se donne $f' \in F'$ tel que :

$$f'(f * w) = f'(f) \times f'(w), \ \forall \ w \in V_q(\psi).$$

Proposer un algorithme qui fournisse la valeur :

$$\lambda_{p,q} = f'(u) \ / \ f'(v)$$

Lorsque la condition du problème P_1 est réalisée.

3. CAS PARTICULIERS

3.0 - Introduction

Les problèmes que nous venons de définir dépendent de plusieurs choix : les suites de fonctions ϕ et ψ, la suite de fonctionnelles σ, la valeur relative de p et q, la fonctionnelle f'.

Afin de mettre en évidence la variété des problèmes dont ce formalisme permet de rendre compte, il est commode de disposer d'un vocabulaire qui traduise ces choix et qui s'accorde, autant que faire se peut, avec la terminologie classique [7]. Précisons ici les grandes lignes de ce vocabulaire.

3.1 - Le choix des suites de fonctions ϕ et ψ

Si ce choix n'est pas explicité, l'interpolation est *généralisée*.

Le cas particulier le plus courant est celui où G est un ouvert de $\mathbb{R} = K$ contenant l'origine. On peut alors proposer deux choix :

. avec $\phi_i(x) = \psi_i(x) = x^i$, $\forall\, x \in G$, $\forall\, i \in \mathbb{N}$, on a le problème d'interpolation *classique*

. avec $\phi_{2i}(x) = \psi_{2i}(x) = \cos(ix)$ et $\phi_{2i-1}(x) = \psi_{2i-1}(x) = \sin(ix)$, $\forall\, x \in G$, $\forall\, i \in \mathbb{N}$, on a le problème d'interpolation *trigonométrique*.

3.2 - Le choix de la suite de fonctionnelles σ

Si ce choix n'est pas explicité, l'interpolation est *fonctionnelle*.

On peut lui opposer le choix classique de l'interpolation *ponctuelle* dans lequel les fonctionnelles σ_k sont de la forme : $\sigma_k(f) = f(x_k)$, où $x_k \in G$.

Entre ces deux cas extrèmes, on peut définir nombre de cas intermédiaires qui supposent davantage d'informations sur G.

Supposons que G soit un intervalle ouvert de $\mathbb{R} = K$ et soit ξ une suite de points de G et ℓ une suite d'entiers non négatifs tels que $i \neq j \Longrightarrow \xi_i \neq \xi_j$ ou $\ell_i \neq \ell_j$. Si F est l'ensemble des fonctions indéfiniment continuement dérivables sur $H = \{\xi_i \mid i \in \mathbb{N}\}$ alors on peut définir la suite de fonctionnelles σ par :

$$\sigma_k(f) = f^{(\ell_k)}(\xi_k),\ \forall\ k \in \mathbb{N}.$$

Le caractère fonctionnel est alors remplacé par le caractère *hermitien*.

En particulier, si $\ell_i = 0$, $\forall\ i \in \mathbb{N}$, on retrouve le problème *ponctuel*, tandis que, si $\xi_i = 0 \in G$, et $\ell_i = i$, $\forall\ i \in \mathbb{N}$, nous dirons que l'interpolation est *Taylorienne*.

3.3 - Les valeurs de p et q

Le cas particulier le plus courant est celui où $q = 0$ dans lequel l'interpolation *polynomiale* remplace l'interpolation *rationnelle*, mais le cas où la différence $|p-q|$ n'excède pas 1 est intéressant en raison des simplifications algorithmiques auxquelles il conduit dans certains cas particuliers.

3.4 - La fonctionnelle f' du problème P_3

Le choix le plus courant est $f'(f) = f(x)$ où $x \in G$.

Toutefois, dans le cas de l'interpolation polynomiale, d'autres choix sont possibles [5].

3.5 - Liaison des problèmes

Convenons de représenter un problème d'interpolation au moyen de 3 initiales :

- la première repèrera le degré : P dans le cas polynomial, R dans le cas rationnel
- la seconde sera relative aux fonctionnelles : F dans le cas fonctionnel, H pour hermitien, P pour ponctuel et T pour Taylorien
- la dernière se rapporte aux fonctions interpolantes :
  C pour l'interpolation classique (polynômes),
  T pour trigonométrique et G pour généralisée.

Nous avons rassemblé quelques uns des problèmes d'interpolation dans le diagramme qui figure en annexe, en matérialisant le fait qu'un problème est cas particulier d'un autre par une flèche dirigée vers le problème particulier.

Nous avons encadré 6 problèmes particuliers en raison de l'intérêt qu'ils présentent :

(1) RFG : le problème d'interpolation rationnelle généralisée est le problème le plus général considéré ici.

(2) RHC : le problème d'interpolation rationnelle d'Hermite classique est très voisin des problèmes considérés par Warner [13] ou Claessens [3].

(3) RPG : le problème d'interpolation rationnelle ponctuelle généralisée a été envisagé dans le contexte particulier de l'extrapolation à l'origine par Brezinski [2].

(4) PHG : le problème d'interpolation polynomiale hermitienne généralisée est une légère extension des problèmes présentés par Davis [7] et résolus par l'algorithme de Mühlbach [11].

(5) PPT : les problèmes d'interpolation polynomiale ponctuelle trigonométri-
(6) PPC que et classique sont les deux problèmes les plus courants qui soient.

4. SOLUTION DES PROBLEMES

4.0 - Notations matricielles

Posons : $N_\ell = \{i \in \mathbb{N} \mid i \leq \ell\}$, $\forall\, \ell \in \mathbb{N}$, avec $N_{-1} = \emptyset$.

Convenons de représenter par A_ℓ^c l'élément de la matrice A dont l'indice de ligne est ℓ et celui de colonne c.

Aux données des problèmes P_0 et P_3, associons les matrices suivantes :

$X : X_k^i = \sigma_k(\phi_i)$, $i \in N_p$, $k \in N_{p+q}$ (p+q+1 lignes, p+1 colonnes)

$Z : Z_k^j = \sigma_k(f * \psi_j)$, $j \in N_q$, $K \in N_{p+q}$ (p+q+1 lignes, q+1 colonnes)

$x : x^i = f'(\phi_i)$, $i \in N_p$ (1 ligne, p+1 colonnes)

$y : y^j = f'(\psi_j)$, $j \in N_q$ (1 ligne, q+1 colonnes)

La solution $(u,v) \in \mathcal{B}^*(\phi,\psi)$ sera associée aux deux vecteurs a (indicé par N_p) et b (indicé par N_q) par :

$$u = \sum_{i=0}^{p} a_i \phi_i \quad \text{et} \quad v = \sum_{j=0}^{q} b_j \psi_j \ ,$$

ces vecteurs étant considérés comme des matrices à 1 colonne.

4.1 - Solution du problème P_1

Proposition 1.- Une condition nécessaire et suffisante pour que le problème P_0 ait une solution unique est que la matrice de Gram (X,Z) associée aux $p+q+1$ fonctionnelles σ_k $(k=0,\ldots,p+q)$ et aux $p+q+2$ fonctions ϕ_i $(i=0,\ldots,p)$ ou $f * \psi_j$ $(j=0,\ldots,q)$ soit de rang maximal $p+q+1$.

Démonstration (esquisse). Notons que le vecteur $\begin{pmatrix} a \\ -b \end{pmatrix}$ se caractérise par son orthogonalité aux lignes de la matrice (X,Z). Son unicité (à un coefficient près que la normalisation θ permet de définir) est étroitement liée au rang de cette matrice.

4.2 - Solution du problème P_3

Proposition 2.- Pourvu que la matrice (X,Z) soit de rang maximal, la solution du problème P_3 est donnée par :

$$\lambda = \frac{f'(u)}{f'(v)} = - \text{dét} \begin{pmatrix} X & Z \\ x & 0 \end{pmatrix} / \text{dét} \begin{pmatrix} X & Z \\ 0 & y \end{pmatrix} ,$$

chacun des deux quotients étant défini dans $\bar{K} = K \cup \{\infty\}$ si et seulement si l'autre l'est.

Démonstration abrégée. La condition $\lambda = f'(u)/f'(v)$ s'écrit $f'(u-\lambda v) = 0$, où (u,v) est la solution du problème P_0. Cette condition s'écrit encore : $(x,\lambda y) \begin{pmatrix} a \\ -b \end{pmatrix} = 0 \in K$. Compte tenu de la CNS du problème P_1, on a alors :

$$\begin{pmatrix} X & Z \\ x & \lambda y \end{pmatrix} \begin{pmatrix} a \\ -b \end{pmatrix} = 0 \in K^{p+q+2}$$

D'où la nullité du déterminant soit :

$$\text{dét} \begin{pmatrix} X & Z \\ x & 0 \end{pmatrix} + \lambda \; \text{dét} \begin{pmatrix} X & Z \\ 0 & y \end{pmatrix} = 0$$

Un raisonnement élémentaire montre encore que :

$$f'(u) = 0 \iff \text{dét} \begin{pmatrix} X & Z \\ x & 0 \end{pmatrix} = 0$$

et

$$f'(v) = 0 \iff \text{dét} \begin{pmatrix} X & Z \\ 0 & y \end{pmatrix} = 0, \text{ d'où le résultat.}$$

4.3 - Solution du problème P_2

Proposition 3.- Pourvu que la matrice (X,Z) soit de rang maximal, la solution du problème P_2 est donnée par :

$$R_{p,q} = -N_{p,q} \,/\, D_{p,q}$$

où

$$N_{p,q} = \sum_{i=0}^{p} \nu_{p,q}^{(i)} \phi_i \quad \text{et} \quad D_{p,q} = \sum_{j=0}^{q} \delta_{p,q}^{(j)} \psi_j$$

où $\nu_{p,q}^{(i)}$ et $\delta_{p,q}^{(j)}$ sont respectivement (au signe près) les mineurs relatifs à x^i et y^j dans la matrice : $\begin{pmatrix} x & y \\ X & Z \end{pmatrix}$.

Démonstration. C'est un corollaire de la proposition 2.

5. ALGORITHMES RECURSIFS

5.0 - Motivation

L'expression $\lambda_{p,q} = -\, \text{dét} \begin{pmatrix} X & Z \\ x & 0 \end{pmatrix} \,/\, \text{dét} \begin{pmatrix} X & Z \\ 0 & y \end{pmatrix}$ peut être utilisée directement si on se contente de la valeur $\lambda_{p,q}$ pour un indice (p,q) fixé. Il suffira alors de calculer chacun des deux déterminants au moyen de la triangularisation de Gauss en profitant du fait que ces déterminants ne diffèrent que par une ligne pour économiser les calculs.

Toutefois, il est fort rare qu'on ait besoin de calculer la valeur ponctuelle d'un interpolant pour un degré fixé à l'avance. Dans la plupart des cas pratiques, on préfère calculer la valeur ponctuelle de plusieurs interpolants de degré distincts et la concordance entre les valeurs obtenues est l'un des critères de choix majeurs du degré de l'interpolant qu'on retient.

Ceci justifie l'intérêt des algorithmes récursifs qui permettent de calculer la valeur ponctuelle d'un ensemble d'interpolants pour un coût qui excède à peine celui d'un seul.

5.1 - Techniques

L'expression de $\lambda_{p,q}$ au moyen d'un quotient de déterminants laisse entrevoir de nombreuses possibilités algorithmiques. S'il est impossible de présenter ici le détail de ces algorithmes, nous pouvons toutefois faire quelques commentaires relatifs à leur élaboration.

Rappelons que l'identité de Sylvester [2,4,8] consiste à particulariser deux lignes et deux colonnes de la matrice dont on calcule le déterminant.

Sans détruire leur structure, on peut l'appliquer aux deux matrices $\begin{pmatrix} X & Z \\ x & 0 \end{pmatrix}$ et $\begin{pmatrix} X & Z \\ 0 & y \end{pmatrix}$ de 3 façons différentes quant à la particularisation des lignes.

- Supprimer la première et la dernière ligne de (X,Z).

- Supprimer la première ligne de (X,Z) et la ligne complémentaire ((x,0) ou (0,y)).

- Supprimer la dernière ligne de (X,Z) et la ligne complémentaire.

Dans le cas particulier où $q = 0$ qui correspond à l'interpolation polynomiale, la première possibilité conduit à une généralisation du schéma de Neville-Aitken tandis que les deux autres conduisent à des généralisations du schéma de Newton [11].

Pour $p,q \geq 0$, on peut particulariser les colonnes qui correspondent respectivement à ϕ_p et ψ_q : ceci permet de définir des algorithmes récursifs à partir de $N_{i,-1}$, $N_{-1,j}$, $D_{i,-1}$, $D_{-1,j}$ et autres quantités similaires. Ces quantités seront initialisées au moyen d'un autre algorithme récursif qui se déduit aisément de celui de Mulhbach [11]. Le détail de tels algorithmes est proposé en [5].

6. CONCLUSION

Ce formalisme permet de rendre compte de nombreux problèmes d'interpolation et il conduit à des algorithmes récurrents très intéressants. Cet intérêt est toutefois tempéré par l'absence de règles singulières permettant de circonvenir les cas où l'on rencontre des sous-matrices (X,Z) qui ne sont pas de rang maximal.

Si ce formalisme ne couvre pas la notion d'opérateur de Padé [6], des modifications mineures permettraient toutefois de prendre en compte les approximants de type Padé [1] ainsi que ceux de Padé-Hermite [8] et par suite ceux de Shafer [12].

RÉFÉRENCES

[1] BREZINSKI C. Rational approximation to formal power series. Journ. Approx. Theory 25 (1979) 295-317.

[2] BREZINSKI C. A general extrapolation algorithm. Numer. Math. 35 (1980) 175-187.

[3] CLAESSENS G. Some aspects of the rational Hermite interpolation table and its applications. Ph. D. Thesis, Univ. of Antwerp, 1976.

[4] CORDELLIER F. Démonstration algébrique de l'extension de l'identité de Wynn aux tables de Padé non normales. dans [15], p. 36-60.

[5] CORDELLIER F. Quelques aspects de l'interpolation rationnelle généralisée. Pub. ANO 30, Univ. de Lille I, 1980.

[6] CUYT A.M. Abstract Padé approximants in operator theory. dans [15], p. 61.87.

[7] DAVIS P.J. Interpolation and approximation. Dover Pub. Inc., New York, 1975.

[8] DELLA DORA Contribution à l'approximation de fonctions de la variable complexe au sens Hermite Padé et de Hardy. Thèse - Grenoble (1980).

[9] GILEWICZ J. Approximants de Padé. Lecture notes in Math. 667, Springer Verlag 1978.

[10] KARLIN S. Total positivity, vol.1. Stanford Univ. Press, Stanford, Cal., 1968.

[11] MUHLBACH G. The general Neville-Aitken-algorithm and some applications.
Numer. Math. 31 (1978) 97-110

[12] SHAFER R.E. On quadratic approximation.
SIAM Journ. N.A. 11 (1974) 447-460.

[13] WARNER D.D. Hermite interpolation with rational functions.
Ph.D. Thesis, Univ. of California, 1974.

[14] WUYTACK L. On the osculatory rational interpolation problem.
Math. Comp. 29 (1975) 837-843.

[15] WUYTACK L. Padé approximation and its applications.
Lecture Notes in Math. 765, Springer Verlag 1979.

ANNEXE

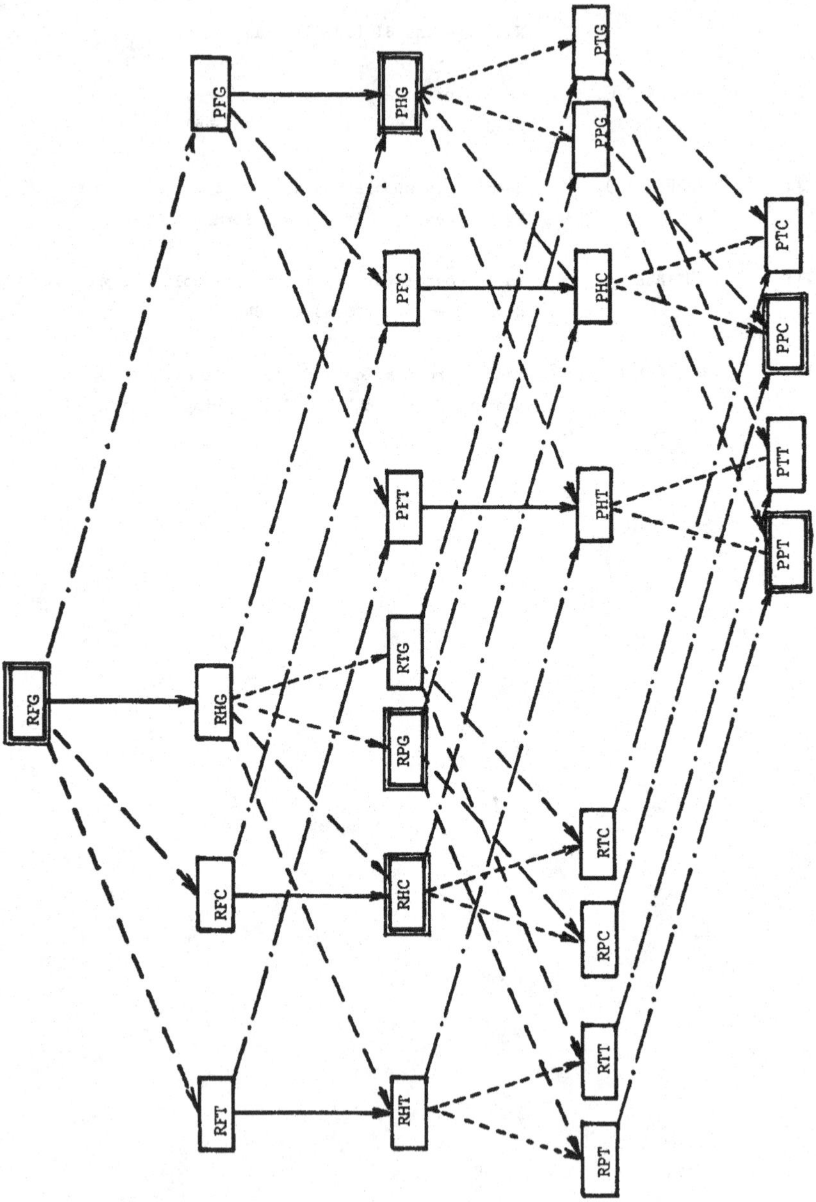
RFG
RHG
RFC
RFT
PFG
PHG
PFC
PFT
RTG
RPG
RHC
RHT
PTG
PPG
PHC
PHT
RTC
RPC
RTT
RPT
PTC
PPC
PTT
PPT

To my dear mother and brother.

NUMERICAL COMPARISON OF ABSTRACT PADE-APPROXIMANTS AND ABSTRACT RATIONAL APPROXIMANTS WITH OTHER GENERALIZATIONS OF THE CLASSICAL PADE-APPROXIMANT **

by

Annie A.M. CUYT

Departement Wiskunde
Universitaire Instelling Antwerpen
Universiteitsplein 1
B-2610 Wilrijk
Belgium

For the introduction of abstract Padé-approximants we refer to (I)*and (II). Now we want to consider an interesting numerical example that can teach us something about the location of zeros and singularities of a nonlinear operator and its different approximations (sections 1-2-3).
Also we shall compare abstract Padé-approximants for a nonlinear operator $\mathbb{R}^2 \to \mathbb{R}$ with other types of 2-variable rational approximants (sections 4-5).

* Roman figures between brackets refer to a work in the reference list.

** This work is in part supported by I.W.O.N.L. (Belgium) and in part by N.F.W.O. (Belgium).

1. Nonlinear operator

$$\text{Let } F: \mathbb{R}^2 \to \mathbb{R}^2 : (x,y) \to \begin{pmatrix} \dfrac{\dfrac{e^x}{2(1-10y)} - \dfrac{1.05-y}{1-10x}}{\sin(\frac{\pi}{2}+0.05+x-y)} \\ \\ \cos(\frac{\pi}{2}-0.05-x+y) \end{pmatrix} = \begin{pmatrix} F_1(x,y) \\ \\ F_2(x,y) \end{pmatrix}$$

The operator F is singular for $x = 0.1$

or $y = 0.1$

or $y = x + (2k+1)\frac{\pi}{2} + 0.05 \quad (k \in \mathbb{Z})$.

The second component F_2 vanishes on $y = x + k\pi + 0.05$ $(k \in \mathbb{Z})$.

For $k = o$: $F = 0$ in $\binom{0}{0.05}$ and $\binom{0.37981434..}{0.42981434..}$.

For $k < o$: the first component F_1 does not vanish on $y = x + k\pi + 0.05$.

For $k > o$: F has two zeros x_1^* and x_2^* on $y = x + k\pi + 0.05$.

On $y = x + k\pi + 0.05$ the operator F has two poles, namely in $x_1^{\blacksquare} = 0.05 - k\pi$ and $x_2^{\blacksquare} = 0.01$.

A characteristic behaviour of F on $y = x + k\pi + 0.05$ for $k>o$ and $k<o$ is respectively shown in F1.1 and F1.2, while F1.3 shows the behaviour of F on $y = x + 0.05$ $(k=o)$.

The fact that for $k>o$: $|x_1^* - x_1^{\blacksquare}|$ decreases for increasing k, complicates the calculation of the root x_1^* of $F(x,y) = 0$.

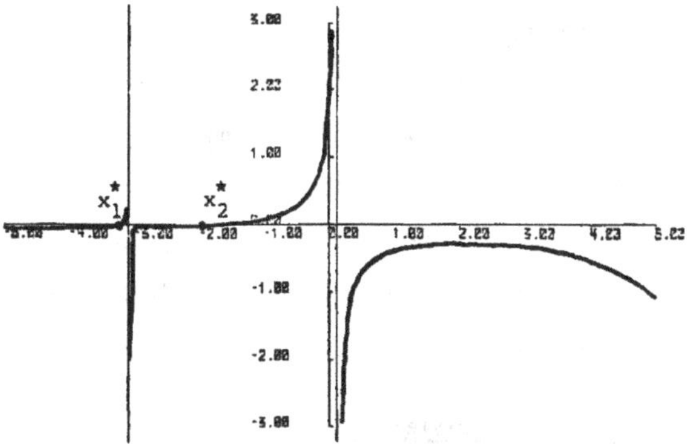

F1.1 (k=1)

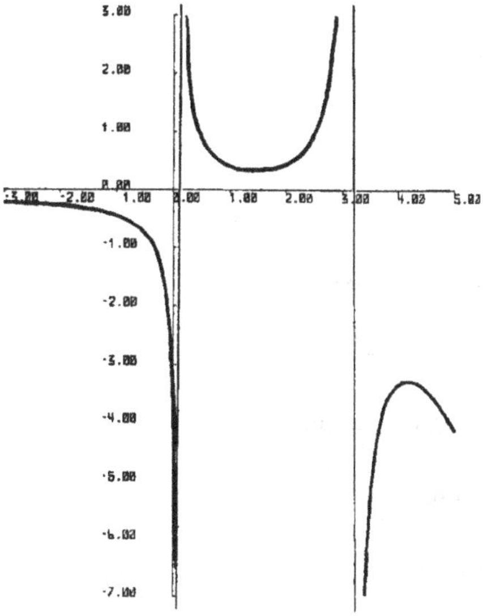

F1.2 (k=-1)

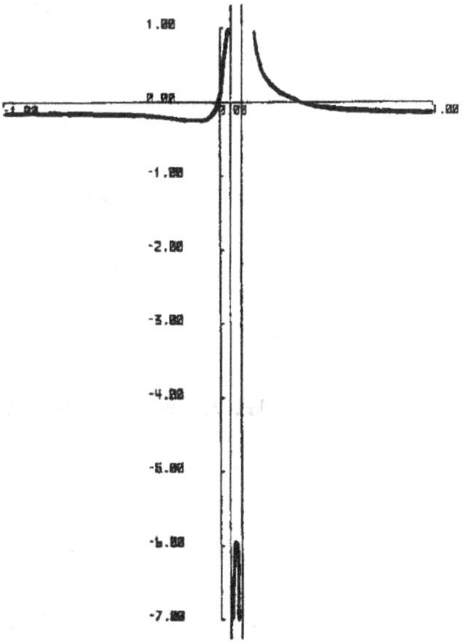

F1.3 (k=o)

2. (1,1) Abstract Padé Approximant (APA)

Let us now approximate F by a rational operator R and study the location of the zeros and the poles of this approximation. We perform the necessary calculations (as described in (I)) to obtain the (1,1)-APA in $\binom{0}{0}$ and have to conclude that its first component is undefined in $\binom{0}{0}$. But the second component is the (1,1) Abstract Padé Approximant to the second component of F.

$$R: \mathbb{R}^2 \to \mathbb{R}^2 : (x,y) \to \begin{pmatrix} \dfrac{ax+by+cx^2+dxy+ey^2}{a'x+b'y+(c'x^2+d'xy+e'y^2)\ \sin(\frac{\pi}{2}+0.05)} \\ \\ \dfrac{\cos(\frac{\pi}{2}-0.05)+(x-y)\ [\sin(\frac{\pi}{2}-0.05)+0.5\mathrm{cotg}(\frac{\pi}{2}-0.05)\cos(\frac{\pi}{2}-0.05)]}{1+0.5\ (x-y)\ \mathrm{cotg}(\frac{\pi}{2}-0.05)} \end{pmatrix}$$

$$= \begin{pmatrix} R_1(x,y) \\ R_2(x,y) \end{pmatrix}$$

with $a = -0.3025\ \mathrm{cotg}(\frac{\pi}{2}+0.05)+5.5$

$b = 0.3025\ \mathrm{cotg}(\frac{\pi}{2}+0.05) - 3.3$

$c = 42.3875 - 5.5\ \mathrm{cotg}(\frac{\pi}{2}+0.05) - 0.3025/\sin^2(\frac{\pi}{2}+0.05)$

$d = -111.75 + 9.6\ \mathrm{cotg}(\frac{\pi}{2}+0.05) + 0.605/\sin^2(\frac{\pi}{2}+0.05)$

$e = 63.5 - 3.3\ \mathrm{cotg}(\frac{\pi}{2}+0.05) - 0.3025/\sin^2(\frac{\pi}{2}+0.05)$

$a' = 0.55\ \cos(\frac{\pi}{2}+0.05) - 10\sin(\frac{\pi}{2}+0.05)$

$b' = -0.55\ \cos(\frac{\pi}{2}+0.05) + 6\sin(\frac{\pi}{2}+0.05)$

$c' = 0.55\ \mathrm{cotg}^2(\frac{\pi}{2}+0.05) - 10\mathrm{cotg}(\frac{\pi}{2}+0.05) + 104.75 + 0.55/\sin^2(\frac{\pi}{2}+0.05)$

$d' = -1.1\mathrm{cotg}^2(\frac{\pi}{2}+0.05) + 16\mathrm{cotg}(\frac{\pi}{2}+0.05) - 15 - 1.1/\sin^2(\frac{\pi}{2}+0.05)$

$e' = 0.55\ \mathrm{cotg}^2(\frac{\pi}{2}+0.05) - 6\mathrm{cotg}(\frac{\pi}{2}+0.05) - 50 + 0.55/\sin^2(\frac{\pi}{2}+0.05)$.

The second component R_2 vanishes on

$$y = x + \frac{\cos(\frac{\pi}{2} - 0.05)}{\sin(\frac{\pi}{2} - 0.05) + 0.5\cot g(\frac{\pi}{2} - 0.05)\cos(\frac{\pi}{2} - 0.05)}$$

$$= x + 0.0499...$$

R has two zeros, near to the zeros of F on $y = x+0.05$ $(k=0)$, namely in $\binom{0.00252235...}{0.05250148...}$ and $\binom{0.49805568...}{0.54803481...}$.

Because numerator and denominator of R are polynomial operators we lose the periodicity of F (no infinite number of zeros). The abstract rational approximant has distributed its poles in a very interesting manner.

Looking at F2.1 and F2.2 which show the poles of F(plotted as 000-lines) and those of R(plotted as XXX-lines) in the considered area, we remark that the dominating direction of the first bisector for the poles of F is somewhat found back in the asymptotic behaviour of the poles of R_1 (hyperbola) and in the situation of the poles of R_2 on $y=x+39.966...$ X-axis and Y-axis are marked by dots (...) as well as the asymptotes for the poles of R_1: $y = 1.305x - 0.015$

$y = -1.649x + 0.138$

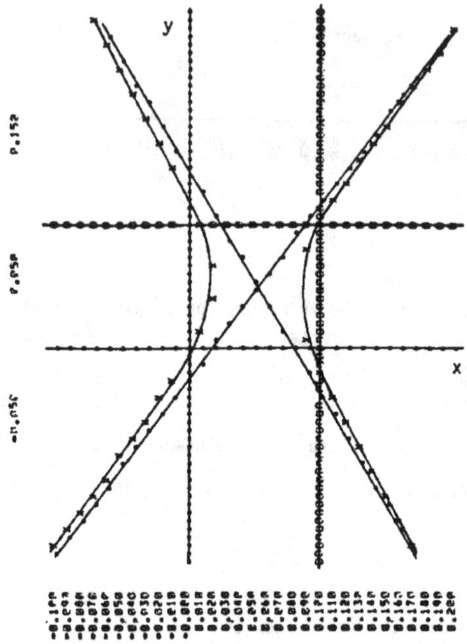

F2.1

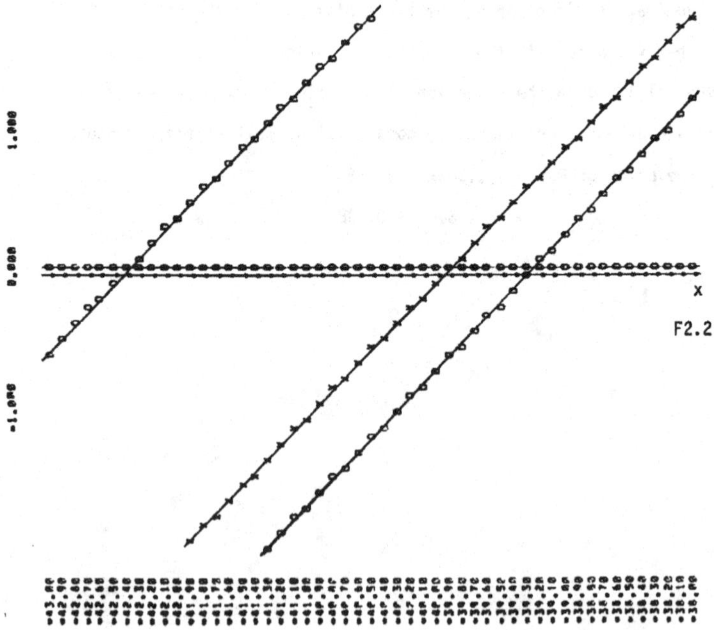

F2.2

3. <u>Taylor series expansion</u>

Since for functions $f : \mathbb{R} \to \mathbb{R}$ we can compare the curve-fitting ability of a rational function of degree n in the numerator and degree m in the denominator with that of a polynomial of degree n+m, we also calculated the Taylor series expansion T in $\binom{0}{0}$ up to and including 2nd order terms.

$$T : \mathbb{R}^2 \to \mathbb{R}^2 : (x,y) \to \begin{pmatrix} \dfrac{-0.55}{\sin(\frac{\pi}{2}+0.05)} + ax + by + cx^2 + dxy + ey^2 \\ \\ \cos(\frac{\pi}{2}-0.05) + (x-y)\sin(\frac{\pi}{2}-0.05) - 0.5\,(x-y)^2\cos(\frac{\pi}{2}-0.05) \end{pmatrix}$$

$$= \begin{pmatrix} T_1(x,y) \\ \\ T_2(x,y) \end{pmatrix}$$

with
$$a = \frac{1}{\sin(\frac{\pi}{2}+0.05)}\,(0.55\,\mathrm{cotg}(\tfrac{\pi}{2}+0.05)-10)$$

$$b = \frac{1}{\sin(\frac{\pi}{2}+0.05)}\,(-0.55\,\mathrm{cotg}(\tfrac{\pi}{2}+0.05)+6)$$

$$c = \frac{1}{\sin(\frac{\pi}{2}+0.05)}\left(-0.55\,\frac{(1+\cos^2(\frac{\pi}{2}+0.05))}{\sin^2(\frac{\pi}{2}+0.05)} + 10\,\mathrm{cotg}(\tfrac{\pi}{2}+0.05) - 104.75\right)$$

$$d = \frac{1}{\sin(\frac{\pi}{2}+0.05)}\left(1.1\,\frac{(1+\cos^2(\frac{\pi}{2}+0.05))}{\sin^2(\frac{\pi}{2}+0.05)} - 16\,\mathrm{cotg}(\tfrac{\pi}{2}+0.05) + 15\right)$$

$$e = \frac{1}{\sin(\frac{\pi}{2}+0.05)}\left(-0.55\,\frac{(1+\cos^2(\frac{\pi}{2}+0.05))}{\sin^2(\frac{\pi}{2}+0.05)} + 6\,\mathrm{cotg}(\tfrac{\pi}{2}+0.05) + 50\right)$$

F3.1 which shows the zeros of T_1 (hyperbola) and T_2 (straight lines) demonstrates that we do not have to look for zeros of T near the origin. The singularities of F are the cause of this bad behaviour of T.

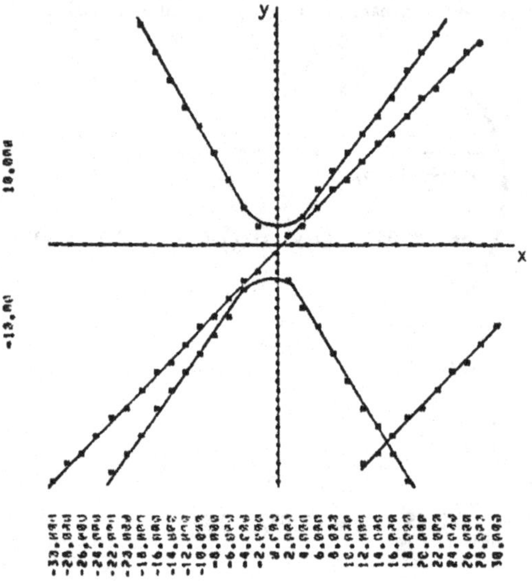

F3.1

But T_2 does also preserve the dominant direction of the first bisector for the zeros of F.

Assume that we got $\binom{0}{0}$, the point in which the approximations were calculated, from a previous iteration-step in a procedure that calculates the root $\binom{0}{0.05}$ of $F(x,y)=0$.

Calculating the approximation R and equating its numerator to zero would supply a good estimate of $\binom{0}{0.05}$, while the approximation T cannot be used to obtain an estimate of the root in $\binom{0}{0.05}$.

4. Different Padé-type 2-variable rational approximants

We are going to compare Abstract Padé Approximants (APA) or Abstract Rational Approximants (ARA) for F with Chisholm diagonal (III) approximants (CA) or Hughes Jones off-diagonal (IV,V) approximants (HJA), Lutterodt (VII)-approximants (LA), Lutterodt-approximants of type R^1 (VIII) (LAR^1), Karlsson and Wallin-approximants (VI) (KWA) and partial sums of the abstract (IX) Taylor-series development (PS), all in $\binom{n}{0}$.

The calculation of each type of approximant $\frac{P}{Q}: \mathbb{R}^2 \to \mathbb{R}$ is based on :

$$(FQ-P)(x,y) = \sum_{i,j=0}^{\infty} d_{ij}\, x^i\, y^j \text{ with } d_{ij}=0 \text{ for } (i,j) \in S \subset \mathbb{N}^2.$$

We call S the interpolationset; the choice of S determines the type of approximant. The KWA is unique when the interpolationset S contains in addition to $\{(i,j)\,|\,i+j < n\}$, as many points as possible in a given enumeration in $\mathbb{N}^2$ (we have used the diagonal enumeration (0,0), (1,0), (0,1),(2,0),(1,1),(0,2),(3,0), ...).

The LA need not be unique with respect to the chosen interpolation set (we shall give the interpolationset together with the calculated approximant). For the CA, HJA and LA we denote by $(n_1,n_2)/(m_1,m_2)$ a rational approximant of degree n_1 in x and n_2 in y in the numerator and of degree m_1 in x and m_2 in y in the denominator. For the APA and KWA we denote by n/m a rational approximant where the sum of the degrees in x and y is at most n in the numerator and at most m in the denominator. The nth partial sum of the Taylor series development is indicated by PSn.

Let N be the amount of (unknown) coefficients in the approximant (for rational approximants 1 coefficient can always be determined by a normalisation).

We consider N-1 to be a measure for the "operator-fitting" ability of the calculated rational approximant, and N to be a measure for the "operator-fitting" ability of the considered partial sum of the Taylor-series development.

For CA, HJA and LA : $N = (n_1+1)(n_2+1)+(m_1+1)(m_2+1)$.

For KWA and APA : $N = \frac{1}{2}(n+1)(n+2) + \frac{1}{2}(m+1)(m+2)$.

For PSn : $N = (n+1)(n+2)/2$.

For increasing N we expect increasing accuracy.

We do remind that for all the types of rational approximants considered, except LAB[1], a lot of classical properties of Padé-approximants for analytic functions: $\mathbb{R} \to \mathbb{R}$ remain valid, now for analytic operators: $\mathbb{R}^2 \to \mathbb{R}$, such as:

a) reciprocal covariance: if $F(0,0) \neq 0$ and P/Q is the Padé-type approximant for F with interpolationset S, then Q/P is the Padé-type approximant for $\frac{1}{F}$ with the same interpolationset.

b) if P/Q is a diagonal Padé-type approximant ($(n_1,n_2) = (m_1,m_2)$ or $n = m$) and for $a,b,c,d \in \mathbb{R}$: $ad-bc \neq 0$, $cF(0,0)+d \neq 0$, then $(aP+bQ)/(cP+dQ)$ is the Padé-type approximant ($(n_1,n_2)/(m_1,m_2)$ or n/m) for $\frac{aF+b}{cF+d}$ with the same interpolationset.

But only the CA, HJA, LA type B[1] and APA have the projection property: equating in the Padé-type approximant P/Q a variable to zero, supplies the Padé-type approximant in the remaining variables.

5. Examples and conclusions

a) Let us consider $F : \mathbb{R}^2 \to \mathbb{R} : (x,y) \to 1 + \dfrac{x}{0.1-y} + \sin(xy)$.

type	approximant	N	exact order of FQ-P
① PS 2	$1+10x+101xy$	6	$O(xy^2)$
② HJA(1,1)/(0,1)	degenerate $\dfrac{1+10x+\alpha y+(101+10\alpha)xy}{1+\alpha y}$	6	$O(xy^2)$ $\alpha = \dfrac{-1000}{101} \Rightarrow O(xy^3)$
③ HJA(1,0)/(1,1)	$\dfrac{1+10x}{1-101xy}$	6	$O(x^2y)$
④ HJA(0,1)/(1,1)	degenerate $\dfrac{1-(\frac{101+\alpha}{10})y}{1-10x-(\frac{101+\alpha}{10})y+\alpha xy}$	6	$O(x^2)\forall\alpha$
⑤ HJA(1,1)/(1,0)	$1+10x+101xy$	6	$O(xy^2)$
⑥ KWA 1/1	$\dfrac{1+10x-10.1y}{1-10.1y}$	6	$O(xy^2)$
⑦ LA(1,1)/(0,1) 	$\dfrac{1+10x+\alpha y+(101+10\alpha)xy}{1+\alpha y}$	6	see ②
⑧ LA(1,0)/(1,1) 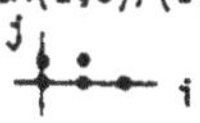	$\dfrac{1+10x}{1-101xy}$	6	$O(x^2y)$
⑨ APA 1/1	$\dfrac{1+10x-10.1y}{1-10.1y}$	6	$O(xy^2)$
⑩ CA(1,1)/(1,1)	degenerate $\dfrac{1+10x+(10-10\alpha)y+\alpha xy}{1+(10-10\alpha)y+(101\alpha-201)xy}$	8	$O(x^2y, xy^2)$ $\alpha = \dfrac{201}{101} \Rightarrow O(xy^3)$

(11) HJA(1,2)/(0,2)	degenerate $\dfrac{1+10x+\alpha y+(101+10\alpha)xy+\beta y^2+(10\beta+101\alpha+1000)xy^2}{1+\alpha y+\beta y^2}$	9	$O(xy^3)$ $\alpha=-10$ and $\beta=0$ $\Rightarrow O(x^3y^3)$
(12) HJA(2,1)/(0,2)	degenerate $\dfrac{1+10x+\alpha y+(101+10\alpha)xy}{1+\alpha y}$	9	$O(xy^2)$ $\alpha=\frac{-1000}{101}\Rightarrow O(xy^3)$
(13) LA(1,2)/(0,2) 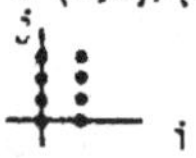	$\dfrac{1+10x+\alpha y+(101+10\alpha)xy+\beta y^2+(10\beta+101\alpha+1000)xy^2}{1+\alpha y+\beta y^2}$ with $\beta=-10^3(10+\alpha)/101$	9	see (11)
(14) LA(2,1)/(0,2) 	$\dfrac{1+10x-\frac{1000}{101}y+\frac{201}{101}xy}{1-\frac{1000}{101}y}$	9	$O(xy^3)$
(15) KWA 1/2	$\dfrac{1+10x-10.1y}{1-10.1y}$	9	$O(xy^2)$
(16) ARA 1/2	$\dfrac{x-1.01y+10y^2+10x^2-20.2xy}{x-1.01y+10y^2-10.1xy+2.01xy^2}$	14	$O(xy^3)$
(17) APA 2/1	$\dfrac{1+10x-\frac{1000}{101}y+\frac{201}{101}xy}{1-\frac{1000}{101}y}$	9	$O(xy^3)$
(18) KWA 2/1	$\dfrac{1+10x-\frac{1000}{101}y+\frac{201}{101}xy}{1-\frac{1000}{101}y}$	9	$O(xy^3)$
(19) PS 3	$1+10x+101xy+1000xy^2$	10	$O(xy^3)$
(20) CA(2,2)/(1,1)	degenerate $\dfrac{1+(10+\alpha)x-10y+(\beta+1)xy+10\alpha x^2-10xy^2+(10\beta+101\alpha)x^2y+(101\beta+1000\alpha)x^2y^2}{1+\alpha x-10y+\beta xy}$	13	$O(x^2y^3)$ $\alpha=0$ and $\beta=0$ $\Rightarrow O(x^3y^3)$

(21) KWA 3/1	$\frac{1+10x-10y+xy-10xy^2}{1-10y}$	13	$0(x^3y^3)$
(22) LA(2,2)/(1,1)	no interpolationset satisfying the description in (VII)	13	
(23) APA 3/1	$\frac{1+10x-10y+xy-10xy^2}{1-10\ y}$	13	$0(x^3y^3)$
(24) PS 4	$1+10x+101xy+1000xy^2+10000xy^3$	15	$0(xy^4)$
(25) $LAB^1(1,1)/(0,1)$	the prescribed interpolationset supplies a system of linearly dependent equations (applies to (25)–(29))	6	
(26) $LAB^1(1,0)/(1,1)$		6	
(27) $LAB^1(1,2)/(0,2)$		9	
(28) $LAB^1(2,1)/(0,2)$		9	
(29) $LAB^1(2,2)/(1,1)$		13	

For each of the mentioned approximations of the first example we have plotted the surface $|F(x,y)\text{-approximation }(x,y)|$ on $A=[-0.09,0.09]\times[-0.09,0.09]$ (F6.1 - F6.10), nearly all from the same viewpoint.

We have also calculated an estimate ε_r of $\dfrac{\sup\limits_A|F(x,y)\text{-approximation}(x,y)|}{\sup\limits_A|F(x,y)|}$

which is a measure for the relative error made by approximating ($\sup\limits_A|F(x,y)|\cong 10$).

We remark that we may call the APA accurate.

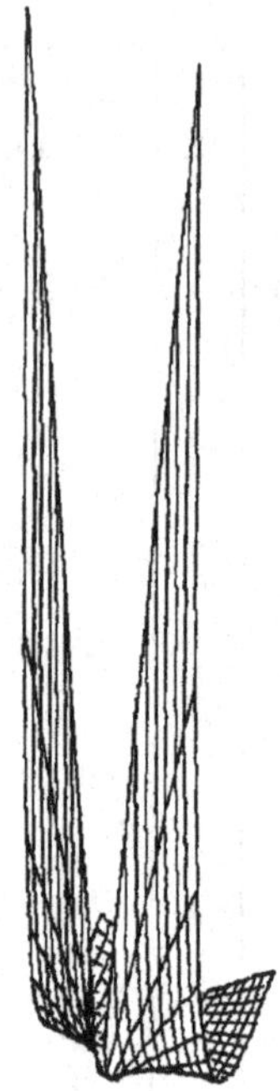

F6.1

$|F(x,y) - (1+10x+101xy)|$

$\epsilon_r = 0.73$

viewpoint (1,2,10)

① ⑤

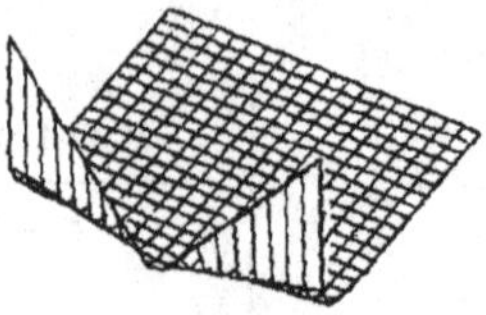

F6.2

$$\left| F(x,y) - \frac{1+10x - \frac{1000}{101} y + \frac{201}{101} xy}{1 - \frac{1000}{101} y} \right|$$

$\epsilon_r = 0.06$

viewpoint (1,2,10) ② ⑩ ⑫ ⑭ ⑰ ⑱

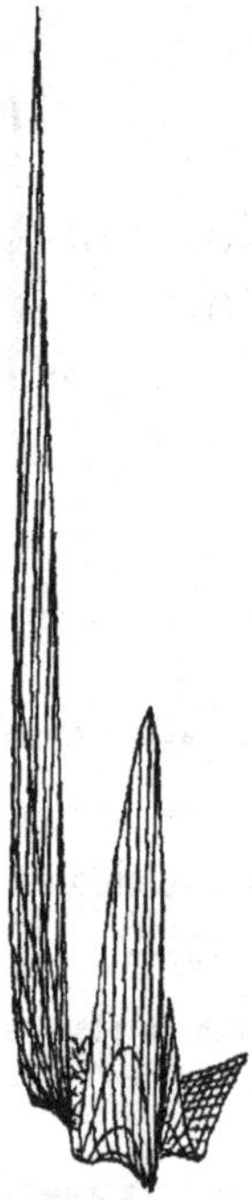

F6.3

$$\left| F(x,y) - \frac{1+10x}{1-101xy} \right|$$

$\epsilon_r = 0.81$

viewpoint (1,2,10)

③ ⑧

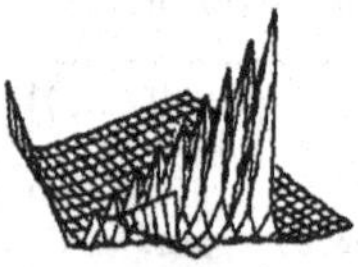

F6.4

$$\left| F(x,y) - \frac{1-10.1y}{1-10x-10.1y} \right|$$

$\epsilon_r = 90.1$

viewpoint (1,1,100) because of steepness

④

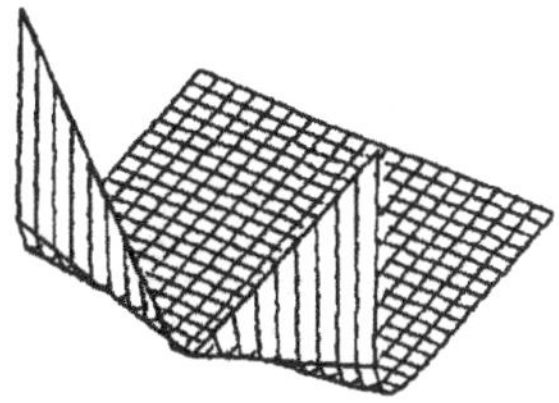

F6.5

$$\left| F(x,y) - \frac{1+10x-10.1y}{1-10.1y} \right|$$

$\epsilon_r = 0.09$

viewpoint (1,2,10)

⑥⑨⑮

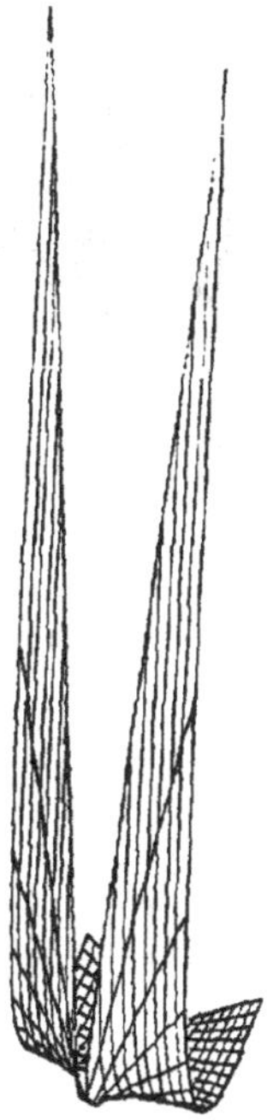

F6.6

$$\left| F(x,y) - \frac{1+10x-\frac{1}{101}y+\frac{10191}{101}xy}{1-\frac{1}{101}y} \right|$$

$\epsilon_r = 0.73$

viewpoint (1,2,10)

⑦

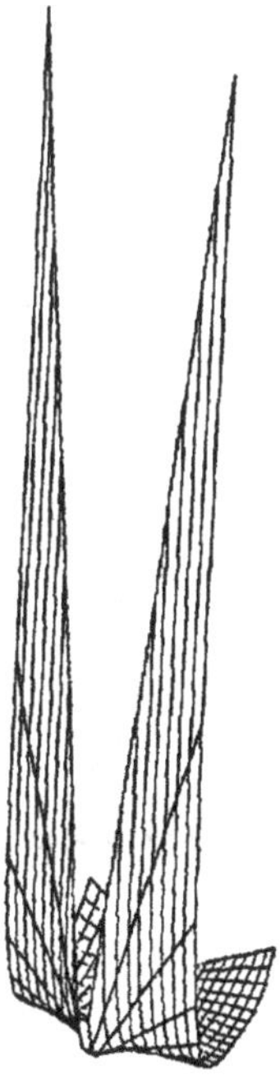

F6.8

$$\left| F(x,y) - (1+10x+101xy+1000xy^2) \right|$$

$\epsilon_r = 0.66$

viewpoint (1,2,10)

⑬ ⑲

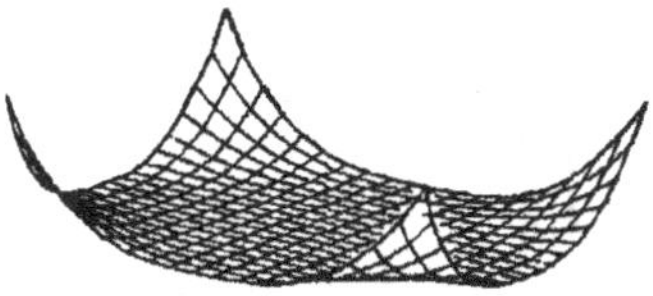

F6.7

$$\left| F(x,y) - \frac{1+10x-10y+xy-10xy^2}{1-10y} \right| * 10^5$$

$\epsilon_r = 0.9 \times 10^{-7}$ for $\left| F(x,y) - \frac{1+10x-10y+xy-10xy^2}{1-10y} \right|$

viewpoint (1,2,1) because of flatness

⑪ ⑳ ㉑ ㉓

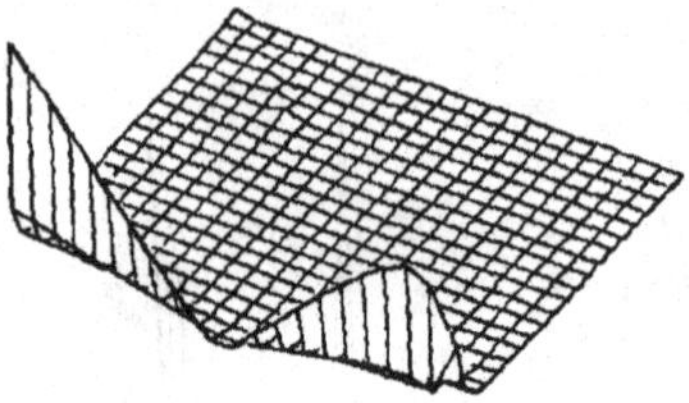

F6.9

$$\left| F(x,y) - \frac{x - 1.01y + 10y^2 + 10x^2 - 20.2xy}{x - 1.01y + 10y^2 - 10.1xy + 2.01x^2y} \right|$$

$\varepsilon_r = 0.07$ (equating $|F - \text{ARA } 1/2|$ $\binom{0}{0}$ to 0)

viewpoint (1,2,10)

(16)

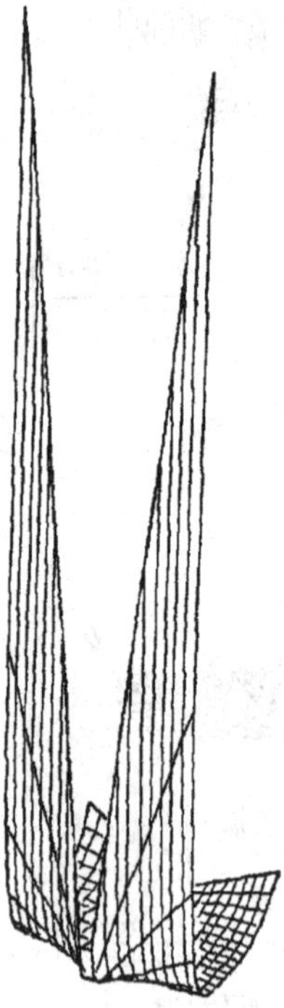

F6.10

$$\left| F(x,y) - (1 + 10x + 101xy + 1000xy^2 + 10000xy^3) \right|$$

$\varepsilon_r = 0.6$

viewpoint (1,2,10)

(24)

We merely have to compare : ε_r for ① - ⑨ and remark that at F6.2 and F6.5 the most accurate approximations are gathered; HJA(1,1)/(0,1) is a bit more accurate than KWA 1/1 and APA 1/1 because a rational function (1,1)/(0,1) fits very well the behaviour of F; however sometimes the approximation cannot be adjusted to F in this way (more complicated operators F) and we can as well at random have chosen worse approximants without knowing it (e.g. (1,0)/(1,1) or (0,1)/(1,1) or (1,1)/(1,0) in this case)

ε_r for ⑩ - ⑲ and remark that F6.2 and F6.9 gather very good approximations; only HJA(1,2)/(0,2) is better, partly because of the very degenerate solution and partly because the denominator $1+\alpha y+\beta y^2$ can fit F very well

ε_r for ⑳ - ㉔ and remark that PS 4 is very bad in comparison with all the rational approximations, what was to be expected.

We compare the different types of approximants on two other examples.

h) Let us consider $F : \mathbb{R}^2 \to \mathbb{R} : (x,y) \to \frac{xe^x - ye^y}{x-y} = \sum_{i,j=0}^{\infty} \frac{1}{(i+j)!} x^i y^j$.

Here we have in the Taylor series expansion of F a term in every power $x^i y^j$.

We compare the function values in some points.

		N					
x			0.05	0.25	0.25	0.65	0.65
y			0.25	0.05	0.45	0.45	0.85
F(x,y)	$\frac{xe^x - ye^y}{x - y}$		1.342	1.342	1.924	2.697	3.718
PS 2	$1+x+y+\frac{1}{2}(x^2+xy+y^2)$	6	1.339	1.339	1.889	2.559	3.349
$LAB^1(1,1)/(1,0)$	$\frac{1+\frac{1}{2}x+y}{1-\frac{1}{2}x}$	6	1.308	1.343	1.800	2.630	3.222
$LAB^1(1,1)/(1,1)$	$\frac{1+\frac{1}{2}(x+y)-\frac{1}{4}xy}{1-\frac{1}{2}(x+y)+\frac{1}{4}xy}$	8	1.328	1.328	2.032	2.109	4.153
ARA 1/1	$\frac{x+y+\frac{1}{2}(x^2+3xy+y^2)}{x+y-\frac{1}{2}(x^2+xy+y^2)}$	10	1.344	1.344	1.958	2.887	4.455
CA(1,1)/(1,1)	$\frac{1+\frac{1}{2}(x+y)-\frac{1}{6}xy}{1-\frac{1}{2}(x+y)+\frac{1}{3}xy}$	8	1.344	1.344	1.936	2.742	3.819
HJA(1,1)/(0,1)	$\frac{1+x+\frac{1}{2}y}{1-\frac{1}{2}y}$	6	1.343	1.308	1.903	2.419	3.609
KWA 1/1	$\frac{1+\frac{1}{2}x+y}{1-\frac{1}{2}x}$	6	1.308	1.343	1.800	2.630	3.222

We see that ARA is good as well for $x>y$ as for $x<y$ (on a not too large neighbourhood), while the other approximations, except CA(1,1)/(1,1), are not. The reason is still the same as in section 5a: (1,1)/(1,0) fits the behaviour of F if $x>y$ and (1,1)/(0,1) fits the behaviour of F if $y>x$. What's more: $F(x,y)=F(y,x)$ and APA and ARA always conserve this property, while the other types of approximants do not.

c) Now consider $F:\{(x,y)\,|\,y \geq -x-1\} \subset \mathbb{R}^2 \to \mathbb{R} : (x,y) \to \sqrt{1+x+y} =$

$$1 + \frac{x+y}{2} + \sum_{k=2}^{\infty} (-1)^{k-1} \frac{(x+y)^k}{k!} \frac{(2k-3)!!}{2^k}$$

where $(2k-3)!! = (2k-3)(2k-1) \ldots 5.3.1$

We calculate some approximants:

APA 1/1	$\dfrac{1 + 0.75\,(x+y)}{1 + 0.25\,(x+y)}$
CA (1,1)/(1,1)	$\dfrac{1 + 0.75\,(x+y) - 0.1875\,xy}{1 + 0.25\,(x+y) - 0.1875\,xy}$
HJA (1,1)/(1,0)	$\dfrac{1 + 0.75x + 0.5y - 0.125\,xy}{1 + 0.25\,x}$
HJA (1,1)/(0,1)	$\dfrac{1 + 0.5x + 0.75y - 0.125\,xy}{1 + 0.25\,y}$
KWA 1/1	$\dfrac{1 + 0.75\,(x+y)}{1 + 0.25\,(x+y)}$
LA (1,1)/(1,1)  or 	$\dfrac{1 + 0.75\,(x+y) - 0.1875\,xy}{1 + 0.25\,(x+y) - 0.1875\,xy}$
LA (1,1)/(1,0) 	$\dfrac{1 + 0.75x + 0.5y - 0.125\,xy}{1 + 0.25\,x}$
LA (1,1)/(0,1) 	$\dfrac{1 + 0.5x + 0.75y - 0.125\,xy}{1 + 0.25\,y}$

The border of the domain of F is nicely simulated by the poles of the APA k/1 : $y = -x - \frac{2k+2}{2k-1}$ with $\lim_{k\to\infty} \frac{-2k-2}{2k-1} = -1$

We also compare the function-values in different points:

	(x,y)=(2,-1)	(x,y)=(-0.4,-0.5)	(x,y)=(2,-2)
F	1.4142	0.3162	1.0000
APA 1/1	1.4000	0.4194	1.0000
CA (1,1)/(1,1)	1.3077	0.3898	1.0000
HJA (1,1)/(1,0)	1.5000	0.4722	1.3333
HJA (1,1)/(0,1)	2.0000	0.4571	2.0000

When we compare the approximations that have the same "operator-fitting" ability (as defined earlier), we see that APA 1/1 and KWA 1/1 are much more accurate than the other types.

References

Sections 1-3 :

(I) Cuyt Annie A.M.
Abstract Padé-approximants in Operator Theory.
Lecture Notes in Mathematics 765: Padé Appr. and its Appl.
(L. Wuytack ed.) pp. 61-87, Springer, Berlin,1979.

(II) Cuyt Annie A.M.
On the properties of abstract rational (1-point) approximants
(ARA).to appear in: Journal of Operator Theory 5(2), spring '81.

Sections 4-5 :

(III) Chisholm J.S.R.
N-Variable Rational Approximants.
in : Saff E.B. and Varga R.S.
Padé and Rational approximations : theory and applications.
Academic Press, London, 1977, pp. 23-42.

(IV) Hughes Jones R.
General Rational Approximants in N-Variables.
Journal of Approximation Theory 16, 1976, pp. 201-233.

(V) Hughes Jones R. and Makinson G.J.
The generation of Chisholm Rational Polynomial Approximants to
Power series in Two Variables.
Journal of the Inst. of Math. and its Appl., 1974, pp. 299-310.

(VI) Karlsson J. and Wallin H.
Rational Approximation by an interpolation procedure in several
variables.
in : Saff E.B. and Varga R.S.
Padé and Rational approximations : theory and applications.
Academic Press, London, 1977, pp. 83-100.

(VII) Lutterodt C.H.
Rational Approximants to Holomorphic Functions in n-Dimensions.
Journal of Mathematical Analysis and Applications 53, 1976,
pp. 89-98.

(VIII) Lutterodt C.H.
A two-dimensional analogue of Padé-approximant theory.
J. Phys. A : Math. Vol. 7 N° 9, 1974, pp. 1027-1037.

(IX) Rall L.B.
Computational Solution of Nonlinear Operator Equations.
John Wiley and Sons Inc., New York, 1969.

CHOIX AUTOMATIQUE ENTRE SUITES DE PARAMETRES

DANS

L'EXTRAPOLATION DE RICHARDSON.

- * -

J.P. DELAHAYE

- * -

U.E.R I.E.E.A - Informatique
University of Lille 1
59655 Villeneuve d'Ascq Cédex
FRANCE

- * -

Le procédé de Richardson [1], [10], [11] qui permet l'extrapolation polynomiale d'une suite est utilisé pour accélérer la convergence des suites (par exemple : méthode de Romberg). Pour le mettre en oeuvre,il faut fixer une suite de paramètres (qui sont les abscisses des points d'interpolation). Si parfois, on a de bonnes raisons de choisir telle suite de paramètres,il arrive souvent que l'on ne sache pas comment déterminer cette suite.

Dans cet article nous proposons des méthodes permettant le choix automatique de la suite des paramètres dans l'extrapolation de Richardson.

Les résultats présentés ici, bien qu'indépendants, complètent ceux déjà énoncés dans [6], [7].

Le point de vue adopté est le suivant : nous considérons les propriétés d'exactitude des transformations de suites obtenues par le procédé de Richardson et nous définissons des méthodes de choix ayant les meilleures propriétés d'exactitudes possibles.

On sait, en accélération de la convergence, qu'une transformation exacte sur une grande famille de suites, accélère efficacement une famille de suites encore plus grande. En conséquence, les méthodes définies ici peuvent être utilisées pour l'accélération de la convergence.

Au paragraphe I nous rappelons diverses définitions, nous donnons une formulation précise du problème de la sélection entre transformations de suites, et nous montrons que parfois ce problème est impossible à résoudre.

Au paragraphe II, nous étudions les propriétés d'exactitude des transformations de suites que l'on peut définir à partir du procédé de Richardson.

Au paragraphe III, nous proposons un procédé de sélection entre les transformations *"K-ième colonne"* (K fixé) des procédés de Richardson correspondant à des suites de paramètres différentes.

Les paragraphes IV et V envisagent les sélections entre transformations *"diagonales descendantes"* et *"diagonales rapides"*.

I. NOTATIONS, DEFINITIONS, PROBLEME DE LA SELECTION, CONTRE-EXEMPLE.

I.1. NOTATIONS, DEFINITIONS.

Soit S un ensemble de suites de nombres réels (ou plus généralement d'éléments d'un espace métriques (E,d)).

Nous appellerons *transformation de suites de S* toute application de S dans S.

Nous dirons qu'une transformation de suites de S est *effectivement calculable* s'il existe un *"algorithme pour suites"* [5], [8] correspondant à cette transformation.

Soit A une transformation de suites de S et soit $(x_m) \in S$. On notera $(A^{(n)}(x_m))_{n \in \mathbb{N}}$ ou plus simplement $(A^{(n)})_{n \in \mathbb{N}}$ la suite transformée de (x_m) par A.

Par exemple si A est le procédé d'Aïtken :

(1) $$A^{(n)}(x_m) = A^{(n)} = (x_{n+2}x_n - x_{n+1}^2)/(x_{n+2} - 2x_{n+1} + x_n)$$

En posant $A = \varepsilon_2$ on retrouve les notations usuelles [1], [2].

La formule (1) montre d'ailleurs que la transformation d'Aïtken est effectivement calculable. Par contre, il est facile d'établir que, par exemple, la transformation, qui à toute suite convergente associe la suite stationnaire égale à sa limite, n'est pas effectivement calculable (d'autres résultats négatifs de ce type sont énoncés dans [5], [8], [9]).

Supposons maintenant que S est une famille de suites convergentes.

Nous dirons que la transformation A est exacte sur S si :

$$\forall (x_m) \in S, \exists n_o \in \mathbb{N}, \forall n \geq n_o : A^{(n)}(x_m) = \lim_{m\to\infty} x_m$$

Si pour une famille S il existe une transformation effectivement calculable exacte sur S nous dirons que S *est exactable*.

La famille des suites convergentes, la famille des suites monotones convergentes, la famille des suites à convergence logarithmique, sont des familles non exactables (il s'agit là de conséquences immédiates des résultats de [8], [9]). Par contre, pour bien des méthodes d'accélération de la convergence, la famille des suites pour laquelle la méthode est exacte, est connue (voir par exemple [1], [2], [4]).

I.2. LE PROBLEME DE LA SELECTION.

Soient ${}^1A, {}^2A, \ldots, {}^{\ell}A$, ℓ transformations de suites effectivement calculables telles que :

$${}^1A \text{ est exacte sur la famille } {}^1S,$$

$${}^2A \text{ est exacte sur la famille } {}^2S,$$

$$\ldots \quad \ldots \quad \ldots \quad \ldots \quad \ldots$$

$${}^{\ell}A \text{ est exacte sur la famille } {}^{\ell}S.$$

Peut-on trouver une transformation de suites A effectivement calculable et exacte sur ${}^1S \cup {}^2S \cup \ldots \cup {}^{\ell}S$?

On parle de "*sélection*" car lorsque cette transformation existe, la plupart du temps elle vérifie :

$$\forall (x_m) \in {}^1S \cup {}^2S \cup \ldots \cup {}^{\ell}S, \forall n \in \mathbb{N} :$$

$$A^{(n)}(x_m) \in \{{}^1A^{(n)}(x_m), {}^2A^{(n)}(x_m), \ldots, {}^{\ell}A^{(n)}(x_m)\}$$

I.3. CONTRE-EXEMPLE.

Il peut arriver que le problème de la sélection n'admette pas de solution, et cela, même dans des cas assez simples comme celui que nous présentons ici :

Soit : $E = \{1/2^p + 1/2^q \mid p \in \mathbb{N},\ q \in \mathbb{N},\ p < q\} \cup \{1/2^p \mid p \in \mathbb{N}\}$.

Soit ${}^1S$ l'ensemble des suites décroissantes de points de E convergentes vers un point de E (les seules limites possibles sont les éléments de $\{1/2^p \mid p \in \mathbb{N}\}$).

Soit ${}^2S$ l'ensemble des suites décroissantes de points de E convergentes vers 0.

${}^1S$ est exactable, car la transformation définie par :

$$\begin{cases} A^{(n)} = 1/2^p \text{ quand } x_n = 1/2^p + 1/2^q,\ p < q, \\ A^{(n)} = 1/2^p \text{ quand } x_n = 1/2^p \end{cases}$$

est une transformation effectivement calculable exacte sur ${}^1S$.

${}^2S$ est exactable, car la transformation définie par :

$$A^{(n)} = 0.$$

est évidemment une transformation effectivement calculable exacte sur ${}^1S$.

La famille ${}^1S \cup {}^2S$ par contre n'est pas exactable car c'est une famille rémanente (voir [8], [9])

II. PROPRIETES D'EXACTITUDE DES TRANSFORMATIONS OBTENUES PAR LE PROCEDE DE RICHARDSON.

Dans ce paragraphe nous étudions quelles sont les différentes transformations (au sens du paragraphe I : à une suite, on fait correspondre *une* suite transformée) que le tableau donné par l'extrapolation de Richardson permet de définir. Nous indiquons à la proposition 1 les propriétés d'exactitude de ces transformations.

Toutes les suites considérées dans le reste du texte sont des suites de nombres réels ou complexes (ou plus généralement d'éléments d'un corps quelconque).

Soit (a_n) une suite de paramètres deux à deux distincts, convergente vers 0.

Pour toute suite donnée (x_n) on considère $P_k^{(n)}$ le polynôme d'interpolation de degré $\leq k$ défini par :

$$P_k^{(n)}(a_n) = x_n, \ldots, P_k^{(n)}(a_{n+k}) = x_{n+k}.$$

Le schéma de Neuville-Aïtken permet le calcul rapide des valeurs de ces polynômes [1], [3] :

$$\begin{cases} P_o^{(n)}(x) = x_n \\[2ex] P_{k+1}^{(n)}(x) = \dfrac{(a_{n+k+1}-x)\, P_k^{(n)}(x) - (a_n - x)\, P_k^{(n+1)}(x)}{a_{n+k+1} - a_n} \end{cases}$$

On pose $T_k^{(n)} = P_k^{(n)}(0)$ et ces quantités peuvent être disposées en tableau de la façon suivante :

$$\begin{array}{ccccccc}
x_o = T_0^{(o)} & & & & & & \\
 & \searrow & & & & & \\
 & & T_1^{(o)} & & & & \\
 & \nearrow & & \searrow & & & \\
x_1 = T_o^{(1)} & & & & T_2^{(o)} & & \\
 & \searrow & & \nearrow & & \searrow & \\
 & & T_1^{(1)} & & & & T_3^{(0)} \quad \ldots \\
 & \nearrow & & \searrow & & \nearrow & \\
x_2 = T_o^{(2)} & & & & T_2^{(1)} & & \\
 & \searrow & & \nearrow & & \searrow & \\
 & & T_1^{(2)} & & & & T_3^{(1)} \quad \ldots \\
 & \nearrow & & \searrow & & \nearrow & \\
x_3 = T_o^{(3)} & & & & T_2^{(2)} & & \\
 & \searrow & & \nearrow & & \searrow & \\
 & & T_1^{(3)} & & & & T_3^{(2)} \quad \ldots \\
\ldots \quad \ldots\ldots & & \ldots\ldots & & \ldots\ldots & & \ldots\ldots
\end{array}$$

Soient $\alpha(n)$ et $\beta(n)$ deux suites d'entiers, nous notons T_α^β la transformation de suites qui à la suite (x_m) fait correspondre la suite $(T_{\alpha(n)}^{(\beta(n))})$.

Si $\alpha(n) = k$ (k entier fixé) et $\beta(n) = n$, on obtient la transformation "*k-ième colonne*" que nous noterons aussi T_k.

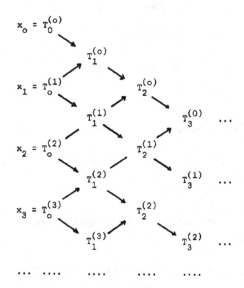

Si $\alpha(n) = n$ et $\beta(n) = k$ (k entier fixé), on obtient la transformation "*k-ième diagonale descendante*" que nous noterons aussi $T^{(k)}$.

Si $\alpha(n) = \beta(n) = n$, on obtient une transformation que l'on peut appeler "*diagonale descendante rapide*" et dont nous allons voir qu'elle présente un certain intérêt, on la notera $T^{()}$.

Considérons les familles de suites suivantes :

$$S_k = \{(x_n) \mid \exists \alpha_o, \alpha_1, \ldots, \alpha_k, \forall n, x_n = \sum_{j=0}^{k} \alpha_j a_n^j\}$$

$$S = \bigcup_{k \in \mathbb{N}} S_k$$

$$C_k = \{(x_n) \mid \exists \alpha_o, \alpha_1, \ldots, \alpha_k, \exists n_o, \forall n \geq n_o : x_n = \sum_{j=0}^{k} \alpha_j a_n^j\}$$

$$C = \bigcup_{k \in \mathbb{N}} C_k$$

Elles sont liées entre elles par les relations d'inclusions suivantes :

$$\begin{array}{ccccc} S_K & \subset & S_{K+1} & \subset & S \\ \cap & & \cap & & \cap \\ C_K & \subset & C_{K+1} & \subset & C \end{array}$$

La proposition suivante est une généralisation du théorème 23 de [1]. C'est une conséquence de ce que les transformations T_α^β sont bâties à partir du polynôme d'interpolation.

Proposition.

1) Si : $\lim_{n\to\infty} \inf \alpha(n) \geq k$ (k entier fixé) *et si* $\lim_{n\to\infty} \beta(n) = +\infty$, *alors :* T_α^β *est une transformation exacte sur* C_k *(en particulier la transformation "k-ième colonne" est exacte sur* C_k*).*

2) Si $\lim_{n\to\infty} \alpha(n) = +\infty$ *alors :* T_α^β *est une transformation exacte sur S (en particulier la transformation "k-ième diagonale descendante" est exacte sur S).*

3) Si $\lim_{n\to\infty} \alpha(n) = \lim_{n\to\infty} \beta(n) = +\infty$ *alors :*

T_α^β est une transformation exacte sur C (en particulier la transformation "diagonale descendante rapide" est exacte sur C).

Remarque.

On voit donc que, en un certain sens, la plus efficace des transformations est la *"diagonale descendante rapide"* ; cela est dû au fait qu'elle interpole sur un nombre de plus en plus grand de points tout en oubliant petit à petit les premiers points de la suite qui n'étaient peut-être pas exacts. Cependant, dans le cas d'une suite régulière dès le début, ce sera la 1ère diagonale descendante (qui n'oublie pas le passé) qui sera la plus efficace.

Démonstration.

Etablissons à titre d'exemple la partie 3°.

Soit $(x_m) \in C$. Il existe $k \in \mathbb{N}$ et $n_o \in \mathbb{N}$ tels que :

$$\forall n \geq n_o : x_n = \sum_{i=0}^{k} \alpha_i a_n^i = P(a_n).$$

Soit $n_1 \geq n_o$ tel que : $\forall n \geq n_1 : \alpha(n) \geq k$, $\beta(n) \geq n_o$.

Pour tout $n \geq n_1$, $P_{\alpha(n)}^{(\beta(n))}$ est un polynôme de degré $\leq \alpha(n)$ tel que :

$$P_{\alpha(n)}^{(\beta(n))}(a_i) = x_i \text{ pour tout } i \in \{\beta(n), \beta(n)+1, \ldots, \beta(n)+\alpha(n)\}.$$

Puisque les (a_i) ont été supposés deux à deux distincts, on a nécessairement $P_{\alpha(n)}^{(\beta(n))} = P$, et donc $T_{\alpha(n)}^{(\beta(n))} = \alpha_o = \lim_{m\to\infty} x_m$.

III. SELECTION ENTRE K-ièmes COLONNES.

Dans ce paragraphe nous définissons une méthode qui permet le choix automatique entre les k-ièmes colonnes (k entier fixé) obtenues à partir de ℓ suites différentes de paramètres (ℓ entier fixé).

Soient k et ℓ deux entiers fixés.

Soient $({}^1a_n)$, $({}^2a_n)$, ..., $({}^\ell a_n)$ ℓ suites de paramètres convergentes vers 0, dans chaque suite les paramètres étant deux à deux distincts.

Pour toute suite donnée (x_n) on considère les polynômes d'interpolation de degrée $\leq k$, ${}^1P_k^{(n)}, {}^2P_k^{(n)}, \ldots, {}^\ell P_k^{(n)}$ définis par :

$$ {}^iP_k^{(n)}({}^ia_n) = x_n, \ldots, {}^iP_k^{(n)}({}^ia_{n+k}) = x_{n+k} $$

Le schéma de Neuville-Aïtken (voir § II) en permet l'utilisation facile.

On pose ${}^iT_k^{(n)} = {}^iP_k^{(n)}(0)$. Les quantités ${}^iT_k^{(n)}$ peuvent être disposées en ℓ tableaux semblables à celui du § II.

Pour tout $i \in \{1,2,\ldots,\ell\}$ on définit :

$$ {}^iS_k = \{(x_n) \mid \exists \alpha_o, \alpha_1, \ldots \alpha_k, \forall n : x_n = \sum_{j=o}^{k} \alpha_j {}^ia_n^j\} $$

$$ {}^iS = \bigcup_{k \in \mathbb{N}} {}^iS_k $$

$$ {}^iC_k = \{(x_n) \mid \exists \alpha_o, \alpha_1, \ldots, \alpha_k, \exists n_o, \forall n \geq n_o \Rightarrow x_n = \sum_{j=o}^{k} \alpha_j {}^ia_n^j\} $$

$$ {}^iC = \bigcup_{k \in \mathbb{N}} {}^iC_k $$

Bien évidemment la transformation k-ième colonne ${}^1T_k$ est exacte sur ${}^1C_k$; de même ${}^2T_k$ sur ${}^2C_k, \ldots, {}^\ell T_k$ ou ${}^\ell C_k$.

La nouvelle transformation $A = S({}^1T_k, {}^2T_k, \ldots, {}^\ell T_k)$ que nous définissons sera exacte sur ${}^1C_k \cup {}^2C_k \cup \ldots \cup {}^\ell C_k$.

Transformation $S({}^1T_k, {}^2T_k, \ldots, {}^\ell T_k) = A$

Etape n.

Pour tout $i \in \{1,2,\ldots,\ell\}$ calculer :
${}^iC^{(n)} = \text{card } \{j \in \{1,2,\ldots,n\} \mid {}^iP_k^{(j)}({}^ia_{j+k+1}) = x_{j+k+1}\}$

Déterminer $i(n) \in \{1,2,\ldots,\ell\}$ tel que :

$$ {}^{i(n)}C^{(n)} = \max_{i \in \{1,2,\ldots \ell\}} {}^iC^{(n)} $$

Poser : $A^{(n)} = {}^{i(n)}T_k^{(n)}$.

Remarques.

1°) L'idée de la méthode, peut-être un peu cachée par le formalisme, est la suivante :

à l'étape n, on détermine la suite de coefficients qui dans les étapes précédentes a permis le plus souvent (à l'aide de (k+1) points consécutifs) de deviner le (k+2)-ième ; puis on propose le transformé par la k-ième colonne correspondant à cette suite de paramètres.

2°) Les ${}^{i}C^{(n)}$ sont calculables par récurrence en posant :

$$ {}^{i}C^{(n)} = {}^{i}C^{(n-1)} + 1 \text{ si } {}^{i}P_k^{(n)}\,({}^{i}a_{n+k+1}) = x_{n+k+1} $$

$$ {}^{i}C^{(n)} = {}^{i}C^{(n-1)} \text{ sinon.} $$

3°) La méthode présentée ici entre dans le cadre général de [6] : les coefficients de décomptes utilisés sont ceux de type 1 (ceux de type 0 ne donneraient rien). Cependant le théorème 1 énoncé plus loin évite l'hypothèse de mutuelle régularité des théorèmes de [6]. Dans le cas du procédé de Richardson, c'est là l'avantage de la méthode de sélection présentée ici sur les méthodes proposées dans [6].

4°) Diverses modifications et généralisations de la méthode de sélection ici décrite sont possibles.

(a) Dans la définition de ${}^{i}C^{(n)}$ on peut remplacer la relation : ${}^{i}P_k^{(j)}({}^{i}a_{j+k+1}) = x_{j+k+1}$ par la relation :

$$ |{}^{i}P_{(k)}^{(j)}({}^{i}a_{j+k+1}) - x_{j+k+1}| = \min_{h\in\{1,2,\dots,\ell\}} |{}^{h}P_k^{(j)}({}^{h}a_{j+k+1}) - x_{j+k+1}| $$

La méthode que l'on obtient ainsi sera mieux adaptée aux problèmes d'accélération (voir les considérations analogues à propos des méthodes C et D dans [6]).

(b) Ici la sélection s'est réalisée à l'aide d'un test sur l'interpolation par le polynôme de Lagrange au point ${}^{i}a_{n+k+1}$. Il est facile d'imaginer des variantes (préservant les résultats du théorème 1) où on utiliserait le point ${}^{i}a_{n+k+2}$, ou bien le point ${}^{i}a_{n-1}$, ou bien même encore plusieurs points (par exemple ${}^{i}a_{n+k+1}$ et ${}^{i}a_{n+k+2}$.)

(c) Moyennant certaines modifications (introduction progressive des diverses transformations) il est possible d'envisager une sélection entre une infinité (dénombrable) de transformations.

(d) Au lieu de considérer des polynômes on peut prendre d'autres fonctions d'interpolation.

Théorème 1.

La transformation $S({}^1T_k, {}^2T_k, \dots, {}^{\ell}T_k)$ *est exacte sur* ${}^1C_k \cup {}^2C_k \cup \dots \cup {}^{\ell}C_k$.

Démonstration.

Soit $(x_n) \in {}^1C_k \cup {}^2C_k \cup \dots \cup {}^{\ell}C_k$. Soit i_o tel que $(x_n) \in {}^{i_o}C_k$; on pose $x = \lim_{n\to\infty} x_n$. D'après la proposition 1, il existe $n_o \in \mathbb{N}$ tel que pour tout $n \geq n_o$: ${}^{i_o}T_k^{(n)} = x$ et ${}^{i_o}P_k^{(n)} = {}^{i_o}P_k$, où ${}^{i_o}P_k$ est le polynôme tel qu'à partir d'un certain rang $x_n = {}^{i_o}P_k({}^{i_o}a_n)$.

Pour tout $n \geq n_o$ on a donc : ${}^{i_o}C^{(n)} > n-n_o$.

Soit maintenant I l'ensemble des entiers $i \in \{1,2,\dots,\ell\}$ tels qu'il existe m_i vérifiant :

$$\forall n \geq m_i \; : \; {}^iP_k^{(n)}\,({}^ia_{n+k+1}) = x_{n+k+1}$$

Soit $j \notin I$; il existe une infinité d'entiers n tels que :

$${}^jP_k^{(n)}({}^ia_{n+k+1}) \neq x_{n+k+1}$$

donc il existe $p_j \geq n_o$ tel que pour tout $n \geq p_j$:

$${}^jC^{(n)} \leq n-n_o$$

Ceci étant vrai pour tout $j \notin I$, il existe $n_1 = \max\{p_j \mid j \notin I\}$ tel que pour tout $n \geq n_1$:

$$(*) \qquad i(n) \in I.$$

Soit $i \in I$; pour tout $n \geq m_i$:

$$\begin{cases} {}^iP_k^{(n)}({}^ia_n) = x_n, \; {}^iP_k^{(n)}({}^ia_{n+1}) = x_{n+1}, \dots, {}^iP_k^{(n)}({}^ia_{n+k+1}) = x_{n+k+1} \\ {}^iP_k^{(n+1)}({}^ia_{n+1}) = x_{n+1}, \; {}^iP_k^{(n+1)}({}^ia_{n+2}) = x_{n+2}, \dots, {}^iP_k^{(n+1)}({}^ia_{n+k+2}) = x_{n+k+2} \end{cases}$$

Les polynômes ${}^{i}P_k^{(n+1)}$ et ${}^{i}P_k^{(n)}$ correspondent donc sur k+1 points, ils sont donc égaux à un polynôme ${}^{i}P_k$ ne dépendant pas de n . Pour $n \geq m_i$ on a :

$$x_n = {}^{i}P_k({}^{i}a_n).$$

En prenant la limite quand n tend vers $+\infty$ on a :

$$x = {}^{i}P_k(0).$$

Avec (*) on obtient que pour tout $n \geq n_1 \cdot n \geq \max\{m_i \mid i \in I\}$

$$ {}^{i(n)}T_k^{(n)} = {}^{i(n)}P_k^{(n)}(0) = {}^{i(n)}P_k(0) = x.$$

IV. SELECTION ENTRE k-ièmes DIAGONALES DESCENDANTES.

Une technique tout à fait semblable à celle du paragraphe précédent est utilisée ici pour permettre le choix automatique entre les k-ièmes diagonales descendantes (k entier fixé) obtenues à partir de ℓ suites différentes de paramètres (ℓ entier fixé).

Soient k et ℓ deux entiers fixés.

Soient $({}^{1}a_n)$, $({}^{2}a_n)$,...,$({}^{\ell}a_n)$ ℓ suites de paramètres comme au paragraphe III.

Pour toute suite donnée (x_n) on considère les polynômes d'interpolation de degré $\leq n$ ${}^{1}P_n^{(k)}$, ${}^{2}P_n^{(k)}$,..., ${}^{\ell}P_n^{(k)}$ définis par :

$${}^{i}P_n^{(k)}({}^{i}a_k) = x_k, \ldots, {}^{i}P_n^{(k)}({}^{i}a_{n+k}) = x_{n+k}.$$

On obtient ℓ transformations k-ièmes diagonales descendantes que, conformément au paragraphe II, on note ${}^{1}T^{(k)}$,..., ${}^{\ell}T^{(k)}$.

D'après la proposition 1 la transformation ${}^{1}T^{(k)}$ est exacte sur ${}^{1}S$; de même ${}^{2}T^{(k)}$ sur ${}^{2}S$,..., ${}^{\ell}T^{(k)}$ sur ${}^{\ell}S$.

La nouvelle transformation $A = S({}^{1}T^{(k)}, {}^{2}T^{(k)}, \ldots, {}^{\ell}T^{(k)})$ que nous définissons sera exacte sur ${}^{1}S \cup {}^{2}S \cup \ldots \cup {}^{\ell}S$.

Transformation $S({}^{1}T^{(k)}, {}^{2}T^{(k)}, \dots, {}^{\ell}T^{(k)}) = A$

Etape n.

Pour tout $i \in \{1,2,\dots,\ell\}$ calculer :

$$ {}^{i}C^{(n)} = \text{card}\,\{j \in \{1,2,\dots,n\} \mid {}^{i}P_j^{(k)}({}^{i}a_{j+k+1}) = x_{j+k+1}\} $$

Déterminer $i(n) \in \{1,2,\dots,\ell\}$ tel que :

$$ {}^{i(n)}C^{(n)} = \max_{i\in\{1,2,\dots,\ell\}} {}^{i}C^{(n)} $$

Poser : $A^{(n)} = {}^{i(n)}T_n^{(k)}$.

Des remarques semblables à celles faites au paragraphe précédent peuvent être répétées.

Théorème 2.

La transformation $S({}^{1}T^{(k)}, {}^{2}T^{(k)}, \dots, {}^{\ell}T^{(k)})$ *est exacte sur* ${}^{1}S \cup {}^{2}S \cup \dots \cup {}^{\ell}S$.

Démonstration.

Soit $(x_n) \in {}^{1}S \cup {}^{2}S \cup \dots \cup {}^{\ell}S$. Soit i_o tel que $(x_n) \in {}^{i_o}S$.; on pose $x = \lim_{n\to\infty} x_n$. D'après la proposition 1 il existe $n_o \in \mathbb{N}$ tel que pour tout $n \geq n_o$: ${}^{i_o}T_n^{(k)} = x$ et ${}^{i_o}P_n^{(k)} = {}^{i_o}P^{(k)}$, où ${}^{i_o}P^{(k)}$ est le polynôme tel que pour tout n $x_n = {}^{i_o}P^{(k)}({}^{i_o}a_n)$.

Pour tout $n \geq n_o$ on a donc : ${}^{i_o}C^{(n)} > n - n_o$.

Comme pour le théorème 1 on établit qu'à partir d'un certain rang n_1 :

$$ i(n) \in I $$

où I désigne l'ensemble des indices i tels qu'il existe m_i vérifiant :

$$ \forall n \geq m_i : {}^{i}P_n^{(k)}({}^{i}a_{n+k+1}) = x_{n+k+1} $$

Soit $i \in I$: pour tout $n \geq m_i$:

$${}^{i}P_n^{(k)}({}^{i}a_k) = x_k,\ {}^{i}P_n^{(k)}({}^{i}a_{k+1}) = a_{k+1},\dots,{}^{i}P_n^{(k)}({}^{i}a_{n+k+1}) = x_{n+k+1}$$

$${}^{i}P_{n+1}^{(k)}({}^{i}a_k) = x_k,\ {}^{i}P_{n+1}^{(k)}({}^{i}a_{k+1}) = a_{k+1},\dots,{}^{i}P_{n+1}^{(k)}({}^{i}a_{n+k+2}) = x_{n+k+2}$$

Le polynôme ${}^{i}P_{n+1}^{(k)}$ (de d° $\leq$ n+1) correspond donc au polynôme ${}^{i}P_n^{(k)}$ (de d° $\leq$ n) sur n+2 points ; il en résulte que ces deux polynômes sont égaux à un polynôme ${}^{i}P^{(k)}$ ne dépendant pas de n.

On conclut comme pour le théorème 1.

V. SELECTION ENTRE DIAGONALES RAPIDES.

En basant le test de choix sur deux points et non plus sur un seul, on définit une méthode de choix automatique entre les diagonales rapides obtenues à partir de ℓ suites différentes de paramètres (ℓ entier fixé).

Comme précédemment on se fixe un entier ℓ, et $({}^{1}a_n)$, $({}^{2}a_n),\dots,({}^{\ell}a_n)$ ℓ suites de paramètres.

Pour toute suite donnée (x_n) on considère les polynômes d'interpolation de d° $\leq$ n, ${}^{1}P_n^{(n)}$, ${}^{2}P_n^{(n)},\dots,{}^{\ell}P_n^{(n)}$ définis par :

$${}^{i}P_n^{(n)}({}^{i}a_n) = x_n,\dots,{}^{i}P_n^{(n)}({}^{i}a_{2n}) = x_{2n}$$

On obtient ℓ transformations diagonales rapides, que conformément au paragraphe II, on note ${}^{1}T(),\dots,{}^{\ell}T()$.

D'après la proposition 1 la transformation ${}^{1}T()$ est exacte sur ${}^{1}C$; de même ${}^{2}T()$ sur ${}^{2}C,\dots,{}^{\ell}T()$ sur ${}^{\ell}C$.

La nouvelle transformation $A = S'({}^{1}T(),\ {}^{2}T(),\dots,\ {}^{\ell}T())$ que nous définissons sera exacte sur ${}^{1}C \cup {}^{2}C \cup \dots \cup {}^{\ell}C$

Transformation $S^{1}({}^{1}T(),\ {}^{2}T(),\dots,\ {}^{\ell}T()) = A.$

Etape n.

Pour tout $i \in \{1,2,\dots,\ell\}$ calculer :

$${}^{i}C^{(n)} = \text{card}\,\{j \in \{1,2,\dots,n\} \mid {}^{i}P_j^{(j)}({}^{i}a_{2j+1}) = x_{2j+1}$$

$$\text{et}\quad {}^{i}P_j^{(j)}({}^{i}a_{2j+2}) = x_{2j+2}\}$$

Déterminer $i(n) \in \{1,2,\dots,\ell\}$ tel que :

$${}^{i(n)}C^{(n)} = \max_{i \in \{1,2,\dots,\ell\}} {}^{i}C^{(n)}$$

Poser $A^{(n)} = {}^{i(n)}T_n^{(n)}$.

Théorème 3.

La transformation $S^1({}^1T^{()}, {}^2T^{()}, \dots, {}^{\ell}T^{()})$ *est exacte sur* ${}^1C \cup {}^2C \cup \dots \cup {}^{\ell}C$.

Démonstration.

Soit $(x_n) \in {}^1C \cup {}^2C \cup \dots \cup {}^{\ell}C$. Soit i_o tel que $(x_n) \in {}^{i_o}C$. On pose $x = \lim_{n\to\infty} x_n$. D'après la proposition 1 il existe n_o tel que pour tout $n \geq n_o$: ${}^{i_o}T^{(n)} = x$ et ${}^{i_o}P_n^{(n)} = {}^{i_o}P$ où ${}^{i_o}P$ est le polynôme tel qu'à partir d'un certain rang $x_n = {}^{i_o}P({}^{i_o}a_n)$.

Pour tout $n \geq n_o$ on a donc : ${}^{i}C^{(n)} > n-n_o$.

Comme pour le théorème 1 on établit qu'à partir d'un certain rang n_1 :

$$i(n) \in I$$

où I désigne l'ensemble des indices i tels qu'il existe m_i vérifiant :

$$\forall n \geq m_i \; : \; {}^{i}P_n^{(n)}({}^{i}a_{2n+1}) = x_{2n+1}, {}^{i}P_n^{(n)}({}^{i}a_{2n+2}) = x_{2n+2}.$$

Soit $i \in I$; pour tout $n \geq m_i$:

$${}^{i}P_n^{(n)}({}^{i}a_n) = x_n, {}^{i}P_n^{(n)}({}^{i}a_{n+1}) = x_{n+1}, \dots, {}^{i}P_n^{(n)}({}^{i}a_{2n+2}) = x_{2n+2}$$

$${}^{i}P_{n+1}^{(n+1)}({}^{i}a_{n+1}) = x_{n+1}, {}^{i}P_{n+1}^{(n+1)}({}^{i}a_{n+2}) = x_{n+2}, \dots, {}^{i}P_{n+1}^{(n+1)}({}^{i}a_{2n+4}) = x_{2n+4}$$

Le polynôme ${}^{i}P_{n+1}^{(n+1)}$ (de d° $\leq n+1$) correspond au polynôme ${}^{i}P_n^{(n)}$ (de d° $\leq n$) sur $n+2$ points ; il en résulte que ces deux polynômes sont égaux à un polynôme ${}^{i}P$ ne dépendant pas de n.

On conclut comme pour le théorème 1.

REFERENCES.

[1] C. BREZINSKI, *Accélération de la convergence en analyse numérique.* Lecture Notes in Mathematics 584 Springer-Verlag, Heidelberg, 1977.

[2] C. BREZINSKI, *Algorithmes d'accélération de la convergence : Etude Numérique.* Technip, Paris, 1978.

[3] C. BREZINSKI, *Analyse Numérique discrète.* Cours polycopié, Lille 1978.

[4] F. CORDELLIER, *Caractérisation des suites que la première étape du θ-algorithme transforme en suites constantes.* C.R. Acad. Sc. Paris t 284 (1977), pp 389-392.

[5] J.P. DELAHAYE, *Algorithmes pour suites non convergentes,* Numer. Math. 34 (1980) pp 333-347.

[6] J.P. DELAHAYE, *Automatic selection between sequence transformations.* A paraître.

[7] J.P. DELAHAYE, *Choix automatique entre transformations de suites destinées à l'accélération de la convergence.* Colloque d'Analyse Numérique de Gouvieux 1980.

[8] J.P. DELAHAYE et B. GERMAIN-BONNE, *Résultats négatifs en accélération de la convergence,* Numer. Math. (à paraître).

[9] J.P. DELAHAYE et B. GERMAIN-BONNE, *Résultats négatifs concernant les algorithmes d'accélération de la convergence.* Séminaire d'Analyse Numérique de Grenoble n° 337, 1980.

[10] P.J. LAURENT, *Etudes des procédés d'extrapolation en analyse numérique.* Thèse de Grenoble 1964.

[11] L.F. RICHARDSON, *The deferred approach to the limit.* Trans. Phil. Roy. Soc. 226 (1927), pp. 261-299.

QUELQUES RESULTATS SUR LA STRUCTURE DES TABLES DE PADE-HERMITE

J. DELLA DORA
Laboratoire IMAG
BP 53X
38041 GRENOBLE CEDEX

§1) RAPPELS ET NOTATIONS

Nous avons introduit dans [1], [2] les définitions suivantes que nous rappelons rapidement

Définition 1 : Soient f_1, f_2, f_3 trois séries formelles, appartenant à $k[[X]]$.
(k étant un corps quelconque).
Soit d'autre part un multi-entier $\bar{n} = (n_1, n_2, n_3)$ appartenant à N^3.
On appelle forme de Padé-Hermite (P-H) associée à (f_1, f_2, f_3) et à $\bar{n}$ tout triple de polynômes $(Y_1^{\bar{n}}, Y_2^{\bar{n}}, Y_3^{\bar{n}})$ qui vérifient :

1) $\partial° Y_i^{\bar{n}} \leq n_i \quad (i = 1,2,3)$

2) $\sum_{i=1}^{3} Y_i^{\bar{n}} f_i = X^{|\bar{n}|+2} \Gamma_{\bar{n}}$

où $|\bar{n}| = \sum_{i=1}^{3} n_i$ et $\Gamma_{\bar{n}}$ appartient à $k[[X]]$

Le problème d'existence d'une telle forme est évident, pour fixer l'unicité il est nécessaire d'imposer quelques conditions (qui naturellement vont poser des problèmes d'existence). Ceci nous a alors conduit à introduire les notations suivantes :

On notera

$$Z_i^{\bar{n}}(X) = \det \left| \begin{array}{ccc|ccc|ccc} f_0^1 & & & f_0^i & & & f_0^3 & & \\ \vdots & \ddots & & \vdots & \ddots & & \vdots & \ddots & \\ \vdots & & f_0^1 & \vdots & & f_0^i & \vdots & & f_0^3 \\ \vdots & & \vdots & \vdots & & \vdots & \vdots & & \vdots \\ f^1_{|\bar{n}|+1} & & f^1_{|\bar{n}|-n_1+1} & f^i_{|\bar{n}|+1} & & f^i_{|\bar{n}|-n_i+1} & f^3_{|\bar{n}|+1} & & f^3_{|\bar{n}|-n_3+1} \\ 0 & \cdots & 0 & 1 & \cdots & X^{n_i} & 0 & \cdots & 0 \end{array} \right|$$

(i = 1,2,3)

et

$\Delta_{\bar{n}}^{j}$ le déterminant obtenu en remplaçant dans le déterminant précédent la dernière ligne par :

$$(f^1_{|\bar{n}|+j+3}, \ldots, f^1_{|\bar{n}|-n_1+j+3} ; f^2_{|\bar{n}|+j+3}, \ldots, f^2_{|\bar{n}|-n_2+j+3} ; f^3_{|\bar{n}|+j+3}, \ldots, f^3_{|\bar{n}|-n_3+j+3})$$

j variant de -1 à + ∞.

On a alors l'expression
$$\sum_{i=1}^{3} f_i\, Z_i^{\bar n} = x^{|\bar n|+2} \left(\sum_{j=0}^{+\infty} \Delta_{\bar n}^{j-1} x^j \right)$$

Nous avons appelé dans [2] forme normalisée de Padé-Hermite d'ordre $\bar n$ associé à (f_1,f_2,f_3) le triple de polynômes ainsi formé.
Nous appelerons enfin table Δ^j la table à 3 dimensions telle qu'au point $\bar n = (n_1,n_2,n_3)$ se trouve le déterminant $\Delta_{\bar n}^j$; de la même façon on appellera table de Padé-Hermite la table à trois dimensions telle qu'au point $\bar n=(n_1,n_2,n_3)$ se trouve le triple $(Z_1^{\bar n}\ ;\ Z_2^{\bar n}\ ;\ Z_3^{\bar n})$.
Le problème de la construction récurrente de ces tables (dans le cas normal : c'est à dire lorsque $\Delta_{\bar n}^{-1} \neq 0$ pour tout $\bar n$ appartenant à N^3) est résolu dans [1], [2].
Afin de construire ces tables dans les cas singuliers il est important de connaître (comme dans le cas de la C-table de Gragg) la structure des blocs possibles dans les diverses tables Δ^j.

§2) STRUCTURE DES ZEROS DANS LES TABLES Δ^j

2.1 Règle du tétraédre

Nous allons exposer cette règle à partir de la table Δ^{-1}, dans les autres tables les résultats s'en déduisent immédiatement.

Le cas intéressant est celui où deux éléments, dans une table Δ^{-1}, adjacents sont nuls. C'est à dire s'il existe un triple $\bar n = (n_1,n_2,n_3)$ et un triple $\tilde n = (n_1+\alpha,\ n_2+\beta,\ n_3+\gamma)$; $(\alpha,\beta,\gamma$ valant un ou zéro et l'un d'entre-eux seulement est non nul) tels que
$$\Delta_{\bar n}^{-1} = \Delta_{\tilde n}^{-1} = 0$$

D'autre part nous devons supposer que $\Delta_{\bar n}^{-1}$ n'est pas adjacent à un point où Δ^{-1} est nul, ce qui nous conduit à supposer tout d'abord
$$H_1 \quad \Delta_{\bar n-e_1}^{-1}\ \Delta_{\bar n-e_2}^{-1}\ \Delta_{\bar n-e_3}^{-1} \neq 0$$

Nous allons maintenant supposer que
$$H \quad \Delta_{\bar n}^{-1} = \Delta_{\bar n+e_3}^{-1} = 0$$

Naturellement nous devons faire sur $\Delta_{\bar n+e_3}^{-1}$ certaines hypothèses nous assurant que nous sommes bien en position générale ; nous supposerons donc
$$H_2 \quad \Delta_{\bar n+e_3-e_2}^{-1} \cdot \Delta_{\bar n+e_3-e_1}^{-1} \neq 0$$

On peut alors prouver le

lemme 1 : Sous les hypothèses H, H_1, H_2 on a l'équivalence
$$\Delta_{\bar n}^{-1} = \Delta_{\bar n+e_3}^{-1} = 0 \qquad \text{et} \qquad \Delta_{\bar n}^{-1} = \Delta_{\bar n}^{0} = 0$$

Cela résulte facilement de la formule suivante (démontrée en [1])

$$\Delta^{-1}_{\bar{n}+e_3}\ \Delta^{-1}_{\bar{n}-e_1} = \Delta^{0}_{\bar{n}+e_3-e_1}\ \Delta^{-1}_{\bar{n}} - \Delta^{0}_{\bar{n}}\ \Delta^{-1}_{\bar{n}+e_3-e_1}$$

Cependant si nous avons $\Delta^{-1}_{\bar{n}} = \Delta^{0}_{\bar{n}} = 0$ nous pouvons espérer d'autres conséquences. En effet considérons les deux formules suivantes :

$$*)\ \Delta^{-1}_{\bar{n}+e_1}\ \Delta^{-1}_{\bar{n}-e_2} = \Delta^{0}_{\bar{n}}\ \Delta^{-1}_{\bar{n}+e_1-e_2} - \Delta^{0}_{\bar{n}+e_1-e_2}\ \Delta^{-1}_{\bar{n}}$$

$$**)\ \Delta^{-1}_{\bar{n}+e_2}\ \Delta^{-1}_{\bar{n}-e_3} = \Delta^{0}_{\bar{n}}\ \Delta^{-1}_{\bar{n}+e_2-e_3} - \Delta^{0}_{\bar{n}+e_2-e_3}\ \Delta^{-1}_{\bar{n}}$$

en utilisant $\Delta^{-1}_{\bar{n}} = \Delta^{0}_{\bar{n}} = 0$ nous avons en utilisant l'hypothèse H_1

<u>Lemme 2</u> : Si $\Delta^{-1}_{\bar{n}} = \Delta^{0}_{\bar{n}} = 0$ et si l'hypothèse H_1 est vérifiée alors

$$\Delta^{-1}_{\bar{n}+e_1} = \Delta^{-1}_{\bar{n}+e_2} = \Delta^{-1}_{\bar{n}+e_3} = 0$$

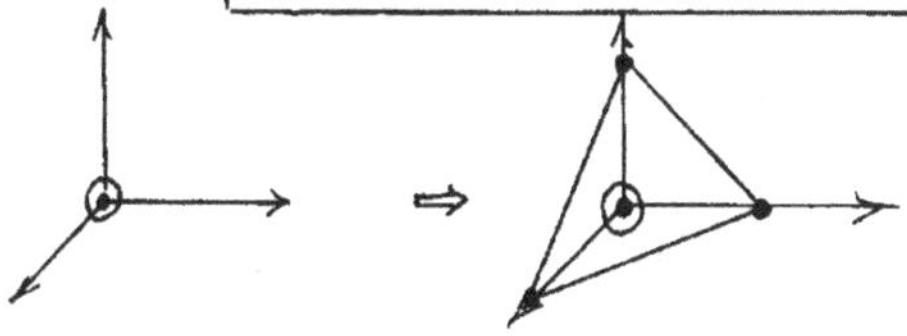

On peut résumer ces deux lemmes en ce que nous appelerons " la <u>règle du tetraédre</u>"

Le problème qui se pose alors est celui de savoir s'il n'existerait pas une structure en bloçs carrés (comme pour la C-table de Gragg). L'exemple suivant montre qu'il n'en est rien :

<u>Exemple 1</u> : Considérons la série formelle dont les premiers termes sont

$$f(z) = 1 + \beta z^2 + \gamma z^3 + \lambda z^4 + \varepsilon z^5 + \dots$$

et considérons le triple $(f,f^2,1)$. Un calcul direct montre que

$$\Delta^{j}_{000} = 0 \quad \text{pour tout } j \geq -1$$

$$\Delta^{-1}_{100} = \Delta^{-1}_{010} = \Delta^{-1}_{001} = 0$$

et pourtant

$$\Delta^{-1}_{111} = \beta^4$$

Comme nous pouvons toujours choisir $\beta \neq 0$ nous avons notre contre exemple.

2.2 <u>Conséquences de la règle du tétraédre</u>

Nous allons utiliser la règle du tétraédre pour étudier la structure des blocs de la table Δ^{-1}. Pour cela nous allons d'abord supposer (à titre d'exemple) que :

$$\Delta^{-1}_{\bar{n}} = \Delta^{-1}_{\bar{n}+e_3} = \Delta^{-1}_{\bar{n}+2e_3} = 0$$

D'autre part nous allons supposer que ces points ne sont pas adjacents à d'autres points eux-mêmes nuls soit

$$\Delta^{-1}_{\bar n-e_1} \cdot \Delta^{-1}_{\bar n-e_2} \cdot \Delta^{-1}_{\bar n-e_3} \neq 0$$

$$\Delta^{-1}_{\bar n+e_3-e_2} \cdot \Delta^{-1}_{\bar n+e_3-e_1} \neq 0$$

De ces deux hypothèses on déduit

1) $\Delta^{0}_{\bar n} = \Delta^{0}_{\bar n+e_3} = 0$

2) D'où l'on a $\Delta^{-1}_{\bar n+e_1} = \Delta^{-1}_{\bar n+e_2} = \Delta^{-1}_{\bar n+e_3} = 0$

La première étape peut donc se résumer à :

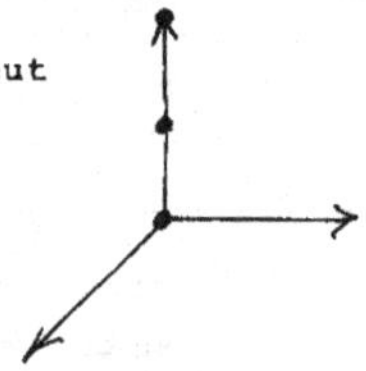

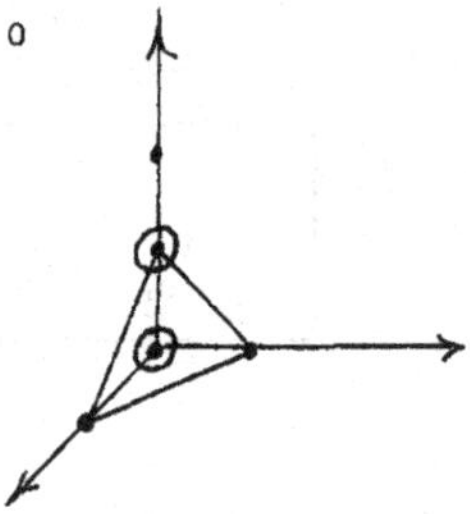

La deuxième étape :

Elle consiste à utiliser la formule (*) du paragraphe précédent en remplaçant $\bar n$ par $\bar n+e_3$:

$$\Delta^{-1}_{\bar n+e_1+e_3} \Delta^{-1}_{\bar n-e_2+e_3} = \Delta^{0}_{\bar n+e_3} \Delta^{-1}_{\bar n+e_1-e_2+e_3} - \Delta^{0}_{\bar n+e_1-e_2+e_3} \Delta^{-1}_{\bar n+e_3}$$

puisque $\Delta^{-1}_{\bar n+e_3} = \Delta^{0}_{\bar n+e_3} = 0$ en utilisant l'hypothèse $\Delta^{-1}_{\bar n+e_3-e_2} \neq 0$

on déduit

$$\Delta^{-1}_{\bar n+e_1+e_3} = 0$$

La formule

$$\Delta^{-1}_{\bar n+e_2+e_3} \Delta^{-1}_{\bar n-e_1+e_3} = \Delta^{0}_{\bar n-e_1+e_2+e_3} \Delta^{-1}_{\bar n+e_3} - \Delta^{-1}_{\bar n-e_1+e_2+e_3} \Delta^{0}_{\bar n+e_3}$$

nous perme -telle de conclure que :

$$\Delta^{-1}_{\bar n+e_2+e_3} = 0$$

Troisième étape :

Nous allons maintenant utiliser la symétrie du problème, pour cela nous ferons l'hypothèse que

$$\Delta^{-1}_{\bar n+e_3+e_1-e_2} \cdot \Delta^{-1}_{\bar n+e_2+e_3-e_1} \neq 0$$

(ce n'est pas la seule possible !) alors

$$\Delta^{-1}_{\bar n+e_2} = \Delta^{-1}_{\bar n+e_2+e_3} = 0 \Rightarrow \Delta^{0}_{\bar n+e_2} = 0$$

$$\Delta^{-1}_{\bar n+e_1} = \Delta^{-1}_{\bar n+e_1+e_3} = 0 \Rightarrow \Delta^{0}_{\bar n+e_1} = 0$$

Quatrième étape

Maintenant reprenons la formule (**) du paragraphe précédent dans laquelle nous supposerons $\bar{n}$ changé en $\bar{n}+e_2$:

$$\Delta^{-1}_{\bar{n}+2e_2} \; \Delta^{-1}_{\bar{n}+e_2-e_3} = \Delta^{0}_{\bar{n}+e_2} \; \Delta^{-1}_{\bar{n}+2e_2-e_3} - \Delta^{0}_{\bar{n}+2e_2-e_3} \; \Delta^{-1}_{\bar{n}+e_2}$$

en supposant $\Delta^{-1}_{\bar{n}+e_2-e_2} \neq 0$ on déduit de $\Delta^{0}_{\bar{n}+e_2} = \Delta^{-1}_{\bar{n}+e_2} = 0$ que

$$\underline{\Delta^{-1}_{\bar{n}+2e_2} = 0}$$

De la même façon si nous supposons $\Delta^{-1}_{\bar{n}+e_1-e_2} \neq 0$ alors en utilisant (*) où l'on remplace $\bar{n}$ par $\bar{n}+e_1$ on déduit

$$\underline{\Delta^{-1}_{\bar{n}+2e_1} = 0}$$

Ceci achève la quatrième étape que nous résumons sur le schéma suivant :

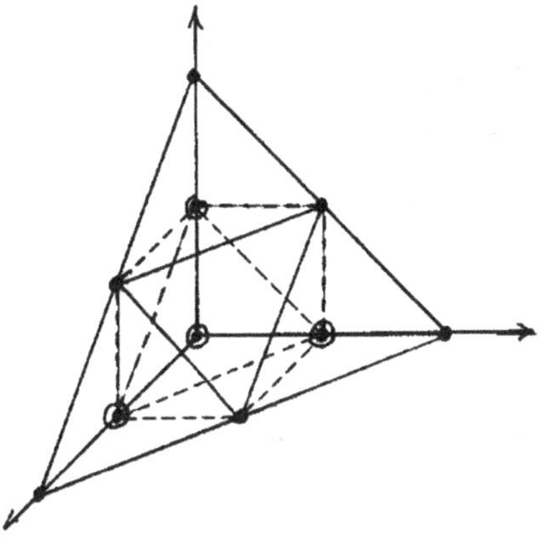

Le cas du point $\bar{n}+e_1+e_2$ reste à régler pour assurer la parfaite symétrie du problème :

Sous l'hypothèse $\Delta^{-1}_{\bar{n}+e_1-e_3} \neq 0$ on déduit de la formule ** en remplaçant $\bar{n}$ par $\bar{n}+e_1$ que $\Delta^{-1}_{\bar{n}+e_1+e_2} = 0$ ce qui complète le tétraédre :

La cinquième étape va prendre en compte les égalités

$$\underline{\Delta^{0}_{\bar{n}} = \Delta^{0}_{\bar{n}+e_1} = \Delta^{0}_{\bar{n}+e_2} = \Delta^{0}_{\bar{n}+e_3} = 0}$$

Nous pouvons alors utiliser une des formules démontrées dans [2] , à savoir

$$\underbrace{\Delta^{0}_{\bar{n}+e_3}}_{=\,0}\ \Delta^{-1}_{\bar{n}-e_2} = \underbrace{\Delta^{-1}_{\bar{n}-e_2+e_3}\ \Delta^{-1}_{\bar{n}}}_{=\,0} - \Delta^{-1}_{\bar{n}-e_2+e_3}\ \Delta^{1}_{\bar{n}}$$

Or nous avons déjà fait l'hypothèse $\Delta^{-1}_{\bar{n}-e_2+e_3} \neq 0$ donc nous pouvons en déduire

$$\underline{\Delta^{1}_{\bar{n}} = 0}$$

Ce qui achève les conséquences de l'application de la règle du tétraédre au cas considéré.

<u>Remarque</u> : On peut alors reprendre le problème à l'envers et considérer que l'on a

$$\Delta^{-1}_{\bar{n}} = \Delta^{0}_{\bar{n}} = \Delta^{1}_{\bar{n}} = 0$$

et, sous des hypothèses génériques convenables, en déduire la structure en grand tétraédre précédent.

Cependant les conséquences de $\Delta^{-1}_{\bar{n}} = \Delta^{0}_{\bar{n}} = \Delta^{1}_{\bar{n}} = 0$ ne s'arrêtent pas là ; en effet nous allons considérer ce qui se passe sur la couche correspondante à une somme des indices $|\bar{n}| + 3$.

<u>Nous rencontrons deux types de points</u> :

a) <u>les points situés sur les bords</u>

(n_1,n_2,n_3+3) ; (n_1,n_2+1,n_3+2) ; (n_1,n_2+2,n_3+1) , (n_1,n_2+3,n_3) et les permutés de ces points.

b) <u>un point central</u>

$(n_1+1,\ n_2+1,\ n_3+1)$

Génériquement il ne se passera rien pour le cas a)

Par contre le point central subit lui de profondes modifications :

Pour cela il nous faut développer $\Delta^{i}_{(n_1+1,n_2+1,n_3+1)}$ d'une manière non habituelle:

$$\text{Soit } \Delta^{-1}_{n_1+1,n_2+1,n_3+1} = \det \left|\begin{array}{cc|cc|cc}
f^1_0 & & f^2_0 & & f^3_0 & \\
 & \ddots & & \ddots & & \ddots \\
 & f^1_0 & & f^2_0 & & \\
 & & & & & f^3_0 \\
f^1_{|\bar{n}|} & & f^2_{|\bar{n}|} & & & \\
f^1_{|\bar{n}|+1} & & f^2_{|\bar{n}|+1} & & & \\
f^1_{|\bar{n}|+2} & f^1_{|\bar{n}|-n_1+2} & f^2_{|\bar{n}|+2} & f^2_{|\bar{n}|-n_2+2} & f^3_{|\bar{n}|+2} & f^3_{|\bar{n}|-n_3+2}
\end{array}\right|$$

(où nous notons $|\hat{n}| = |\bar{n}| + 3$) nous allons le développer suivant la règle de Sylvester en supprimant les lignes $|\hat{n}|+1$ et $|\hat{n}|+2$ et les colonnes n_1+2 et n_1+n_2+4. Nous obtenons alors

$$\Delta^{-1}_{n_1+1,n_2+1,n_3+1} \cdot \Delta^{1}_{n_1,n_2,n_3+1} = A^{-1}_{n_1,n_2+1,n_3+1} \; \Delta^{-1}_{n_1+1,n_2,n_3+1}$$

$$- \; A^{-1}_{n_1+1,n_2,n_3+1} \; \Delta^{-1}_{n_1,n_2,n_3+1}$$

les déterminants A jouant un rôle secondaire nous ne les préciserons pas plus. Cependant puisque $\Delta^{1}_{n_1,n_2,n_3+1} \neq 0$ nous en déduisons que $\underline{\Delta^{-1}_{n_1+1,n_2+1,n_3+1} = 0}$

Si maintenant nous considérons plus généralement $\Delta^{i}_{n_1+1,n_2+1,n_3+1}$ $i \geq -1$

par un raisonnement du même type que celui que nous venons de faire nous pouvons déduire le

<u>Théorème</u> : Si $\Delta^{-1}_{\bar{n}} = \Delta^{0}_{\bar{n}} = \Delta^{-1}_{\bar{n}} = 0$ et si les conditions génériques habituelles sont vérifiées, alors : $\Delta^{j}_{\bar{n}+e_1+e_2+e_3} = 0 \quad \forall j \geq -1$

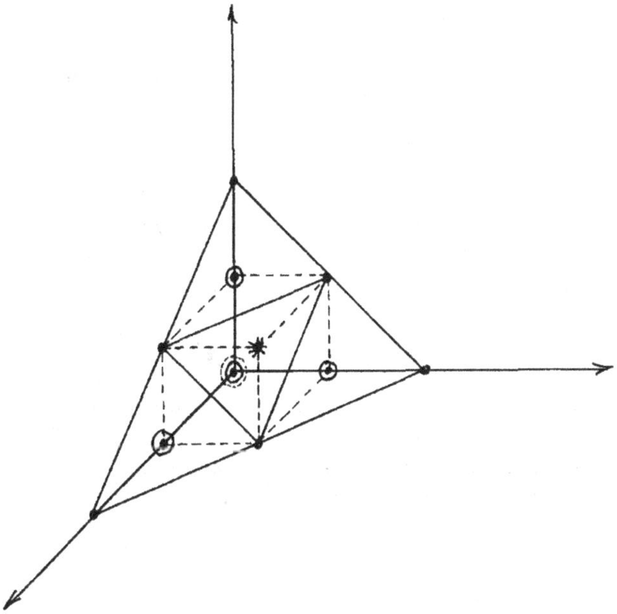

§ 3) CONSEQUENCES DE LA REGLE DU TETRAEDRE POUR LA TABLE P-H

Nous allons nous placer dans une situation "générique" où l'on a pour le triple $\bar{n}$

$$\Delta^{-1}_{\bar{n}} = \Delta^{-1}_{\bar{n}+e_i} = 0 \quad \text{(pour tout } i = 1,2,3)$$

et donc $\Delta^{0}_{\bar{n}} = 0$

Nous noterons $\bar{n} = (n_1, n_2, n_3)$

Commençons par démontrer les lemmes suivants :

lemme 1 : Sous les hypothèses précédentes : $\partial^\circ(Z_1^{\bar{n}+e_i}) \le n_1$; $\partial^\circ(Z_2^{\bar{n}+e_i}) = n_2$; $\partial^\circ(Z_3^{\bar{n}+e_i}) = n_3$

La démonstration repose sur le fait que, par exemple,

$$Z_1^{\bar{n}+e_i}(X) = \varepsilon \, . \, \Delta^{-1}_{\bar{n}} \, X^{n_1+1} + \ldots .$$

Sur un raisonnement identique nous avons le

lemme 2 : Sous les hypothèses précédentes : $\partial^\circ(Z_1^{\bar{n}+e_i+e_j}) \le n_1$; $\partial^\circ(Z_2^{\bar{n}+e_i+e_j}) \le n_2$ et $\partial^\circ(Z_3^{\bar{n}+e_i+e_j}) \le n_3$

De ceci nous allons déduire

lemme 3 : Sous les hypothèses précédentes les triples

$(Z_i^{\bar{n}})_i$; $\{(Z_i^{\bar{n}+e_j})_i\}_j$; $\{(Z_i^{\bar{n}+e_j+e_k})_i\}_{j,k}$

donnent tous le même ordre $|\bar{n}|+4$ et sont tous proportionels à $(Z_i^{\bar{n}})_i$

Cela résulte des définitions même des triples et de l'existence d'une solution non nulle.

Conclusion :

Nous sommes dans un cas où la table est stationnaire, sur le schéma suivant nous avons porté les points où les formes sont proportionelles.

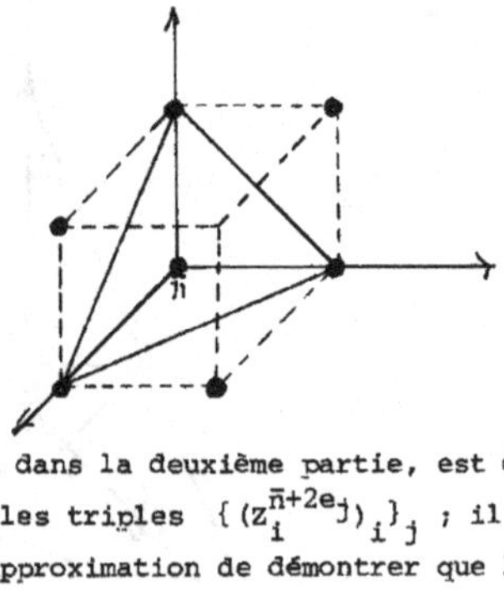

L'étape suivante, en procédant toujours comme dans la deuxième partie, est de considérer la conséquence des hypothèses sur les triples $\{(Z_i^{\bar{n}+2e_j})_i\}_j$; il est alors aisé par des considérations d'ordre d'approximation de démontrer que les triples considérés sont proportionnels à $(Z_i^{\bar{n}})_i$.

La difficulté est en fait localisée au point $\hat{n} = \bar{n} + e_1 + e_2 + e_3$

On commence par démontrer très facilement que $\partial(Z_1^{\hat{n}}) = n_i + 1$, puis si l'on étudie par exemple directement le terme constant de $Z_1^{\hat{n}}$:

Au signe près, il va pouvoir s'écrire :

$(-1)^{n_1+1} f_0^2 \; \xi_1 + (-1)^{n_1+n_2} f_0^3 \; \xi_2$ dans cette expression ξ_1 et ξ_2 sont deux déterminants obtenus de manière évidente dont nous allons parler :

$$\xi_1 = \det \left| \begin{array}{ccc:ccc:c:ccc} f_0^1 & & & f_0^2 & & & f_1^3 & f_0^3 & & \\ & \ddots & & & \ddots & & & & \ddots & \\ & & f_0^1 & & & f_0^2 & & & & f_0^3 \\ & & & & & & & & & \\ f^1_{|\bar{n}|+3} & & f^1_{|\bar{n}|+3-n_1} & f^2_{|\bar{n}|+3} & & f^2_{|\bar{n}|+3-n_2} & f^3_{|\bar{n}|+4} & f^3_{|\bar{n}|+3} & & f^3_{|\bar{n}|-n_3+3} \end{array} \right|$$

$$\longleftarrow n_1+1 \longrightarrow \longleftarrow n_2+1 \longrightarrow \longleftarrow n_3+2 \longrightarrow$$

Nous allons développer ce déterminant grâce à la règle de Sylvester en supprimant par exemple les colonnes n_1+1 et n_1+n_2+3 et les deux dernières lignes :

$$\xi_1 = \frac{\gamma_1 \, \Delta^{-1}_{n_1,n_2,n_3} - \gamma_2 \, \Delta^{0}_{n_1,n_2,n_3}}{\Delta^{-1}_{n_1-1,n_2,n_3}}$$

de cette expression on déduit $\xi_1 = 0$ et de la même manière $\xi_2 = 0$; Ceci nous montre que

$$Z_1^{\hat{n}} = X \, P_1^{\hat{n}}$$

$$\partial^\circ(P_1^{\bar{n}}) \leq n_1$$

(même démonstration pour $Z_2^{\hat{n}}$ et $Z_3^{\hat{n}}$). En revenant alors à la définition même on a

$$X \, P_1^{\hat{n}} \, f_1 + X \, P_2^{\hat{n}} \, f_2 + X \, P_3^{\hat{n}} \, f_3 = X^{|\hat{n}|+4} \; \Gamma_{\hat{n}}$$

ce qui démontre qu'il n'existe pas de forme de Padé-Hermite <u>irréductible</u> associée à (f_1,f_2,f_3) et $\hat{n}$. Soit :

<u>Théorème de structure</u> : Si $\Delta^{-1}_{\bar{n}} = \Delta^{0}_{\bar{n}} = 0$ et si les conditions "génériques" sont vérifiées, alors :

. les formes de P-H associées à

$\bar{n}$, $\bar{n}+e_1$, $\bar{n}+2e_1$

$\bar{n}+e_2$, $\bar{n}+2e_2$

$\bar{n}+e_3$, $\bar{n}+2e_3$

$\bar{n}+e_1+e_2$, $\bar{n}+e_1+e_3$, $\bar{n}+e_2+e_3$

existent et sont proportionelles, ces formes peuvent être choisies irréductibles

. on peut associer à $\bar{n}+e_1+e_2+e_3$ une forme de P-H non irréductible proportionelle à celle associée à $\bar{n}$. Cependant, il n'existe pas de forme de P-H irréductible associée à

$\bar{n}+e_1+e_2+e_3$

§ 4) <u>GENERALISATION DU THEOREME DE STRUCTURE</u>

Afin de na pas rendre "indigeste" l'exposé nous allons simplement supposer

$$\boxed{\Delta_{\bar{n}}^{-1} = \Delta_{\bar{n}}^{0} = \Delta_{\bar{n}}^{1} = 0}$$

(plus les éternelles conditions génériques).

Nous avons étudié au §3) les points associés à l'ordre d'approximation $|\bar{n}|+4$; passons maintenant à la couche $|\bar{n}|+5$.

Sur cette couche seuls les points "sur les bords" sont à considérer : prenons par exemple (n_1, n_2+1, n_3+2)

En utilisant une formule du type

$$z_i^{\bar{n}+e_2+2e_3} = \frac{\Delta^{-1}_{\hat{n}+e_2+e_3}}{\Delta^{-1}_{\bar{n}-e_1+e_2+e_3}} \; z_i^{\bar{n}-e_1+e_2+e_3} - \frac{\Delta^{-1}_{\bar{n}-e_1+e_2+2e_3}}{\Delta^{-1}_{\bar{n}-e_1+e_2+e_3}} \; z_i^{\bar{n}+e_2+e_3}$$

compte tenu de $\Delta^{-1}_{\bar{n}+e_2+e_3} = 0$ on démontre facilement que le triple $(z_i^{\bar{n}+e_2+2e_3})_i$ est toujours proportionel à $(z_i^{\bar{n}})_i$. La table <u>stationne toujours</u> (sauf naturellement en $\hat{n}$).

Cependant il est très intéressant de revenir au triple

$$(z_1^{\hat{n}} , z_2^{\hat{n}} , z_3^{\hat{n}})$$

nous pouvons démontrer le

<u>Théorème</u> : Sous les conditions de ce paragraphe

$$z_i^{\hat{n}}(X) \equiv 0 \quad (i = 1,2,3)$$

On peut procéder de plusieurs manières, la plus courte étant d'utiliser un résultat du paragraphe deux où les $\Delta^j_{\bar{n}} = 0$ pour tout j.

Ceci étant nous pouvons passer à la couche $\bar{n}+6$: Sur cette couche les points intérieurs :

$$\bar{n} + e_1 + e_2 + e_3 \ , \ \bar{n} + 2e_1 + e_2 + e_3 \ , \ \bar{n} + e_1 + 2e_2 + e_3$$

sont tels qu'en ces points il n'existe pas de forme P-H réduite. Cette situation n'entrainant rien sur les points frontières.

C'est tout ce que nous savons déduire des hypothèses initiales.

BIBLIOGRAPHIE

[1] J. DELLA DORA

Contribution à l'approximation de fonctions de la variable complexe au sens de Hermite-Padé et de Hardy.
Thèse de Doctorat es Sciences Mathématiques, Université Scientifique et Médicale de Grenoble (1980).

[2] J. DELLA DORA et C. DI CRESCENZO

Approximation de Padé-Hermite.
In "Padé Approximation and its applications" (ed. L. Wuytack), Lecture Notes in Mathematics 765, Springer Verlag (1979).

[3] J. GILEWICZ

Approximants de Padé.

Lecture Notes in Mathematics 667, Springer Verlag (1978).

APPROXIMANTS OF EXPONENTIAL TYPE
GENERAL ORTHOGONAL POLYNOMIALS

André DRAUX

1. APPROXIMANTS OF EXPONENTIAL TYPE

1.1 - Definition

Let f be a formal power series in one variable.

$$f(x) = \sum_{i=0}^{\infty} \frac{c_i}{i!} x^i, \quad c_i \in \mathbb{R}, \ \forall\, i \in \mathbb{N}.$$

Problem (1)

We search the families of exponential type

$$F_p(x) = \sum_{j=0}^{\ell_p} S_j(x)\, e^{m_{j,p} x}$$

where the $S_j(x)$ are polynomials of degree $q_{j,p}-1$, for $j \in \mathbb{N}$, $0 \le j \le \ell_p$

and $\displaystyle\sum_{j=0}^{\ell_p} q_{j,p} = p,$

$$S_j(x) = \sum_{i=i_j}^{i_{j+1}-1} k_{i,p}\, x^{i-i_j} \quad \text{for } j \in \mathbb{N},\ 0 \le j \le \ell_p,$$

with
$$\begin{cases} i_0 = 0, \\ i_j = \displaystyle\sum_{s=0}^{j-1} q_{s,p} \text{ for } j \in \mathbb{N},\ 1 \le j \le \ell_p + 1, \end{cases}$$

$m_{j,p}$ and $k_{i,p}$ are unknown.

We want the expansion of $F_p(x)$ in ascending powers of x to coincide with the expansion of f up to the degree 2p-1, that is to say :

$$\text{ord}\,(f - F_p) \geq 2p.$$

Problem (2)

We search the solutions $F_p(x)$ of the linear homogeneous differential equation (E) with constant coefficients, i.e.

$$\sum_{i=0}^{p} \lambda_{i,p}\, y^{(p-i)} = 0,$$

where the $\lambda_{i,p}$ are the solutions of the linear system (M_p) :

$$(M_p) \quad \begin{pmatrix} c_0 & c_1 & \cdots & c_{p-1} \\ c_1 & c_2 & \cdots & c_p \\ \vdots & & & \vdots \\ c_{p-1} & c_p & \cdots & c_{2p-2} \end{pmatrix} \begin{pmatrix} \lambda_{p,p} \\ \lambda_{p-1,p} \\ \vdots \\ \lambda_{1,p} \end{pmatrix} = - \begin{pmatrix} c_p \\ c_{p+1} \\ \vdots \\ c_{2p-1} \end{pmatrix} \text{ and } \lambda_{0,p} = 1.$$

We also require $\text{ord}\,(f - F_p) \geq 2p$.

Property 1 :

The problems (1) and (2) are equivalent.

To determine the coefficients $m_{i,p}$ *we have*

$$P_p(z) = \sum_{i=0}^{p} \lambda_{i,p}\, z^{p-i} = \prod_{j=0}^{\ell_p} (z - m_{j,p})^{q_{j,p}}$$

and the $k_{i,p}$*'s are solutions of the linear system* (K_p)

$$(K_p) \quad \begin{pmatrix} 1 & 0 & \cdots & 1 & \cdots \\ m_{0,p} & 1 & \cdots & m_{1,p} & \cdots \\ m_{0,p}^2 & 2m_{0,p} & \cdots & m_{1,p}^2 & \cdots \\ \vdots & \vdots & & & \\ m_{0,p}^{p-1} & (p-1)\, m_{0,p}^{p-2} & \cdots & m_{1,p}^{p-1} & \cdots \end{pmatrix} \begin{pmatrix} k_{0,p} \\ \vdots \\ k_{p-1,p} \end{pmatrix} = \begin{pmatrix} c_0 \\ \vdots \\ c_{p-1} \end{pmatrix}$$

1.2 - Existence and Unicity of F_p

We have proved that :

i) If (M_p) has the rank p, F_p exists and is unique.

ii) If (M_p) has the rank $r < p$ and is consistent, $F_p \equiv F_r$ and is unique.

iii) If (M_p) has the rank $r < p$ and is inconsistent, there is obviously no solution.

The part ii) is the consequence of the following fundamental theorem.

Theorem 1 :

If (M_p) has the rank $r < p$ and is consistent and if $P_p(x) = \sum_{i=0}^{p} \lambda_{i,p} x^{p-i}$ where the $\lambda_{i,p}$'s are solutions of (M_p) and $P_r(x) = \sum_{i=0}^{r} \lambda_{i,r} x^{r-i}$ where the $\lambda_{i,r}$'s are solutions of (M_r), the set of zeros of $P_r(x)$ is included in the set of zeros of $P_p(x)$.

The last $(p-r)$ zeros of $P_p(x)$ are arbitrary. □

We shall write : $P_p(x) = w_{p-r}(x)\, P_r(x)$, where $w_{p-r}(x)$ is an arbitrary polynomial of degree $(p-r)$.

2. GENERAL ORTHOGONAL POLYNOMIALS

2.1 - Definition

If we define a linear functional acting on the space P of real polynomials by $c(x^i) = c_i$, $\forall\, i \in \mathbb{N}$, then the preceding polynomials $P_p(x)$ are orthogonal with respect to the functional c.

Let

$$H_k^{(i)} = \begin{vmatrix} c_i & c_{i+1} & \cdots & c_{i+k-1} \\ c_{i+1} & c_{i+2} & \cdots & c_{i+k} \\ \vdots & & & \vdots \\ c_{i+k-1} & c_{i+k} & \cdots & c_{i+2k-2} \end{vmatrix}$$

We place these determinants in a two dimensional array H.

$$\begin{array}{lll} H_1^{(0)} & & \\ H_1^{(1)} & H_2^{(0)} & \\ H_1^{(2)} & H_2^{(1)} & H_3^{(0)} \\ \vdots & \vdots & \vdots \end{array}$$

We have the important following property for the determination of orthogonal polynomials.

<u>*Property 2*</u> :

Let j *be the integer part of* $\dfrac{p_{\ell+1} + 1 - h_{\ell+1}}{2}$.

If $\begin{cases} H_i^{(0)} \neq 0 \text{ for } i \in \mathbb{N},\ p_\ell + 1 \leq i \leq h_{\ell+1} + 1, \\ H_i^{(0)} = 0 \text{ for } i \in \mathbb{N},\ h_{\ell+1} + 2 \leq i \leq p_{\ell+1}, \\ H_{p_{\ell+1}+1}^{(0)} \neq 0 , \end{cases}$

(the index $\ell \in \mathbb{N}$ *denotes the successive blocks of zeros along the main diagonal* $(H^{(0)})$ *of the array* H *of Hankel determinants).*

Then :

i) *For* $i \in \mathbb{N}$, $p_\ell + 1 \leq i \leq h_{\ell+1} + 1$ *the orthogonal polynomial* $P_i(x)$ *with respect to* c *exists. It is unique, if the coefficient of* x^i *is fixed.*

ii) *For* $i \in \mathbb{N}$, $h_{\ell+1} + 2 \leq i \leq p_{\ell+1} - j$, $P_i(x)$ *exists and* $P_i(x) = P_{h_{\ell+1}+1}(x)\ w_{i-h_{\ell+1}-1}(x)$, *where* $w_{i-h_{\ell+1}-1}(x)$ *is an arbitrary polynomial of degree* $i - h_{\ell+1} - 1$.

iii) $P_i(x)$ *does not exist for* $i \in \mathbb{N}$, $p_{\ell+1} + 2 - j \leq i \leq p_{\ell+1}$.

iv) $P_{p_{\ell+1}+1}(x)$ *exists. It is unique if the coefficient of* $x^{p_{\ell+1}+1}$ *is fixed.*

v) $P_{p_{\ell+1}+1-j}(x)$ *exists if* $(p_{\ell+1} - h_{\ell+1})$ *is odd and is equal to* $P_{h_{\ell+1}}(x)\ w_{j-1}(x)$ *where* $w_{j-1}(x)$ *is an arbitrary polynomial of degree* j-1.
It does not exist if $(p_{\ell+1} - h_{\ell+1})$ *is even.*

Definition

The orthogonal polynomial $P_p(x)$ is said to be regular if $H_p^{(0)} \neq 0$.

The orthogonal polynomial $P_p(x)$ is said to be singular if $H_p^{(0)} = 0$.

We define a basis of the vector space $\mathcal{P}$ of real polynomials with the set of regular orthogonal polynomials and the polynomials $P_i(x) = w_{i-h_{\ell+1}-1}(x)\ P_{h_{\ell+1}+1}(x)$, for $i \in \mathbb{N}$, $h_{\ell+1}+2 \le i \le p_{\ell+1}$, where $w_{i-h_{\ell+1}-1}(x)$ is an arbitrary polynomial of degree $(i - h_{\ell+1} - 1)$.

2.2 - Recurrence formula

We have a recurrence formula with three regular orthogonal polynomials :

$$P_k(x) = (A_k \times \omega_{k-1-pr(k)}(x) + B_k)\ P_{pr(k)}(x) + C_k\ P_{pr(pr(k))}(x),$$

where $A_k \neq 0$ and $C_k \neq 0$.

$P_{pr(k)}(x)$ (resp. $P_{pr(pr(k))}(x)$) is the regular orthogonal polynomial that precedes $P_k(x)$ (resp. $P_{pr(k)}(x)$) ; A_k, B_k and C_k do not depend on x.

$\omega_{k-1-pr(k)}(x)$ is a polynomial of degree $k-1-pr(k)$; its coefficients are determined by a regular triangular linear system. We take $P_0(x)$ = an arbitrary non zero constant and $P_{-1}(x) = 0$.

We introduce the associated polynomials $Q_k(t) = c\left(\dfrac{P_k(x) - P_k(t)}{x - t}\right)$.

They satisfy the same recurrence relation as the polynomials $P_k(x)$ with $Q_{-1}(x)=-1$ and $Q_0(x) = 0$.

Properties of the zeros

Lemma 1 :

If $k \in \mathbb{N}$, $p_\ell \le k \le h_{\ell+1}$, then

$$P_{k+1}(x)\, Q_{pr(k+1)}(x) - P_{pr(k+1)}(x)\, Q_{k+1}(x) = - A_{k+1}\, c(P_k\, P_{pr(k+1)})$$

where $A_{k+1}\, c(P_k\, P_{pr(k+1)})$ is different from zero.

Property 3 :

i) $P_k(x)$ *and* $P_{pr(k)}(x)$ *have no common zero.*

ii) $Q_k(x)$ *and* $Q_{pr(k)}(x)$ *have no common zero.*

iii) $P_k(x)$ *and* $Q_k(x)$ *have no common zero.*

Proof

We use the relation of the lemma 1.

If $P_k(x)$ and $P_{pr(k)}(x)$ had a common zero, we should have :

$$0 = - A_{k+1}\, c(P_k\, P_{pr(k+1)})$$

which is impossible since $A_{k+1}\, c(P_k\, P_{pr(k+1)})$ is different from zero. The proof is the same for the two other pairs of polynomials. □

The following property is particularly useful for the detection of blocks of zeros in the table H.

Property 4 :

$H^{(0)}_{h_\ell+1} \neq 0,\ H^{(0)}_i = 0$ *for* $i \in \mathbb{N},\ h_\ell+2 \leq i \leq p_\ell$ *and* $H^{(0)}_{p_\ell+1} \neq 0$

if and only if

$$\begin{cases} P_{h_\ell+1} \text{ is a regular orthogonal polynomial ,} \\ c(x^j\, P_{h_\ell+1}) = 0 \text{ for } j \in \mathbb{N},\ 0 \leq j \leq p_\ell - 1 , \\ c(x^{p_\ell} P_{h_\ell+1}) \neq 0. \end{cases}$$

Detection of blocks of zeros

If $P_k(x)$ is a regular orthogonal polynomial, we compute $c(x^k\, P_k(x))$.

If this quantity is equal to zero, then $H^{(0)}_{k+1} = 0$ and $P_{k+1}(x)$ is not a regular orthogonal polynomial, else $H^{(0)}_{k+1} \neq 0$ and $P_{k+1}(x)$ is a regular orthogonal polynomial.

If $c(x^k\, P_k(x)) = 0$, we compute $c(x^{k+1}\, P_k(x))$. If $c(x^{k+1}\, P_k(x)) = 0$, then $H^{(0)}_{k+2} = 0$ and $P_{k+2}(x)$ is not a regular orthogonal polynomial, else $H^{(0)}_{k+2} \neq 0$ and $P_{k+2}(x)$ is a regular orthogonal polynomial. And so on.

2.3 - Adjacent systems of orthogonal polynomials

We define the linear functional $c^{(i)}$, for $i \in \mathbb{N}$, by $c^{(i)}(x^k) = c_{k+i}$, $\forall\, k \in \mathbb{N}$.

Then, we take the orthogonal polynomials $P_k^{(i)}(x)$ with respect to $c^{(i)}$; we display these polynomials in a table P.

$$\begin{array}{llll} P_{-1}^{(0)}(x) & & & \\ P_{-1}^{(1)}(x) & P_0^{(0)}(x) & & \\ P_{-1}^{(2)}(x) & P_0^{(1)}(x) & P_1^{(0)}(x) & \\ \cdot & \cdot & \cdot & \cdot \\ \cdot & \cdot & \cdot & \cdot \end{array}$$

In this table we have square blocks of polynomials that are not regular.

We denote these blocks by P, and its sides and corners by a geographical mark

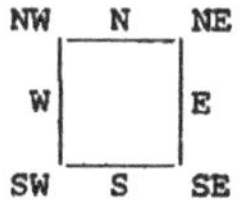

(North, South, East and West for the sides, North-West, North-East, South-West and South-East for the corners).

We suppose that the orthogonal polynomials are monic.

Property 5 :

If $P_k(x)$ is a regular orthogonal polynomial in the North-West corner, then :

i) *all the polynomials on the West side are identical.*

ii) *a polynomial on the North side is equal to the preceding multiplied by the x polynomial.*

We now denote the regular orthogonal polynomial that precedes $P_k^{(n)}(x)$ on the nth diagonal, by $P_{pr(k,n)}^{(n)}(x)$.

Property 6 :

A regular orthogonal polynomial $P_i^{(n+1)}$ satisfies the following relations for $i \in \mathbb{N}$, $p_\ell \le i \le h_{\ell+1} + 1$.

$$P_i^{(n+1)}(x) = x^{-1}\left(P_{i+1}^{(n)}(x) + q_{i+1,\ell}^{(n)}\, P_{pr(i+1,n)}^{(n)}(x)\right) .$$

$$P_i^{(n+1)}(x) = \omega_{i-pr(i+1,n)}^{(n)}(x)\, P_{pr(i+1,n)}^{(n)}(x) + E_{i+1}^{(n)}\, P_{pr(pr(i+1,n),n+1)}^{(n+1)}(x).$$

If $i = p_\ell$ *we have* $\begin{cases} B^{(n)}_{p_\ell+1} = - q^{(n)}_{p_\ell+1,\ell} - e^{(n)}_{h_\ell+1,\ell} \\ C^{(n)}_{p_\ell+1} = - q^{(n)}_{h_\ell+1,\ell-1}\, e^{(n)}_{h_\ell+1,\ell}. \end{cases}$

In the other cases we have $\begin{cases} B^{(n)}_{i+1} = - q^{(n)}_{i+1,\ell} - e^{(n)}_{pr(i+1,n),\ell} \\ C^{(n)}_{i+1} = - q^{(n)}_{pr(i+1,n),\ell}\, e^{(n)}_{pr(i+1,n),\ell}\,. \end{cases}$

$B^{(n)}_{i+1}$ *and* $C^{(n)}_{i+1}$ *are the coefficients that are used in the recurrence formula with the orthogonal polynomial* $P^{(n)}_{i+1}(x)$.

If the coefficients $q^{(n)}_{pr(i+1,n),\ell}$ *and* $e^{(n)}_{pr(i+1,n),\ell}$ *are defined, then* $E^{(n)}_{i+1} = - e^{(n)}_{pr(i+1,n),\ell}$, *else* $E^{(n)}_{i+1} = C^{(n)}_{i+1}$.

The complete text of this property contains some other relations and gives many particular cases, but owing to restrictions of space we cannot give complete results.

Property 7 :

If $P^{(n+1)}_i$ *is a regular orthogonal polynomial, then the associated polynomial* $Q^{(n+1)}_i(t)$ *satisfies the two following relations for* $i \in \mathbb{N}$, $p_\ell \le i \le h_{\ell+1} + 1$,

$$Q^{(n+1)}_i(t) = Q^{(n)}_{i+1}(t) + q^{(n)}_{i+1,\ell}\, Q^{(n)}_{pr(i+1,n)}(t) - c_n\, P^{(n+1)}_i(t),$$

$$Q^{(n+1)}_i(t) = \omega^{(n)}_{i-pr(i+1,n)}(t)\left(t\, Q^{(n)}_{pr(i+1,n)}(t) - c_n\, P^{(n)}_{pr(i+1,n)}(t)\right) + E^{(n)}_{i+1}\, Q^{(n+1)}_{pr(pr(i+1,n),n+1)}(t).$$

We have proved a "q-d" algorithm. The set of given relations allows to settle again the relations of Claessens and Wuytack. Their relations are only valid for the particular case where a block P is surrounded by two rows of regular orthogonal polynomials. Our relations are valid for all the cases, but they are too numerous to include them in this paper.

They are contained in our internal report of the University of Lille with many other properties on the approximants of exponential type and general orthogonal polynomials. With these relations we have proved the formula of Gilewicz and Froissart to compute the Hankel determinants behind a block H.

2.4 - Relations in all the directions in the table P

Let $\tilde{P}_k^{(n)}(x) = x^k P_k^{(n)}(x^{-1})$ and $\tilde{H}_k^{(n)}(x) = H_k^{(n)} \tilde{P}_k^{(n)}(x)$.

The recurrence formula and the relations of the property 6 allow us to obtain relations between polynomials $\tilde{H}(x)$ along a diagonal or two adjacent diagonals.

We have also obtained relations along an antidiagonal or two adjacent diagonals with the introduction of new orthogonal relations. Let $\gamma^{(m)}$ a linear functional acting on the space of real polynomials by $\gamma^{(m)}(x^i) = c_{m-i}$, $\forall\, i \in \mathbb{N}$, $i \leq m$.

We want to find the polynomials $W^{(m)}(x)$ that are orthogonal with respect to the functional $\gamma^{(m)}$.

The polynomials $W_i^{(m)}(x)$ are displayed in a table W.

$$
\begin{array}{llll}
W_1^{(1)} & & & \\
W_1^{(2)} & W_2^{(3)} & & \\
W_1^{(3)} & W_2^{(4)} & W_3^{(5)} & \\
W_1^{(4)} & W_2^{(5)} & W_3^{(6)} & W_4^{(7)} \\
\vdots & \vdots & \vdots & \vdots
\end{array}
$$

The polynomials $\tilde{P}_k^{(n)}(x)$ are also displayed in a table $\tilde{P}$.

$$
\begin{array}{lll}
\tilde{P}_1^{(0)} & & \\
\tilde{P}_1^{(1)} & \tilde{P}_2^{(0)} & \\
\tilde{P}_1^{(2)} & \tilde{P}_2^{(1)} & \tilde{P}_3^{(0)} \\
\vdots & \vdots & \vdots
\end{array}
$$

We have showed that the polynomials $W_i^{(m)}(x)$ satisfy a recurrence formula with three regular orthogonal polynomials along an antidiagonal.

We have similar relations to the ones of property 6 between two adjacent systems of orthogonal polynomials.

At last we have the following property between the polynomials $W_k^{(m)}(x)$ and $\tilde{P}_k^{(n)}(x)$.

Property 8 :

$P_k^{(m+1-2k)}(x)$ *is orthogonal regular with respect to* $c^{(m+1-2k)}$ *and* $P_k^{(m+1-2k)}(0) \neq 0$ *if and only if* $\tilde{P}_k^{(m+1-2k)}(x)$ *is orthogonal regular with respect to* $\gamma^{(m)}$ *and* $\tilde{P}_k^{(m+1-2k)}(0) \neq 0$.

In this case, we have $\tilde{P}_k^{(m+1-2k)}(x) = (-1)^k \dfrac{H_k^{(m+2-2k)}}{H_k^{(m+1-2k)}} W_k^{(m)}(x)$ *where* $W_k^{(m)}(x)$ *is orthogonal regular with respect to* $\gamma^{(m)}$.

Proof

Let $\tilde{P}_k^{(m+1-2k)}(x) = \sum_{i=0}^{k} \tilde{\lambda}_{k-i,m+1-2k}\, x^i$

and $P_k^{(m+1-2k)}(x) = \sum_{i=0}^{k} \lambda_{k-i,m+1-2k}\, x^i$

with $\lambda_{0,m+1-2k} = 1$.

Then $\lambda_{j,m+1-2k} = \tilde{\lambda}_{k-j,m+1-2k}$ for $j \in \mathbb{N}$, $0 \le j \le k$.

If $P_k^{(m+1-2k)}(x)$ is orthogonally regular with respect to the linear functional $c^{(m+1-2k)}$, we have the following system (O_1) of orthogonality :

$$(O_1) \quad \sum_{i=0}^{k} \lambda_{k-i,m+1-2k}\, c_{m+1-2k+i+j} = 0 \text{ for } j \in \mathbb{N},\ 0 \le j \le k-1.$$

If $P_k^{(m+1-2k)}(0) \neq 0$, then $H_k^{(m+2-2k)} \neq 0$ and $\tilde{\lambda}_{0,m+1-2k} \neq 0$. We have the system (O_2) of orthogonality that is deduced from (O_1) :

$$(O_2) \quad \sum_{i=0}^{k} \tilde{\lambda}_{k-i,m+1-2k}\, c_{m-i-j} = 0 \text{ for } j \in \mathbb{N},\ 0 \le j \le k-1.$$

Therefore the polynomial $\tilde{P}_k^{(m+1-2k)}(x)$ is orthogonal with respect to $\gamma^{(m)}$.
Further $\tilde{P}_k^{(m+1-2k)}(0) = \tilde{\lambda}_{k,m+1-2k} = 1$.

Thus it is different from zero.

This completes the proof of the necessary condition.

The sufficient condition can be proved by the same way.

At last the system (O_2) is also satisfied by the polynomial $W_k^{(m)}(x)$ that is orthogonal regular, since $H_k^{(m+2-2k)} \neq 0$.

Then it is proportional to $\tilde{P}_k^{(m+1-2k)}(x)$.

$$\tilde{P}_k^{(m+1-2k)}(x) = \sum_{i=0}^{k} \lambda_{i,m+1-2k}\, x^i .$$

It is easy to see that :

$$\lambda_{k,m+1-2k} = (-1)^k \frac{H_k^{(m+2-2k)}}{H_k^{(m+1-2k)}}$$

and therefore

$$\tilde{P}_k^{(m+1-2k)}(x) = (-1)^k \frac{H_k^{(m+2-2k)}}{H_k^{(m+1-2k)}} W_k^{(m)}(x) . \square$$

Then we have obtained relations along an antidiagonal or two adjacent anti-diagonals.

With these relations and those along the diagonal it is possible to derive relations along a column or a row of the table $\tilde{P}$.

To end we give an expression of the error between the formal powers series f and the approximant $F_p(x)$ of exponential type.

$$f(x) - F_p(x) = \frac{H_{p+1}^{(0)}}{H_p^{(0)}} \frac{x^{2p}}{(2p)!} + O(x^{2p+1}).$$

One will find in our internal report some properties on the derivation and the integration of $f(x)$ and $F_p(x)$, on the connexion between the approximants of the derivation or the integration of $f(x)$ and the adjacent systems of orthogonal polynomials, and on two-point Padé approximants.

BIBLIOGRAPHY

[1] C. BREZINSKI. *"Padé-type approximation and general orthogonal polynomials"*. Birkhaüser. (1980). ISNM 50.

[2] G. CLAESSENS and L. WUYTACK. *"On the computation of non normal Padé approximants"*. Journal of computational and Applied Mathematics. Vol. : 5 n°4 (1979). p.283-289.

[3] A. DRAUX. *"Approximants de type exponentiel. Polynômes orthogonaux généraux"*. Publication ANO 27. Equipe d'Analyse Numérique et d'Optimisation. Université des Sciences et Techniques de Lille I, UER d'IEEA.

[4] F.R. GANTMACHER. *"The theory of matrices"*. New-York (1959).

[5] J. GILEWICZ. *"Approximants de Padé"*. Lecture Notes in Mathematics 667. Springer-Verlag. Heidelberg (1978).

[6] L. WEISS and R.N. MC. DONOUGH. *"Prony's method, Z-transform and Padé approximation"*. Science Review. Vol. : 5 n°2 (1963). p.145-149.

[7] L. WUYTACK. (Edited by). *"Padé approximation and its applications - Proceedings. Antwerp 1979"*. Lecture Notes in Mathematics 765. Springer-Verlag. Heidelberg (1979).

André DRAUX

UER d'I.E.E.A.
Informatique
Université des Sciences et
Techniques de Lille I

59655 VILLENEUVE D'ASCQ CEDEX
- France -

MULTIPOINT PADÉ APPROXIMANTS CONVERGING TO FUNCTIONS OF STIELTJES' TYPE

J.K. Gelfgren
Department of Mathematics
University of Umeå
S-901 87 Umeå
Sweden

Abstract.

A function of Stieltjes type can be written $f(z) = \int_a^b \frac{d\alpha(t)}{z-t}$ where a,b are extended real numbers and $\alpha(t)$ is a bounded, non-decreasing real function. In recent years some people have studied the convergence of Padé approximants to such functions. In this paper we show the geometric convergence to f for multipoint Padé approximants.

1. Introduction.

Let $f(z) = \int_a^b \frac{d\alpha(t)}{z-t}$ where a, b are extended real numbers and α is a bounded, non-decreasing real function. Let $x_1, x_2, \dots, x_{k_1} \in \mathbb{R}\setminus[a,b]$ and $z_1, \bar{z}_1, z_2, \bar{z}_2, \dots, z_{k_2}, \bar{z}_{k_2} \in \mathbb{C}\setminus\mathbb{R}$. Let $R_{n-1,n}(z) = P_{n-1}(z)/Q_n(z)$ where P_{n-1} is a polynomial of degree $\leq n-1$ and $Q_n(z)$ is a polynomial of degree $\leq n$. We are interested in the rational function $R_{n-1,n}$ which satisfies

$$\begin{cases} f(z)Q_n(z)-P_{n-1}(z) = A(z)(z-x_1)(z-x_2)\dots(z-x_{k_1})(z-z_1)(z-\bar{z}_1)\dots(z-z_{k_2})(z-\bar{z}_{k_2}), \\ f(z)Q_n(z)-P_{n-1}(z) = B(z)\cdot z^{n-k_3-1}, \quad k_1+2k_2+k_3 = 2n; \end{cases} \tag{1.1}$$

$B(z)$ is bounded at infinity and $A(z)$ is bounded at $x_1,\dots,x_{k_1}, z_1, \bar{z}_1,\dots,\bar{z}_{k_2}$. We see that if $k_1=k_2=0$ we get the usual Padé approximant. The reader interested in Padé approximation of functions of the type $f(z) = \int_a^b \frac{d\alpha(t)}{z-t}$ may read [5, 6, 7, 8, 10].

2. Results.

Theorem 1: *If a,b are finite numbers, then for* $z \in \mathbb{C}\setminus[a,b]$

$$|f(z)-R_{n-1,n}(z)| \leq C_\varepsilon \cdot \frac{1}{\sqrt{|z-a||z-b|}} \cdot \frac{1}{(1-|\psi(z)|)^{1+\varepsilon}} \cdot \prod_1^{2n} |G_k(z)|$$

where $G_k(z) = \dfrac{\psi(z)-\psi(\gamma_k)}{1-\overline{\psi(\gamma_k)}\psi(z)}$, γ_k *is an interpolation point,* ψ *is a conformal mapping of* $\mathbb{C}\setminus[a,b]$ *onto the interior of the unit circle and* C_ε *is a constant depending on* ε *but not on* n *or* z.

Theorem 2: If $a = -\infty$ and b is a finite point, the interpolation points γ_k must be finite points belonging to $\mathbb{C} \setminus (-\infty,b]$. Then for $z \in \mathbb{C} \setminus (-\infty,b]$

$$|f(z)-R_{n-1,n}(z)| \le C_\varepsilon \frac{1}{\sqrt{|z-b|}} \cdot \frac{1}{(1-|\psi(z)|)^{2+\varepsilon}} \cdot \prod_1^{2n} |G_k(z)|.$$

Theorem 3: If $a = -\infty$, $b = \infty$ the interpolation points belong to $\mathbb{C} \setminus \mathbb{R}$. Then, if $\mathrm{Im}\, z > 0$, we have

$$|f(z)-R_{n-1,n}(z)| \le C_\varepsilon \frac{1}{(1-|\psi(z)|)^{1+\varepsilon}} \cdot \prod_1^{n} |G_k(z)|,$$

where

$$G_k(z) = \frac{\psi(z)-\psi(z_k)}{1-\overline{\psi(z_k)}\psi(z)}, \quad \mathrm{Im}\, z_k > 0.$$

Theorem 4: The rational function $R_{n-1,n}$ in (1.1) can be written

$$R_{n-1,n}(z) = \sum_1^n \frac{\lambda_k}{z-\alpha_k} \text{ where } \lambda_k > 0 \text{ and } \alpha_k \in (a,b) \text{ for all } k.$$

From theorem 1 it is clear, that if the distance from the interpolation points $\{\gamma_k\}$ to the interval $[a,b]$ is not less than a fixed number, then $|G_k(z)| \le \Delta_F < 1$ on each compact set $F \subset C \setminus [a,b]$. From this we get

$$|f(z)-R_{n-1,n}(z)| \le C_\varepsilon (\Delta_F)^{2n}, \quad z \in F,$$

which is geometric convergence on F.

On the other hand we know from the theory of Blaschke products that if $\lim_{n\to\infty} \sum_{k=1}^{n} (1-|\psi(\gamma_k)|) = \infty$ then $R_{n-1,n}$ converges uniformly to f on compact subsets of $\mathbb{C} \setminus [a,b]$.

3. Characterization of the Class $M(a,b)$.

Let a, b be extended real numbers.

Definition: $f(z) \in M(a,b) \Leftrightarrow f(z) = \int_a^b \frac{d\alpha(t)}{z-t}$ for a non-decreasing real function α of bounded variation.

Characterization theorem: $f \in M(a,b) \Leftrightarrow f$ satisfies (i) - (iii),

(i) f is holomorphic in $\tau_+ = \{z \mid \mathrm{Im}\, z > 0\}$ and maps τ_+ into $\overline{\tau}_- = \{\omega \mid \mathrm{Im}\, \omega \le 0\}$.

(ii) f satisfies the inequality $\sup_{y \ge 1} |y \cdot f(iy)| \le C < \infty$ $(z = x+iy)$.

(iii) f is continuous on $\mathbb{R} \setminus [a,b]$.

Proof of ⇐: From Achiezer [1, p. 93] we see that (i) and (ii) imply $f(z) = \int_{-\infty}^{\infty} \frac{d\alpha(t)}{z-t}$ for some non-decreasing real function α of bounded variation. Together with Stieltjes-Perron's inversion formula ([10, p. 72-75], [9, p. 188-90]) it follows from (iii) that α is constant outside $[a,b]$.

Proof of ⇒: The calculation in Achiezer shows (ii) and since (i) and (iii) follow easily from the definition of f the proof is complete.

4. A continued fraction expansion of f used for showing that $R_{n-1,n} \in M(a,b)$.

Lemma 1: Let $f \in M(a,b)$ and let x_1, x_2 be two finite real points outside $[a,b]$ or two complex-conjugated points. If we define $g_1(z)$ from the relation

$$f(z) = \frac{f(x_1)f(x_2)}{f(x_2) + (z-x_1)\, b_1 - (z-x_1)(z-x_2)\, g_1(z)}, \quad b_1 = \frac{f(x_1) - f(x_2)}{x_2 - x_1},$$

then $g_1 \in M(a,b)$. (The lemma is proved in section 6.)

Since $g_1 \in M(a,b)$ we define g_2 from the relation

$$g_1(z) = \frac{g_1(x_3)g_1(x_4)}{g_1(x_4) + (z-x_3)\, b_2 - (z-x_3)(z-x_4)\, g_2(z)}, \quad b_2 = \frac{g_1(x_3) - g_1(x_4)}{x_4 - x_3}.$$

From lemma 1 we conclude that $g_2 \in M(a,b)$.

We now go on with the procedure (compare Baker [2, p. 246])

$$f(z) = \cfrac{f(x_1)f(x_2)}{f(x_2)+(z-x_1)b_1-\cfrac{(z-x_1)(z-x_2)g_1(x_3)g_1(x_4)}{g_1(x_4)+(z-x_3)b_2-\cfrac{(z-x_3)(z-x_4)g_2(x_5)g_2(x_6)}{\ddots \; -\cfrac{(z-x_{2n-3})(z-x_{2n-2})g_{n-1}(x_{2n-1})g_{n-1}(x_{2n})}{g_{n-1}(x_{2n})+(z-x_{2n-1})b_n-(z-x_{2n-1})(z-x_{2n})g_n(z)}}}} \tag{4.1}$$

where $g_k \in M(a,b)$, $b_k = \frac{g_{k-1}(x_{2k-1}) - g_{k-1}(x_{2k})}{x_{2k} - x_{2k-1}}$, $g_0=f$, $k = 1,2,\ldots, n$.

From (4.1) we construct the rational function

$$R_{n-1,n}(z) = \cfrac{f(x_1)f(x_2)}{f(x_2)+(z-x_1)b_1-\cfrac{(z-x_1)(z-x_2)g_1(x_3)g_1(x_4)}{g_1(x_4)+(z-x_3)b_2-\cfrac{(z-x_3)(z-x_4)g_2(x_5)g_2(x_6)}{\ddots \; -\cfrac{(z-x_{2n-3})(z-x_{2n-2})g_{n-1}(x_{2n-1})g_{n-1}(x_{2n})}{g_{n-1}(x_{2n})+(z-x_{2n-1})b_n}}}} \tag{4.2}$$

We now define $R_{0,1}(z) = \dfrac{g_{n-1}(x_{2n-1})g_{n-1}(x_{2n})}{g_{n-1}(x_{2n}) + (z-x_{2n-1})b_n}$. Simple calculations show that $R_{0,1} \in M(a,b)$. Then we define $R_{k,k+1}(z) =$

$$= \frac{g_{n-k}(x_{2n-2k-1})g_{n-k}(x_{2n-2k})}{g_{n-k}(x_{2n-2k})+(z-x_{2n-2k-1})b_{n-k}-(z-x_{2n-2k-1})(z-x_{2n-2k})R_{k-1,k}(z)}.$$

If $R_{k-1,k} \in M(a,b)$ we show in lemma 2 (section 7) that $R_{k,k+1} \in M(a,b)$. Starting with $R_{0,1}$ and iterating it is clear that $R_{n-1,n} \in M(a,b)$. When we interpolate at infinity we need some lemmas similar to lemma 1 and lemma 2 in order to construct a rational function $R_{n,n+1} \in M(a,b)$. In section 8 we formulate these lemmas.

When we interpolate both at infinity and at finite points we glue together the expansion at infinity with (4.1) and using the lemmas 1-4 we get a rational function $R_{n-1,n} \in M(a,b)$.

5. Proof of the Convergence Theorems.

Let a, b be finite numbers and $f(z) = \int_a^b \frac{d\alpha(t)}{z-t}$.

$$\text{Now}\quad (z-a)f(z) = \int_a^b \frac{x^2 + y^2 - ax + at - iy(t-a)}{|z-t|^2}\, d\alpha(t) \tag{5.1}$$

$$\text{and}\quad (z-b)f(z) = \int_a^b \frac{x^2 + y^2 - bx + bt + iy(b-t)}{|z-t|^2}\, d\alpha(t). \tag{5.2}$$

Since $(x-a)f(x)>0$, and $(x-b)f(x) > 0$ if $x \in \mathbb{R}\setminus[a,b]$, $(z-a)f(z)$ and $(z-b)f(z)$ both map $\mathbb{C}\setminus[a,b]$ into $\mathbb{C}\setminus(-\infty,0]$. From (5.1) we see that $\mathrm{Im}(z-a)f(z) \leq 0$ and from (5.2) it is clear that $\mathrm{Im}(z-b)f(z) \geq 0$. Therefore $(z-a)f(z)(z-b)f(z)$ maps $\mathbb{C}\setminus[a,b]$ into $\mathbb{C}\setminus(-\infty,0]$. If $\omega \in (-\infty,0]$ then $-\pi < \arg < \pi$. From this follows $\mathrm{Re}\sqrt{(z-a)f(z)(z-b)f(z)} > 0$ when $z \in \mathbb{C}\setminus[a,b]$. We now define

$\psi_v(z) = \dfrac{\sqrt{(\frac{z-a}{z-b})} - \sqrt{(\frac{v-b}{v-a})}}{\sqrt{(\frac{z-a}{z-b})} + \sqrt{(\frac{v-b}{v-a})}}$, where v is an interpolation point. This is a conformal mapping of $\mathbb{C}\setminus[a,b]$ onto the interior U of the circle and $\psi_v(v) = 0$. We also define $\phi(\omega) = \psi_v^{-1}(\omega)$, $\omega \in U$. We see that $\phi(0) = v$.

Next we define

$$g(\omega) = \frac{\sqrt{(\phi(\omega)-a)f(\phi(\omega))(\phi(\omega)-b)f(\phi(\omega))} - i\,\mathrm{Im}\sqrt{(v-a)f(v)(v-b)f(v)}}{\mathrm{Re}\sqrt{(v-a)f(v)(v-b)f(v)}}, \tag{5.3}$$

$\omega \in U$.

We now need the

Theorem (Duren [3, p.34]): Every analytic function g with positive real part in $|\omega|<1$ is of class H^p for all $p<1$.

Since $g(0) = 1$ the proof of the theorem tells us that

$$\int_0^{2\pi} |g(re^{it})|^p \, dt \le \int_0^{2\pi} \left|\frac{1+re^{it}}{1-re^{it}}\right|^p dt \le K_p < \infty. \tag{5.4}$$

From (5.3) and $|a+b|^p \le |a|^p + |b|^p$ if $p<1$, we get

$$\int_0^{2\pi} (|\sqrt{\phi(re^{it})-a}|\,|\sqrt{\phi(re^{it})-b}|\,|f(\phi(re^{it}))|)^p \, dt \le K_{p,v} < \infty,$$

$$\text{if} \quad 0<p<1. \tag{5.5}$$

Now $R_{n-1,n} \in M(a,b)$ and interpolates to f at $z=v$. Thus $R_{n-1,n}$ satisfies (5.5) and using $|a+b|^p \le |a|^p + |b|^p$ we get

$$\int_0^{2\pi} |\sqrt{(\phi(re^{it})-a)(\phi(re^{it})-b)}\; [f(\phi(re^{it}))-R_{n-1,n}(\phi(re^{it}))]|^p \, dt \le 2K_{p,v},$$

$$0<p<1. \tag{5.6}$$

Hence $\sqrt{(\phi(\omega)-a)(\phi(\omega)-b)}\; [f(\phi(\omega))-R_{n-1,n}(\phi(\omega))] \in H^p, \quad 0<p<1.$

Using the canonical factorization theorem (Duren [3, p. 24]) it is possible to write

$$\sqrt{(\phi(\omega)-a)(\phi(\omega)-b)}\; [f(\phi(\omega))-R_{n-1,n}(\phi(\omega))] = B_n(\omega)S_n(\omega)F_n(\omega), \tag{5.7}$$

where B_n is a Blaschke product, S_n is a singular function and F_n is an outer function for the class H^p, $0<p<1$.

Following Duren we have $0 < |S_n(\omega)| \le 1$ and

$$F_n(\omega) = e^{i\gamma} \exp\{\frac{1}{2\pi} \int_0^{2\pi} \frac{e^{it}+\omega}{e^{it}-\omega} \ln|\sqrt{(\phi(e^{it})-a)(\phi(e^{it})-b)}\; [f(\phi(e^{it}))-R_{n-1,n}(\phi(e^{it}))]|dt\},$$

where γ is a real number

$$|F_n(\omega)| \le [\exp\{\int_0^{2\pi} \ln^+|\sqrt{(\phi(e^{it})-a)(\phi(e^{it})-b)}\; [f(\phi(e^{it}))-R_{n-1,n}(\phi(e^{it}))]^p \, d\mu_{r,\theta}(t)\}]^{1/p}$$

where $d\mu_{r,\theta}(t) = \frac{1}{2\pi} \operatorname{Re} \frac{e^{it}+\omega}{e^{it}-\omega}$, $\omega = re^{i\theta}$ and $\ln^+|\omega| = \begin{cases} 0, & |\omega| < 1 \\ \ln|\omega|, & |\omega| \ge 1 \end{cases}$.

Since $\int_0^{2\pi} d\mu_{r,\theta}(t) = 1$ we can use Jensen's inequality and (5.6) to obtain

$$|F_n(\omega)| \le [\int_0^{2\pi} |\sqrt{(\phi(e^{it})-a)(\phi(e^{it})-b)}|^p|\, f(\phi(e^{it}))-R_{n-1,n}(\phi(e^{it}))|^p+1 d\mu_{r,\theta}(t)]^{1/p} \le$$

$$\le (\frac{2}{1-r}\frac{1}{2\pi} \int_0^{2\pi} |\sqrt{(\phi(e^{it})-a)(\phi(e^{it})-b)}\; [f(\phi(e^{it}))-R_{n-1,n}(\phi(e^{it}))]|^p + 1 \, dt)^{1/p} \le$$

$$\leq [(\frac{2}{1-r})(\frac{K_{p,v}}{\pi} + 1)]^{1/p}, \quad 0<p<1.$$

Thus $|F_n(\omega)| \leq C_\varepsilon \dfrac{1}{(1-r)^{1+\varepsilon}}$ where $\varepsilon>0$ and C_ε is a constant depending on ε but not on n or ω. Using (5.7) we finally get

$$|f(z)-R_{n-1,n}(z)| \leq C_\varepsilon \cdot \frac{1}{\sqrt{|z-a|\,|z-b|}} \cdot \frac{1}{(1-|\psi_v(z)|)^{1+\varepsilon}} \cdot \prod_{k=1}^{2n} |G_k(z)|,$$

where $G_k(z) = \dfrac{\psi_v(z) - \psi_v(\gamma_k)}{1 - \overline{\psi_v(\gamma_k)}\psi_v(z)}$ and γ_k is one of the interpolation points.

Thus we have proved theorem 1. Theorem 2 and 3 can be proved in a similar way.

6. Proof of Lemma 1.

Lemma 1: Let x_1, x_2 be finite real points outside $[a,b]$ and let g be defined from the relation

$$f(z) = \frac{f(x_1)f(x_2)}{f(x_2) + (z-x_1)b_1 - (z-x_1)(z-x_2)g(z)}, \quad b_1 = \frac{f(x_1)-f(x_2)}{x_2-x_1}.$$

If $f \in M(a,b)$ then $g \in M(a,b)$.

Proof: Let $x_3 = \dfrac{x_1+x_2}{2}$ and define $f_1(z) = f(z+x_3)$, $g(z) = g(z+x_3)$.

It is easy to see that $f_1 \in M(a-x_3,\ b-x_3)$ and also that it is enough to prove that $g_1 \in M(a-x_3,\ b-x_3)$.

Let $x_4 = \dfrac{x_1-x_2}{2}$. Since $f(x_1) = f_1(x_4)$ and $f(x_2) = f_1(-x_4)$ we get

$$f_1(z) = \frac{f_1(x_4)f_1(-x_4)}{f_1(-x_4) + (z-x_4)b_1 - (z-x_4)(z+x_4)g_1(z)}.$$

From this we can show that

$$g_1(z) = \frac{1}{(z-x_4)(z+x_4)f_1(z)} [(z-x_4)b_1f_1(z) - f_1(-x_4)(f_1(x_4) - f_1(z))]. \tag{6.1}$$

Since $f_1 \in M(a-x_3,b-x_3)$,

$$\frac{f_1(x_4) - f_1(z)}{(z-x_4)} = \int_{a-x_3}^{b-x_3} \frac{d\alpha(t)}{(x_4-t)(z-t)} \quad \text{and}$$

$$b_1 = \frac{f_1(x_4) - f_1(-x_4)}{-2x_4} = -\int_{a-x_3}^{b-x_3} \frac{d\alpha(t)}{x_4^2 - t^2}.$$

Using these expressions in (6.1) we get

$$g_1(z) = \frac{1}{f_1(z)} [\int \frac{t^2\, d\alpha(t)}{(z-t)(x_4^2-t^2)} \int \frac{d\alpha(t)}{(z-t)(x_4^2-t^2)} - (\int \frac{t\, d\alpha(t)}{(z-t)(x_4^2-t^2)})^2]. \qquad (6.2)$$

We now have to show that g_1 satisfies the three conditions of the characterization theorem in section 3.

$$\mathrm{Im}\, g_1(z) = \frac{-\mathrm{Im}\, z}{|f_1(z)|^2} \cdot \frac{1}{\int \frac{d\alpha(t)}{|z-t|^2}} \cdot \Big[[(|\bar z| \cdot \int \frac{d\alpha(t)}{|z-t|^2})^2 - 2x \int \frac{t\, d\alpha(t)}{|z-t|^2} \int \frac{d\alpha(t)}{|z-t|^2} +$$

$$+ (\int \frac{t\, d\alpha(t)}{|z-t|^2})^2] \cdot (\int t^2\, d\sigma(t) \int d\sigma(t) - (\int t\, d\sigma(t))^2) - (\int \frac{d\alpha(t)}{|z-t|^2})^2 \cdot (\int t^2 d\sigma(t) \int d\sigma(t) -$$

$$- (\int t\, d\sigma(t))^2) - (\int \frac{d\alpha(t)}{|z-t|^2})^2 (\int t^3\, d\sigma(t) \int t\, d\sigma(t) - (\int t^2\, d\sigma(t))^2) +$$

$$+ \int \frac{t\, d\alpha(t)}{|z-t|^2} \int \frac{d\alpha(t)}{|z-t|^2} \cdot (\int t^3\, d\sigma(t) \int d\sigma(t) - \int t^2\, d\sigma(t) \int t\, d\sigma(t))]$$

where $d\sigma(t) = \dfrac{d\alpha(t)}{|z-t|^2(x_4^2-t^2)}$.

Since $d\sigma(t)$ has constant sign for $t \in [a-x_3, b-x_3]$ we know from Hölder's inequality that

$$\int t^2\, d\sigma(t) \int d\sigma(t) - (\int t\, d\sigma(t))^2 \geq 0 \quad \text{and thus}$$

$$[(|\bar z| \int \frac{d\alpha(t)}{|z-t|^2})^2 - 2x \int \frac{t\, d\alpha(t)}{|z-t|^2} \int \frac{d\alpha(t)}{|z-t|^2} + (\int \frac{t d\alpha(t)}{|z-t|^2})^2][\int t^2 d\alpha(t) \int d\sigma(t) - (\int t d\sigma(t))^2] \geq$$

$$\geq (|z| \int \frac{d\alpha(t)}{|z-t|^2} - |\int \frac{t\, d\alpha(t)}{|z-t|^2}|)^2 \cdot (\int t^2\, d\alpha(t) \int d\sigma(t) - (\int t d\alpha(t))^2) \geq 0.$$

$$\mathrm{Im}\, g_1(z) \leq \frac{-\mathrm{Im}\, z}{|f_1(z)|^2} [\int \frac{t\, d\alpha(t)}{|z-t|^2} \int \frac{d\alpha(t)}{|z-t|^2} (\int t^3\, d\sigma(t) \int d\sigma(t) - \int t^2\, d\sigma(t) \int t\, d\sigma(t) -$$

$$- (\int \frac{d\alpha(t)}{|z-t|^2})^2 \cdot (\int t^3\, d\sigma(t) \int t\, d\sigma(t) - (\int t^2\, d\sigma(t))^2) -$$

$$- (\int \frac{t d\alpha(t)}{|z-t|^2})^2 (\int t^2\, d\sigma(t) \int d\sigma(t) - (\int t\, d\sigma(t))^2).$$

Remembering $d\sigma(t) = \dfrac{d\alpha(t)}{|z-t|^2(x_4^2-t^2)}$ we finally get

$$\mathrm{Im}\, g_1(z) \leq \frac{-\mathrm{Im}\, z}{|f_1(z)|^2 \int \frac{d\alpha(t)}{|z-t|^2}} \Big[[x_4^2 \cdot (\int t^2\, d\sigma(t) \int d\sigma(t) - (\int t\, d\sigma(t))^2) -$$

$$- (\int t^3 d\sigma(t) \int t\, d\sigma(t) - (\int t^2 d\sigma(t))^2)]^2 + x_4^2 (\int t^3 d\sigma(t) \int d\sigma(t) - \int t^2 d\sigma(t) \int t\, d\sigma(t))^2 \Big].$$

From this we can conclude that $\operatorname{Im} g_1(z) \leq 0$ when $\operatorname{Im} z > 0$. From (6.2) we see that g_1 is continuous on $\mathbb{R} \setminus [a-x_3, b-x_3]$ and also that

$$y \cdot g_1(iy) = \frac{y^2}{y \cdot f_1(iy)} \left[\int \frac{t^2 d\alpha(t)}{(iy-t)(x_4^2-t^2)} \int \frac{d\alpha(t)}{(iy-t)(x_4^2-t^2)} - (\int \frac{t\, d\alpha(t)}{(iy-t)(x_4^2-t^2)})^2 \right].$$

Since $|y \cdot f_1(iy)| = \left| i \int_{a-x_3}^{b-x_3} \frac{y^2 d\alpha(t)}{y^2+t^2} - \int_{a-x_3}^{b-x_3} \frac{y+d\alpha(t)}{y^2+t^2} \right| \geq C_1 > 0$, $y \geq 1$ and $\sup_{y \geq 1} |y \cdot f_1(iy)| \leq C_2 < \infty$ it is easy to show that

$$\left| \int \frac{y\, t^2\, d\alpha(t)}{(iy-t)(x_4^2-t^2)} \right|, \quad \left| \int \frac{y\, d\alpha(t)}{(iy-t)(x_4^2-t^2)} \right|, \quad \left| \int \frac{y\, t\, d\alpha(t)}{(iy-t)(x_4^2-t^2)} \right|$$ are bounded when $y \geq 1$.

This implies that $\sup_{y \geq 1} |y g_1(iy)| \leq C < \infty$. We have now proved that $g_1 \in M(a-x_3, b-x_3)$ and thus that $g \in M(\overline{a}, b)$.

7. Proof of Lemma 2.

Lemma 2: Let

$$g_1(z) = \frac{g_1(x_1) g_1(x_2)}{g_1(x_2) + (z-x_1) b_1 - (z-x_1)(z-x_2) g_2(z)}, \tag{7.1}$$

$$R_{k,k+1}(z) = \frac{g_1(x_1) g_1(x_2)}{g_1(x_2) + (z-x_1) b_1 - (z-x_1)(z-x_2) R_{k-1,k}(z)}, \tag{7.2}$$

where $g_1, g_2, R_{k-1,k} \in M(a,b)$, $b_1 = \dfrac{g_1(x_1) - g_1(x_2)}{x_2 - x_1}$, $x_1, x_2 \in (b, \infty)$ (or $x_1, x_2 \in (-\infty, a)$).

Let $R_{k-1,k}$ be a rational function with numerator of degree exactly $k-1$ and denominator of degree exactly k satisfying the inequalities

$$0 < \lim_{y \to \infty} iy\, R_{k-1,k}(iy) \leq \lim_{y \to \infty} iy \cdot g_2(iy) < \infty, \tag{7.3}$$

$$0 < R_{k-1,k}(b) < g_2(b), \text{ if } b \text{ is finite} \tag{7.4}$$

$$0 > g_2(a) > R_{k-1,k}(a) \text{ if } a \text{ is finite}. \tag{7.5}$$

Then $R_{k,k+1} \in M(a,b)$ and $R_{k,k+1}$ is a rational function with numerator of degree k and denominator of degree $k+1$ satisfying the inequalities

$$0 < \lim_{y \to \infty} iy\, R_{k,k+1}(iy) \leq \lim_{y \to \infty} iy\, g_1(iy) < \infty, \tag{7.6}$$

$$0 < R_{k,k+1}(b) \overline{<} g_1(b), \text{ if } b \text{ is finite}, \tag{7.7}$$

$$0 > g_1(a) > R_{k,k+1}(a), \quad \underline{\text{if}} \quad a \quad \underline{\text{is finite}}. \tag{7.8}$$

Proof: Since $g_1 \in M(a,b)$, $g_1(z) = \int_a^b \frac{d\alpha_1(t)}{z-t}$ and it is easy to show

$$0 < \lim_{y\to\infty} iy\, g_1(iy) < \infty, \tag{7.9}$$

and from (7.1) we then have

$$0 < \lim_{y\to\infty} iy\, g_1(iy) = \frac{g_1(x_1)g_1(x_2)}{b_1 - \lim\limits_{y\to\infty} iy\, g_2(iy)} < \infty.$$

Since both x_1 and $x_2 \in (-\infty,a)$ (or (b,∞)) $g_1(x_1)g_1(x_2) > 0$ and (7.2), (7.3) now show

$$0 < b_1 - \lim_{y\to\infty} iy\, g_2(iy) \le b_1 - \lim_{y\to\infty} iy\, R_{k-1,k}(iy).$$

From comparing (7.1) with (7.2) we see that

$$0 < \lim_{y\to\infty} iy\, R_{k,k+1}(iy) \le \lim_{y\to\infty} iy\, g_1(iy) < \infty. \tag{7.10}$$

Since $g_1 \in M(a,b)$, $g_1(b) > 0$ when b is finite. Thus from (7.1), (7.2) and (7.4) we get

$$0 < g_1(x_2) + (b-x_1)b_1 - (b-x_1)(b-x_2)g_2(b) < g_1(x_2) + (b-x_1) - (b-x_1)(b-x_2)R_{k-1,k}(b)$$

which implies

$$0 < R_{k,k+1}(b) < g_1(b), \quad \text{if } b \text{ is finite}. \tag{7.11}$$

A similar estimate at $z=a$ gives

$$0 > g_1(a) > R_{k,k+1}(a) \quad \text{if } a \text{ is finite}.$$

We know that $R_{k-1,k}(z) = \frac{p_{k-1}(z)}{q_k(z)}$ where p_{k-1} and q_k are polynomials of degree exactly $k-1$ and k, respectively.

From (7.2) we see that

$$R_{k,k+1}(z) = \frac{g_1(x_1)g_1(x_2)\cdot q_k(z)}{q_k(z)(g_1(x_2) + (z-x_1)b_1) - (z-x_1)(z-x_2)\cdot p_{k-1}(z)} \tag{7.12}$$

and from (7.10) we can conclude that

$$q_{k+1}(z) = q_k(z)(g_1(x_2) + (z-x_1)b_1) - (z-x_1)(z-x_2)p_{k-1}(z)$$

has degree $k+1$ and that the leading coefficient of $q_{k+1}(z)$ has the same sign as that of $q_k(z)$. From (7.3) we see that the leading coefficient of $q_k(z)$ has the same sign as that of $p_{k-1}(z)$. Thus the leading coefficient of $q_{k+1}(z)$ has the same sign as that of $p_{k-1}(z)$.

Now we used the same method as in Freud [4, p. 20] to prove that the zeros of $q_k(z)$ interlace those of $q_{k+1}(z)$. That all the zeros of q_{k+1} belong to (a,b) is a consequence of (7.11) and (7.12).

Thus $R_{k,k+1}(z) = \sum_{k=1}^{n} \frac{\lambda_k}{z-\alpha_k}$ where $\lambda_k > 0$ and $\alpha_k \in (a,b)$ and from this and section 2 it is clear that $R_{k,k+1} \in M(a,b)$. We have now shown lemma 2. If x_1 and x_2 are two complex-conjugated points the proof is almost identical. If x_1 and x_2 belong to different subintervals of $\mathbb{R} \setminus [a,b]$ some small changes are necessary.

8. Some Lemmas Concerning Interpolation at Infinity.

Lemma 3: *Let $h_1 \in M(a,b)$ where a,b are finite real numbers and let h_2 be defined from the relation*

$$h_1(z) = \frac{c_1}{z+d_1 - h_2(z)} \quad \text{where } c_1 = \lim_{z\to\infty} zh_1(z)$$

and

$$d_1 = \lim_{z\to\infty} \frac{c_1 - zh_1(z)}{h_1(z)} .$$

Then $h_2 \in M(a,b)$.

Lemma 4: *Let* $h_1(z) = \dfrac{c_1}{z+d_1 - h_2(z)}$,

$$R_{k,k+1}(z) = \frac{c_1}{z+d_1 - R_{k-1,k}(z)} ,$$

where $h_1, h_2, R_{k-1,k} \in M(a,b)$, $c_1 = \lim_{z\to\infty} zh_1(z)$, $d_1 = \lim_{z\to\infty} \dfrac{c_1 - zh_1(z)}{h_1(z)}$. *Let $R_{k-1,k}$ be a rational function with numerator of degree k-1 and denominator of degree k satisfying*

$$0 < R_{k-1,k}(b) < h_2(b),$$

$$0 > g_2(a) > R_{k-1,k}(a).$$

Then $R_{k,k+1} \in M(a,b)$ and $R_{k,k+1}$ is a rational function with numerator of degree k and denominator of degree k+1 satisfying

$$0 < R_{k,k+1}(b) < h_1(b),$$

$$0 > h_1(a) > R_{k,k+1}(a).$$

It is clear that $R_{k,k+1}$ satisfies $0 < \lim_{y\to\infty} iy\, R_{k,k+1}(iy) = \lim_{y\to\infty} iy\, h_1(iy)$. From lemma 4 we see that $R_{k,k+1}$ satisfies the conditions of lemma 2 and therefore it is possible to join an expression at infinity with an expansion at some finite points. The proofs of lemma 3 and lemma 4 are similar to the proofs of lemma 1 and lemma 2 and are omitted.

References.

[1] Achiezer, N.I. "The Classical Moment Problem", Oliver and Boyd, Edinburgh 1965.

[2] Baker, G.A. Jr. "Essentials of Padé Approximants", Academic Press, New York 1975.

[3] Duren, P.L. "Theory of H^p-spaces", Academic Press, New York 1970.

[4] Freud, G. "Orthogonale Polynome" Birkhäuser Verlag, Basel und Stuttgart 1969.

[5] Goncar, A.A. "On Convergence for Padé approximants to some classes of meromorphic functions" (Russian), Mat.Sb., 97(139)(1975), 607-629.

[6] Goncar, A.A., Lopez, G. "On Markov's theorem for multipoint Padé approximation" (Russian), Mat.Sb., 105(147), (1978), 512-524.

[7] Karlsson, J., von Sydow B. "The convergence of Padé approximants to series of Stieltjes", Ark.Mat., 14(1976), 43-53.

[8] Lopez, G. "Conditions for convergence of multipoint Padé approximants to functions of Stieltjes' type" (Russian), Mat.Sb., 107(149), (1978), 69-83.

[9] Perron, O. "Die Lehre von den Kettenbrüchen", band II, 3rd ed., Teubner, Stuttgart 1957.

[10] Stieltjes, T.J. "Recherces sur les fractions continues", Ann.Fac.Sci. Toulouse, 8(1894), J, 1-122; 9(1895), A, 1-47.

PADE APPROXIMANT INEQUALITIES FOR THE FUNCTIONS OF THE CLASS S

J. GILEWICZ
Centre de Physique Théorique, CNRS, Marseille
and Université de Toulon

and

E. LEOPOLD
Presently at the Centre de Physique Théorique, CNRS, Marseille

ABSTRACT : We prove the inequalities allowing to determine the best Padé approximant in the finite rectangular set of approximants in the Padé table.

INTRODUCTION

The main result of this work is to prove the existence of some small compact in the neighbourhood of the origin in the complex plane where both the function of class S and its Padé approximants have no poles and no zeros. For this compact we can determine the best Padé approximant in the finite rectangular set of approximants and thus watch over the convergent sequence in the Padé table. It is well known that the Padé approximants are the best local rational approximations [2] of analytic functions. In particular, Padé approximants converge (asymptotically in the Padé table) to the functions of the class S in all compacts containing no poles of the function [1], [2]. Our result covers the intermediate step between these two properties. We define the functions of the class S as follows :

$$\forall z \in \mathbb{C} \setminus \{\beta_i\} : \qquad z \mapsto f(z) = e^{-\gamma z} \frac{\prod_{i=1}^{\infty} (1 - \alpha_i z)}{\prod_{i=1}^{\infty} (1 + \beta_i z)} \tag{1}$$

$$\gamma \geq 0, \sum (\alpha_i + \beta_i) < \infty, \alpha_1 \geq \alpha_2 \geq \ldots \geq 0, \beta_1 \geq \beta_2 \geq \ldots \geq 0.$$

(to obtain the classical definition [2] put $z \to -z$).
The corresponding power series is defined by :

$$f(z) = \sum_{n=0}^{\infty} c_n (-z)^n \tag{2}$$

$$c_0 = 1; \quad n > 0: \quad c_n = \frac{1}{n} \left\{ \sum_{k=1}^{n} c_{n-k} \left[\sum_{j=1}^{\infty} \left(\beta_j^k - (-\alpha_j)^k \right) + \gamma \delta_{n-k,n-1} \right] \right\} \tag{3}$$

The sequence (c_n) is totally positive [1], [2]. Denote by S^* the subclass of normal (no-rational) elements of S. In particular in S^* all Toeplitz determinants C_n^m defined later in (17) are strictly positive.

In the case of Padé approximants to the Stieltjes functions one obtains the set of inequalities on the real axis and can define the best Padé approximant [3] owing to the exceptional possibility of control of the positions of poles and zeros of Padé approximants. In the case of the class S we cannot control in general these positions. However, many numerical experiments with the functions of class S show us the existence of some small (with respect to the interval $\left[-\frac{1}{\beta_1}, \frac{1}{\alpha_1}\right]$) region in the complex plane, free of poles and zeros of Padé approximants. This observation has no contradiction with the Froissart, Saff and Varga results [2] giving the asymptotic distribution of poles and zeros of Padé approximants to the exponential element of the class S^*.

Encouraged by numerical experiments and, in some way, using the method employed for the Stieltjes case, we establish the inequalities occurring for some part of the real axis contained in the observed region mentioned before. Working in the following with the class S^* we eliminate the trivial case of rational functions.

INEQUALITIES

We use the notations $[m/n]_f = P_f^{m/n}/Q_f^{m/n}$ for the Padé approximant to f, but in obvious cases the index f can be omitted. The exact degrees of the numerator $P^{m/n}$ and the denominator $Q^{m/n}$ are respectively m and n , because we are limited by normal cases. In the tables the index " m" represent the row index and the index "n" the column one. We accompany the relations between the elements of a table by diagrams of its positions in the table. We prove, in order, the inequalities for P, Q, P/Q and finally for the errors f - P/Q. The limits, for instance $\lim_{m\to\infty} Q^{m/n}$ shall be denoted by $Q^{\infty/n}$.

The following result of uniform convergence in all compacts in $\mathbb{C}$ for the functions of the class S^* is due to Arms and Edrei [1], [2] :

$$Q^{\infty/n}(z) = \prod_{j=1}^{n} (1 + \beta_j z) \tag{4}$$

$$P^{m/\infty}(z) = \prod_{j=1}^{m} (1 - \alpha_j z) \quad . \tag{5}$$

We complete this by :

Theorem 1

Let f belong to S^*, then for all m (resp. n) :

$$\forall z \in D_\xi = \{|z| \leq \xi < 1/\alpha_{m+1}\}: \qquad Q^{m/\infty}(z) = e^{\gamma z} \prod_{j=1}^{\infty} (1+\beta_j z) \Big/ \prod_{j=m+1}^{\infty} (1-\alpha_j z) \tag{6}$$

$$\forall z \in D_\eta = \{|z| \leq \eta < 1/\beta_{n+1}\}: \qquad P^{\infty/n}(z) = e^{-\gamma z} \prod_{j=1}^{\infty} (1-\alpha_j z) \Big/ \prod_{j=n+1}^{\infty} (1+\beta_j z) \tag{7}$$

where the convergence is uniform.

Proof

Consider the expansions :

$$Q^{m/\infty}(z) = \sum_{k=0}^{\infty} q_k^{m/\infty} z^k$$

$$Q^{m/n}(z) = \sum_{k=0}^{n} q_k^{m/n} z^k$$

Following [1] we have :

$$\forall k,n \,|\, 0 \leq k \leq n : \qquad 0 < q_k^{m/n} \leq q_k^{m/n+1} \leq \dots \leq q_k^{m/\infty} \tag{8}$$

$$\lim_{n \to \infty} q_k^{m/n} = q_k^{m/\infty} \tag{9}$$

Then :

$$\forall \varepsilon > 0, \exists N_\xi \,|\, \forall z \in D_\xi, \forall n \geq N_\xi : \sum_{k=n+1}^{\infty} q_k^{m/\infty} |z|^k \leq \frac{\varepsilon}{2}$$

$$\forall k \,|\, 1 \leq k \leq N_\xi, \forall \varepsilon, \exists N_k \geq k \,|\, \forall n \geq N_k : \quad 0 \leq q_k^{m/\infty} - q_k^{m/n} \leq \frac{\varepsilon e^{-\xi}}{2\, k!}$$

Let $N = \max_{1 \leq k \leq N_\xi} \{N_k, N_\xi\}$, then by (8) $\forall n > N$:

$$|Q^{m/n}(z) - Q^{m/\infty}(z)| \leq$$

$$\leq \sum_{k=1}^{N_\xi} (q_k^{m/\infty} - q_k^{m/n})|z|^k + \sum_{k=N_\xi+1}^{n} (q_k^{m/\infty} - q_k^{m/n})|z|^k + \sum_{k=n+1}^{\infty} q_k^{m/\infty}|z|^k \leq$$

$$\leq \sum_{k=1}^{N_\xi} (q_k^{m/\infty} - q_k^{m/n})|z|^k + \sum_{k=N_\xi+1}^{\infty} q_k^{m/\infty}|z|^k \leq \frac{\varepsilon e^{-\xi}}{2} \sum_{k=1}^{N_\xi} |z|^k/k! + \frac{\varepsilon}{2} \leq \varepsilon$$

which proves (6).

If f belongs to S^*, then h :

$$h(z) = f^{-1}(z) = \sum_{j=0}^{\infty} h_j (-z)^j \quad ; \quad h_0 = 1 \tag{10}$$

also belongs to S^*. Applying to h the identities :

$$P_f^{m/n}(z) = Q_h^{n/m}(-z) \quad , \quad Q_f^{m/n}(z) = P_h^{n/m}(-z) \tag{11}$$

we obtain (7) from (6). Q.E.D.

Theorem 2

Let f belong to S^*, then :

$x > 0$:

$$m \geq 0: \qquad 1 = Q^{m/0} < Q^{m/1} < \dots < Q^{m/n} < \dots < Q^{m/\infty} \tag{12}$$

$$n > 0: \qquad Q^{0/n} > Q^{1/n} > \dots > Q^{\infty/n} \geq 1 \tag{13}$$

$-\frac{1}{C_1} \leq x < 0$:

$$m = 0 : \qquad 1 = Q^{0/0} > Q^{0/2} > \dots > \frac{1}{f} = Q^{0/\infty} > \dots > Q^{0/3} > Q^{0/1} \geq 0 \tag{14}$$

$m > 0: \quad 1 = Q^{m/0} > Q^{m/1} > \ldots > Q^{m/\infty} > 0$ (15)

except $Q^{1/1}(-1/c_1) = Q^{1/2}(-1/c_1)$

$n > 0: \quad 0 < Q^{0/n} < Q^{1/n} < \ldots < Q^{\infty/n}$ (16)

except $Q^{0/1}(-1/c_1) = 0$ and $Q^{0/2}(-1/c_1) = Q^{1/2}(-1/c_1)$

<u>Proof</u>

We start from the well known relations [1, 3] :

$$Q^{m/n}(x) = \frac{1}{C_n^m} \begin{vmatrix} 1 & (-x) & \ldots & (-x)^n \\ c_{m+1} & c_m & \ldots & c_{m-n+1} \\ \cdot & \cdot & \cdot & \cdot \\ c_{m+n} & \ldots & \ldots & c_m \end{vmatrix} \quad \text{where} \quad C_n^m = \begin{vmatrix} c_m & c_{m-1} & \ldots & c_{m-n+1} \\ c_{m+1} & \cdot & \cdot & \cdot \\ \cdot & \cdot & \cdot & \cdot \\ c_{m+n-1} & \ldots & \ldots & c_m \end{vmatrix} \tag{17}$$

$$Q^{m/n}(x) = Q^{m/n-1}(x) + x\,Q^{m-1/n-1}(x) \cdot C_n^{m+1} C_{n-1}^{m-1} / (C_n^m C_{n-1}^m) \tag{18}$$

$$Q^{m/n}(x) = Q^{m-1/n}(x) - x\,Q^{m-1/n-1}(x) \cdot C_{n+1}^m C_{n-1}^{m-1} / (C_n^m C_n^{m-1}) \tag{19}$$

The total positivity of the sequence (c_m) implying that all coefficients of the polynomial (17) are strictly positive, $Q^{m/n}$ is positive for x positive. Now, according to (18) and (19), we obtain (12) and (13) respectively. We complete these inequalities by the limits (6) and (4).

The sequence (h_j) of (10) is totally positive ; in particular $\begin{vmatrix} h_1 & h_0 \\ h_{n+1} & h_n \end{vmatrix} > 0$ which gives $h_n/h_{n+1} > h_0/h_1 = 1/c_1$. Then $Q^{0/n+1} - Q^{0/n-1} = (h_n + h_{n+1}x)x^n$, obtained from (11), have the sign depending of n for $-1/c_1 \leqslant x < 0$. Thus, with (6), we obtain (14).

Writing $Q^{0/n}(x) = (h_0 + h_1 x) + (h_2 + h_3 x)x^2 + \ldots$ we see that all $Q^{0/n} > 0$ for $x \geqslant -1/c_1$ except the value $Q^{0/1}(-1/c_1) = 0$. Using this in (18) we obtain (15) for $m = 1$. By induction we obtain (15) for all m.

Putting $Q^{n/0} = 1$ in (19) we obtain (16) for $n=1$ and by induction : for all n.

<u>**Q.E.D.**</u>

According to (11) we can change Q by P in the theorem 2, which gives the corollary :

<u>Theorem 3</u>

Let f belong to S^*, then :

<u>$x < 0$:</u>

$n \geqslant 0: \quad 1 = P^{0/n} < P^{1/n} < \ldots < P^{m/n} < P^{(m+1)/n} < \ldots$ (20)

$m > 0: \quad P^{m/0} > P^{m/1} > \ldots > P^{m/\infty} \geqslant 1$ (21)

$0 < x \leq \frac{1}{c_1}$:

$n = 0$: $\quad 1 = P^{0/0} > P^{2/0} > \ldots > f = P^{\infty/0} > \ldots > P^{3/0} > P^{1/0} \geq 0$  (22)

$n > 0$: $\quad 1 = P^{0/n} > P^{1/n} > \ldots > P^{\infty/n} > 0$  (23)

except $\quad P^{1/1}(1/c_1) = P^{2/1}(1/c_1)$

$m > 0$: $\quad 0 < P^{m/0} < P^{m/1} < \ldots < P^{m/\infty}$  (24)

except $\quad P^{1/0}(1/c_1) = 0$ and $P^{2/0}(1/c_1) = P^{2/1}(1/c_1)$.

In the Padé table the following inequalities occur :

Theorem 4

Let f belong to S^*, then for all m and n :

$x > 0$:

$$(-1)^m\,[m/n] > (-1)^m\,[m/n+1] \tag{25}$$

$$(-1)^m\,[m/n] > (-1)^m\,[m+1/n] \tag{26}$$

$$(-1)^m\,[m/n] > (-1)^m\,[m+1/n+1] \tag{27}$$

$$(-1)^m\,[m/n+1] > (-1)^m\,[m+1/n] \tag{28}$$

$-\frac{1}{c_1} < x < 0$:

$$(-1)^m\,[m/n] < (-1)^m\,[m/n+1] \tag{29}$$

$$(-1)^m\,[m/n] < (-1)^m\,[m+1/n] \tag{30}$$

$$(-1)^m\,[m/n] < (-1)^m\,[m+1/n+1] \tag{31}$$

$$(-1)^m\,[m+1/n] < (-1)^m\,[m/n+1] \tag{32}$$

$$[m/n] \geq 1 \tag{33}$$

$$(-1)^m\,[m/n] < (-1)^m f \tag{34}$$

$0 \leq x \leq \frac{1}{c_1}$: $\quad 0 < [m/n] \leq 1 \quad$ except $\quad [1/0](-1/c_1) = 0$ (35)

$0 < x < \frac{1}{\alpha_{m+1}}$: $\quad (-1)^m\,[m/n] > (-1)^m f$ (36)

Proof

We can easily estimate the sign of the left hand members of the following classical identities :

$$[m/n](x) - [m+1/n](x) = (-1)^m x^{m+n+1} C_{n+1}^{m+1} \left[C_n^m Q^{m/n}(x) Q^{m+1/n}(x)\right]^{-1} \tag{37}$$

$$[m/n](x) - [m/n+1](x) = (-1)^m x^{m+n+1} C_{n+1}^{m+1} \left[C_n^m Q^{m/n}(x) Q^{m/n+1}(x)\right]^{-1} \tag{38}$$

because, according to theorem 2, all Padé denominators are positive in mentioned intervals. Thus we obtain all inequalities from (25) to (32). Theorems 2 and 3 both lead to (33) and (35). Going to the limit in the row (25) (resp. in the column (30)) we obtain (34) according to (5) and (6) (resp.(36) according to (4) and (7)).

Q.E.D.

The following theorem comes from (35). Define some intervals where by convention we put $\alpha_0^{-1} \equiv 0$ and $\beta_0^{-1} \equiv 0$:

$$I_m = \bigcup_{k=0}^{m} [\alpha_{2k}^{-1}, \alpha_{2k+1}^{-1}] - \{\alpha_{2m+1}^{-1}\} \; ; \quad \hat{I}_m = \bigcup_{k=0}^{m} [\alpha_{2k+1}^{-1}, \alpha_{2k+2}^{-1}] - \{\alpha_{2m+2}^{-1}\}$$

$$J_n = \bigcup_{k=0}^{n} [-\beta_{2k+1}^{-1}, -\beta_{2k}^{-1}] - \{-\beta_{2n+1}^{-1}\} \; ; \quad \hat{J}_n = \bigcup_{k=0}^{n} [-\beta_{2k+2}^{-1}, -\beta_{2k+1}^{-1}] - \{-\beta_{2n+2}^{-1}\}$$

$$J_n^* = \bigcup_{k=0}^{n}]-\beta_{2k+1}^{-1}, -\beta_{2k}^{-1}[\; ; \quad \hat{J}_n^* = \bigcup_{k=0}^{n}]-\beta_{2k+2}^{-1}, -\beta_{2k+1}^{-1}[$$

Theorem 5

Let f belong to S^*, then for all m, n and x in the following intervals, we have :

I_m	:	$P^{2m/m} > 0$		(39)
$I_m - \{0\}$	:	$P^{2m+1/0} < P^{2m+1/1} < \ldots < P^{2m+1/\infty}$	$\circ\!\leftarrow\!\circ$	(40)
		$P^{2m+1/m} < P^{2m/m}$	$\downarrow$	(41)
$\hat{I}_m$	:	$P^{2m+1/m} < 0$		(42)
		$P^{2m+2/0} > P^{2m+2/1} > \ldots > P^{2m+2/\infty}$	$\circ\!\rightarrow\!\circ$	(43)
		$P^{2m+2/m} > P^{2m+1/m}$	$\uparrow$	(44)
J_n	:	$Q^{m/2n} > 0$		(45)
$J_n - \{0\}$	:	$Q^{0/2n+1} < Q^{1/2n+1} < \ldots < Q^{\infty/2n+1}$	$\uparrow$	(46)
		$Q^{m/2n+1} < Q^{m/2n+2}$	$\circ\!\leftarrow\!\circ$	(47)
$\hat{J}_n$	:	$Q^{m/2n+1} < 0$		(48)
		$Q^{0/2n+2} > Q^{1/2n+2} > \ldots > Q^{\infty/2n+2}$	$\downarrow$	(49)
		$Q^{m/2n+2} > Q^{m/2n+1}$	$\circ\!\leftarrow\!\circ$	(50)
J_n^*	:	$f > [m/2n]$		(51)
$\hat{J}_n^*$	:	$f < [m/2n+1]$		(52)

Proof

Multiplying (36) by $Q^{m/m}$ and looking at (1), we find (39) and (42) according to the positivity of $fQ^{m/m}$. Now (11) leads to (45) and (48). Applying (36) to the function h (see (10)) we obtain for $-1/\beta_{m+1} < x < 0$:

$$(-1)^m \left([m/m]_f^{-1}(x) - f^{-1}(x) \right) > 0 \; .$$

Hence, multiplying this inequality by $[m/n]\cdot f$, the sign of which is known by (45) and (48), we obtain (51) and (52). The inequalities (46), (47), (49), (50) are the consequences of the inequalities (45) and (48) used in the formulae (18) and (19). The four inequalities (40), (41), (43), (44) for numerators clearly follow from the four corresponding inequalities for denominators, according to (11).

Q.E.D.

We prepare the theorem for the Padé errors by :

Theorem 6

Let f belong to S^*, then for all m and n :

$$-\frac{1}{c_1}<x<0 : \quad (-1)^n\left(f-[m/n+1]\right)+(-1)^n\left(f-[m/n]\right)=\frac{|x|^{m+n+1}C_{n+1}^{m+1}/C_n^m}{Q^{m/n+1}(x)\,Q^{m/n}(x)}>0 \qquad (53)$$

$$(-1)^n\left(f-[m/n]\right)-(-1)^n\left(f-[m+1/n]\right)=\frac{|x|^{m+n+1}C_{n+1}^{m+1}/C_n^m}{Q^{m/n}(x)\,Q^{m+1/n}(x)}>0 \qquad (54)$$

$$x\in J_m^* : \quad \left(f-[m/2n]\right)-\left(f-[m+1/2n]\right)=\frac{|x|^{m+2n+1}C_{2n+1}^{m+1}/C_{2n}^m}{Q^{m/2n}(x)\,Q^{m+1/2n}(x)}>0 \qquad (55)$$

$$x\in \hat{J}_m^* : \quad \left([m/2n+1]-f\right)-\left([m+1/2n+1]-f\right)=\frac{|x|^{m+2n+2}C_{2n+2}^{m+1}/C_{2n+1}^m}{Q^{m/2n+1}(x)\,Q^{m+1/2n+1}(x)}>0 \qquad (56)$$

$$0<x<\frac{1}{\alpha_{m+1}} : \quad (-1)^m\left([m/n]-f\right)+(-1)^{m+1}\left([m+1/n]-f\right)=\frac{x^{m+n+1}C_{n+1}^{m+1}/C_n^m}{Q^{m/n}(x)\,Q^{m+1/n}(x)}>0 \qquad (57)$$

$$(-1)^m\left([m/n]-f\right)-(-1)^m\left([m/n+1]-f\right)=\frac{x^{m+n+1}C_{n+1}^{m+1}/C_n^m}{Q^{m/n}(x)\,Q^{m/n+1}(x)}>0 \qquad (58)$$

Remark :

Both terms in the left hand members are positive according to theorem 4. so that all their terms can be replaced by the moduli $|f-[M/N]|$.

Proof

According to theorems 2 and 5 all $Q^{m/n}$ have the same sign in each identity, so that their products are positive in all mentioned intervals. Consequently (38) with (34) leads to (53), (37) with (34) leads to (54), (37) with (51) leads to (55), (37) with (52) leads to (56), (37) with (36) leads to (57) and (38) with (36) leads to (58).

Q.E.D.

For the next theorem we define the following bounding functions :

For negative x specified later :

$$M_1(m,n,x)=|x|^{m+n+1}C_{n+1}^{m+1}\left[C_m^m Q^{m/n}(x)\,Q^{m+1/n}(x)\right]^{-1}\left[1+\left(Q^{m+1/n}(x)-Q^{m+1/n+1}(x)\right)/Q^{m+2/n}(x)\right]$$

$$M_2(m,n,x)=\min\left\{|x|\,C_{n+1}^{m+1}\left[C_n^m Q^{m/n+1}(x)\right]^{-1},\ C_m^{m+1}\left[C_{n-1}^m Q^{m/n-1}(x)\right]^{-1}\right\}\cdot|x|^{m+n}/Q^{m/n}(x)$$

in particular we define : $M_2(m,0,x) = c_{m+1}|x|^{m+1}(1+x\,c_{m+1}/c_m)^{-1}$

$$M_1^*(m,n,x) = \max\left\{ M_1(m,n,x),\ \tfrac{1}{2}|x|^{m+n+1}\, C^{m+1}_{n+1}\left[C^m_n\, Q^{m/n}(x)\, Q^{m/n+1}(x)\right]^{-1}\right\}$$

$$M_2^*(m,n,x) = \min\left\{ |x|\, C^{m+1}_{n+1}\left[C^m_n\, Q^{m/n+1}(x)\right]^{-1},\ \tfrac{1}{2} C^{m+1}_{n}\left[C^{m}_{n-1}\, Q^{m/n-1}(x)\right]^{-1}\right\}\cdot |x|^{m+n}/Q^{m/n}(x)$$

in particular we define : $M_2^*(m,0,x) = M_2(m,0,x)$

<u>For positive x specified later</u> :

$$M_3(m,n,x) = x^{m+n+1}\, C^{m+1}_{n+1}\left[C^m_n\, Q^{m/n}(x)\, Q^{m/n+1}(x)\right]^{-1}\left[1+\left(Q^{m/n+1}(x) - Q^{m+1/n+1}(x)\right)/Q^{m/n+2}(x)\right]$$

$$M_4(m,n,x) = \min\left\{ x\, C^{m+1}_{n+1}\left[C^m_n\, Q^{m+1/n}(x)\right]^{-1},\ C^{m}_{n+1}\left[C^{m-1}_{n}\, Q^{m-1/n}(x)\right]^{-1}\right\}\cdot x^{m+n}/Q^{m/n}(x)$$

in particular we define : $M_4(0,n,x) = x^{n+1}\, C^{1}_{n+1}\left[Q^{1/n}(x)\, Q^{0/n}(x)\right]^{-1}$

$$M_3^*(m,n,x) = \max\left\{ M_3(m,n,x),\ \tfrac{1}{2} x^{m+n+1}\, C^{m+1}_{n+1}\left[C^m_n\, Q^{m/n}(x)\, Q^{m+1/n}(x)\right]^{-1}\right\}$$

$$M_4^*(m,n,x) = \min\left\{ x\, C^{m+1}_{n+1}\left[C^m_n\, Q^{m+1/n}(x)\right]^{-1},\ \tfrac{1}{2} C^{m}_{n+1}\left[C^{m-1}_{m}\, Q^{m-1/n}(x)\right]^{-1}\right\}\cdot x^{m+n}/Q^{m/n}(x)$$

in particular we define : $M_4^*(0,n,x) = M_4(0,n,x)$

<u>Theorem 7</u>

Let f belong to S^*, then for all m and n :

$-\frac{1}{c_1} < x < 0$:

$$0 < (-1)^m\left(f - [m+1/n]\right) < (-1)^m\left(f - [m/n]\right) \qquad (59)$$

$$0 < M_1(m,n,x) < (-1)^m\left(f - [m/n]\right) < M_2(m,n,x) \qquad (60)$$

$x \in J_m^*$:

$$0 < f - [m+1/2m] < f - [m/2m] \qquad (61)$$

$$0 < M_1(m,2m,x) < f - [m/2m] \qquad (62)$$

$x \in \hat{J}_m^*$:

$$0 < [m+1/2m+1] - f < [m/2m+1] - f \qquad (63)$$

$$0 < M_1(m,2m+1,x) < [m/2m+1] - f \qquad (64)$$

$-\frac{1}{2c_1} \leq x < 0$:

$$0 < (-1)^{m+1}\left(f - [m/n+1]\right) < (-1)^m\left(f - [m/n]\right) \qquad (65)$$

$$0 < (-1)^{n+1}\left(f - [m+1/n+1]\right) < (-1)^m\left(f - [m/n]\right) \qquad (66)$$

$$0 < M_1^*(m,n,x) < (-1)^m\left(f - [m/n]\right) < M_2^*(m,n,x) \qquad (67)$$

$0 < x < \frac{1}{\alpha_2}$:

$$0 < (-1)^m\left([m/n+1] - f\right) < (-1)^m\left([m/n] - f\right) \qquad (68)$$

$$0 < M_3(m,n,x) < (-1)^m([m/n]-f) < M_4(m,n,x) \tag{69}$$

<u>$0 \leq x \leq B$</u> where $B = \frac{1}{c_1}$ for $m \geq n$ and $B = \frac{1}{2c_1}$ for $m < n$:

$$0 < (-1)^{m+1}([m+1/n]-f) < (-1)^m([m/n]-f) \tag{70}$$

$$0 < (-1)^{m+1}([m+1/n+1]-f) < (-1)^m([m/n]-f) \tag{71}$$

$$0 < M_3^*(m,n,x) < (-1)^m([m/n]-f) < M_4^*(m,n,x) \tag{72}$$

<u>Proof</u>

The inequalities (59), (61), (63), (68) are the same as (54), (55), (56) and (58), respectively. The bound M_1 in (60) is obtained in the following way : denoting by R_n^m the right hand member of (54) we have $|f-[m/n]| > R_n^m$, thus $|f-[m/n]| = R_n^m + |f-[m+1/n]| > R_n^m + R_n^{m+1} = M_1(m,n,x)$, the last equality being given by (18). The bound M_2 in (60) follows from (53). In a similar way we obtain:(62) where M_1 follows from (55) and (18), (64) where M_1 follows from (56) and (18). We detail the proof of (65) : according to (53) the inequality $|f-[m/n+1]| < |f-[m/n]|$ is true if and only if $|f-[m/n]| > \frac{1}{2}T_n^m$, where T_n^m denotes the right hand member of (53). But, according to (54) we have $|f-[m/n]| > R_n^m$, then for (65) it is sufficient to have $\frac{1}{2}T_n^m \leq R_n^m$, that is $Q^{m+1/n} \leq 2\,Q^{m/n+1}$. Now by (14) $Q^{m+1/n} < 1$ and by (14) and (16) $Q^{m/n+1} > Q^{0/1}$, then the sufficient condition for (65) becomes $1 \leq 2\,Q^{0/1}$, thus finally $x \geq -1/2c_1$. (66) follows from (59) and (65). The bound M_1^* in (67) is deduced from (53), (54) and (65), the bound M_2^* by (53) and (65). The bound M_3 in (69) is given by (58) and (19) ; M_4 by (57). (70) is obtained from (57) and (58) in the same way as the proof of (65), but the improvement of the bound to $1/c_1$ is more complicated. (71) is the consequence of (68) and (70). The bounds M_3^* and M_4^* follow from (57), (58) and (70).

<u>Q.E.D.</u>

<u>Remark</u> : One of the authors (E.L.) has recently improved the bound $-1/2c$ in (65), (66) and (67) to :

$$A = \left[(c_1)^2 - c_2 - \sqrt{(c_1)^4 + (c_2)^2}\,\right] / (2\,c_1 c_2) \tag{73}$$

and the bound B in (70), (71) and (72) to :

$$B = \left[\frac{c_1}{2} - \frac{c_2}{c_1} + \sqrt{(c_1 - c_2/c_1)^2 + (c_1/2)^2}\,\right] / \left[(c_1)^2 - c_2\right] . \tag{74}$$

As a consequence of theorem 7, we obtain the final theorem on the best Padé approximant in the "rectangular" set in the Padé table :

<u>Theorem 8</u>

Let f belong to S^*, then for all m, n, M, N such that $m < M$ and $n < N$ the following inequalities are true :

$\underline{A \leqslant x < 0}$: $$0 < (-1)^N (f - [M/N]) < (-1)^m (f - [m/n]) \quad (75)$$

$$0 < M_1^*(M,N,x) < (-1)^N (f - [M/N]) < M_2^*(M,N,x) \quad (76)$$

$\underline{0 < x \leqslant B}$: $$0 < (-1)^M ([M/N] - f) < (-1)^m ([m/n] - f) \quad (77)$$

$$0 < M_3^*(M,N,x) < (-1)^M ([M/N] - f) < M_4^*(M,N,x) \quad (78)$$

CONCLUSIONS

This work is carried on in the following way :

(i) improvement of the interval $[A,B]$ to the optimal interval ;

(ii) extension of the inequalities to the complex plane ;

(iii) comparison of results with the valleys algorithm [3] .

SOME NUMERICAL EXPERIMENTS

In the following we reproduce only the significant (for understanding the result) digits of numbers and we use, if necessary, the decimal exponent notation.

(i) Table of Padé approximants $[m/n]$

$$f(x) = e^{-\frac{x}{4}} \prod_{k=1}^{2} \left(1 - \frac{x}{2k}\right) \Big/ \prod_{k=1}^{5} \left(1 + \frac{x}{2^{k+1}}\right)$$

$$-\frac{1}{C_1} \simeq -.68$$

m \ n	0	1	2	3	4	5	6
0	1.	49.	1.8	3.	2.4	2.6	2.55
1	2.	2.7	2.54	2.57	2.566	2.567	
2	2.4	2.58	2.565	2.5669	2.566789		
3	2.5	2.568	2.566733	2.566797			
4	2.560	2.5669	2.566791				
5	2.566	2.5668					
6	2.567						

Values for $x = -.66$

We observe the agreement with (14) and (15) giving the positivity of P and Q, thus of $[m/n]$. We also see that the values of Padé approximants converge rapidly to $f(-.66) \simeq 2.5667947$.

(ii) Table of errors $\varepsilon_{m/m} = [m/m] - f$ for the same function

m \ n	0	1	2	3	4	5	6
0	-7.17	-8.99	-7.51	-9.1	-7.26	-9.76	-6.73
1	-4.95	-866.	-2.58	1.85	-.53	.21	
2	-2.72	1.4	-.25	.05	.007		
3	-1.28	.22	-.02	.002			
4	-.55	.04	-.003				
5	-.22	.008					
6	-.08						

Values for x = -1.5

Here $x \in J_0^* =]-4,0[$ and we observe the agreement with (61), i.e. $0 < -\varepsilon_{m+1/2n} < -\varepsilon_{m/2n}$. We also observe the following property proved by one of the authors (E.L., thesis) : if it exists an M such that $|\varepsilon_{M-1/2n+1}| < |\varepsilon_{M/2n+1}|$ and $|\varepsilon_{M/2n+1}| > |\varepsilon_{M+1/2n+1}|$, then for all $m \leq M$ we have $|\varepsilon_{m-1/2n+1}| < |\varepsilon_{m/2n+1}|$ and for all $m \geq M$ we have $|\varepsilon_{m/2n+1}| > |\varepsilon_{m+1/2n+1}|$ (convergence of columns).

(iii) Table of errors $\varepsilon_{m/m} = [m/m] - f$ for the same function

m \ n	0	1	2	3	4	5	6
0	678.	677.2	677.44	677.37	677.4	677.38	677.39
1	685.	675.	679.	676.	678.	677.	
2	705.	664.	692.	656.	728.	568.	
3	744.	631.	773.	516.	-32632.	132.	
4	804.	564.	1526.	214.	-72.4		
5	883.	459.	-588.	48.9			
6	981.	337.	-101.				

Values for x = -4.5

Here $x \in \hat{J}_0^* =]-8,-4[$ and we observe the agreement with (63), i.e. $0 < \varepsilon_{m+1/2n+1} < \varepsilon_{m/2n+1}$. For the even columns we get exactly the same property as in the (ii) case for the odd columns.

(iv) Table of errors $\varepsilon_{m/n} = [m/n] - f$

$$f(x) = \prod_{k=1}^{\infty} \left(1 - \frac{x}{k^2}\right) \quad ; \quad -\frac{1}{c_1} \simeq -.608$$

m \ n	0	1	2	3	4	5	6
0	-.68	.664	-.158	.064	-.021	.0075	-.0026
1	-.11	.012	-.0012	.00012	-.11 -4	.96-6	
2	-.85-2	.32 -3	-.14 - 4	.62-6	-.27-7		
3	-.4 -3	.7-5	-.16-6				
4	-.12-4	.12-6					
5	-.27-6						
6	-.44-8						

Values for x = -.3486

Here $-\frac{1}{c_1} < x$ and therefore we observe the agreement with (59). In spite of the fact that x is outside the interval $\left[-\frac{1}{2c_1}, 0\right[$ we can see that the inequality (65) is still satisfied. In fact, we could improve the bound $-\frac{1}{2c_1}$ and we have found numerically the optimal bound $\delta = -.353 < A \simeq -.35$ (see (73)).

REFERENCES

[1] ARMS, R.J., EDREI, A., "The Padé Tables and Continued Fractions Generated by Totally Positive Sequences", in "Math.Essays dedicated to A.J. Macintyre", Ohio University Press (1970).

[2] GILEWICZ, J., "Approximants de Padé", L.N.M. 667, Springer-Verlag (1978).

[3] GILEWICZ, J., MAGNUS, A., "Valleys in c-Table", in "Padé Approximation and its Applications", L.N.M. 765, L. Wuytack Ed., Springer-Verlag (1979).

ACCELERATION OF CONVERGENCE OF POWER ITERATIVE PROCESS

W. Guziński
Institute of Nuclear Research
Computing Centre CYFRONET
05-400 Świerk-Otwock, Poland

Introduction

Let A be a square matrix, $\lambda_o, \lambda_1 \ldots \lambda_{M-1}$ its eigenvalues arranged according to decreasing absolute values and $v_o, v_1, \ldots, v_{M-1}$ the corresponding eigenvectors.

The simplest form of the power method is:
choose the initial vector $x^{(o)}$ such that $(x^{(o)}, v_o) \neq 0$ and construct the sequence of vectors $x^{(n)}$ by means of iterations $x^{(n)} = Ax^{(n-1)}$ for $n = 1,2,\ldots$

Setting:

$$s_n = \left(x^{(n)}, y\right), \quad \lambda^{(n)} = s_n / s_{n-1}, \quad v^{(n)} = x^{(n)} / s_n ,$$

where y is an arbitrary nonzero vector, one can easily prove that if the matrix A has a dominating eigenvalue λ_o then:

$$\lim_{n\to\infty} \lambda^{(n)} = \lambda_o \qquad \text{and} \qquad \lim_{n\to\infty} v^{(n)} = v_o$$

The rate of the convergence is given by

$$\left\| v^{(n)} - v_o \right\| = O\left(\frac{\lambda_1}{\lambda_o}\right)^n$$

To accelerate the convergence we will construct the family of sequences $V_m^{(n)}$ by means of the transformation:

$$V_{m+1}^{(n-1)} = V_m^{(n)} + \omega_m^{(n)} \Delta V_m^{(n)}, \quad \Delta V_m^{(n)} = V_m^{(n+1)} - V_m^{(n)}$$

For these sequences we expect to reach the rate of convergence given by

$$\left\| V_m^{(n)} - v_o \right\| = O\left(\frac{\lambda_{m+1}}{\lambda_o}\right)^n$$

1. F-transformation

Let $x^{(n)} = A^n x^{(o)}$ be a given sequence of vectors, λ_i and v_i for $i = 0,1,\dots,M-1$ the eigenvalues and corresponding eigenvectors of the matrix A. Setting:

$$s_n = \left(x^{(n)}, y\right), \quad c_n = s_{n+1} - \lambda_o s_n$$

we can define for $m = 1,2,\dots,M-1$ the transformation of the sequence of vectors $\left\{x^{(n)}\right\}$:

$$X_m^{(n)} = F_m\left[x^{(n)}\right] = \mathrm{Det}\begin{vmatrix} c_n & c_{n+1} & \cdots & c_{n+m} \\ c_{n+1} & c_{n+2} & \cdots & c_{n+m+1} \\ \vdots & \vdots & & \vdots \\ c_{n+m-1} & c_{n+m} & \cdots & c_{n+2m-1} \\ x^{(n+m)} & x^{(n+m+1)} & \cdots & x^{(n+2m)} \end{vmatrix} \qquad (1)$$

The determinant in (1) should be interpreted as a formal expansion with respect the last row.

Setting

$$S_m^{(n)} = \left(X_m^{(n)}, y\right), \quad V_m^{(n)} = X_m^{(n)} / S_m^{(n)}$$

and using the well known Sylvester's determinant identity we can easily obtain the following recursive formulae:

$$X_{m+1}^{(n-1)} = \frac{C_m^{(n-1)} X_m^{(n+1)} - C_m^{(n)} X_m^{(n)}}{C_{m-1}^{(n+1)}}, \quad X_o^{(n)} = x^{(n)} \qquad (2)$$

$$S_{m+1}^{(n-1)} = \frac{C_m^{(n-1)} S_m^{(n+1)} - C_m^{(n)} S_m^{(n)}}{C_{m-1}^{(n+1)}}, \quad S_o^{(n)} = s_n \qquad (2')$$

$$C_{m+1}^{(n-1)} = \frac{C_m^{(n-1)} C_m^{(n+1)} - C_m^{(n)^2}}{C_{m-1}^{(n+1)}}, \quad C_o^{(n)} = c_n, \; C_{-1}^{(n)} = 1 \qquad (2'')$$

$$V_{m+1}^{(n-1)} = \frac{C_m^{(n-1)} S_m^{(n+1)} V_m^{(n+1)} - C_m^{(n)} S_m^{(n)} V_m^{(n)}}{C_m^{(n-1)} S_m^{(n+1)} - C_m^{(n)} S_m^{(n)}}, \quad V_o^{(n)} = v^{(n)} = x^{(n)}/s_n \qquad (2''')$$

The last equation can be transformed to the simplest form:

$$V_{m+1}^{(n-1)} = V_m^{(n)} + \omega_m^{(n)} \Delta V_m^{(n)}, \quad \omega_m^{(n)} = \frac{1}{1 - \frac{S_m^{(n)} C_m^{(n)}}{S_m^{(n+1)} C_m^{(n-1)}}} \qquad (3)$$

2. F-algorithm

Let v_o be the eigenvector of the given matrix A, $x^{(o)}$ an arbitrary real vector of R^M such that $(x^{(o)}, v_o) \neq 0$ and let us construct the sequence of vectors $\{x^{(n)}\}$: $x^{(n)} = Ax^{(n-1)}$ for $n = 1,2,\ldots$.
Now we shall study the application of the F-algorithm defined by $(2')$, $(2'')$ and $(2''')$ or (3) where:

$$s_n = (x^{(n)}, y), \quad c_n = s_{n+1} - \lambda_o s_n, \quad v^{(n)} = x^{(n)}/s_n \quad \text{and } \lambda_o$$

is the eigenvalue corresponding to the eigenvector v_o.
For large problems cost of this algorithm is relatively small and proportional to $M^2 \cdot N$ where M is the order of the algorithm and N is the dimension of the problem.
For the sequences of vectors:

$$V_m^{(n)} = \frac{F_m[x^{(n)}]}{(F_m[x^{(n)}], y)}$$

the following theorem is valid:

<u>Theorem 1</u> If the λ and v are the eigenvalue and corresponding eigenvector of the given matrix A and eigenvector v belongs to the subspace generated by initial vector $x^{(o)}$ then there exists a number $M \leqslant \text{Dim}(A)$ such that

$$V_{M-1}^{(n)} = v \quad \text{for } n = 0,1,\ldots$$

Proof Let l_m be an m - dimensional linear subspace generated by the vector $x^{(o)}$:

$$l_m = l\left(x^{(o)}, x^{(1)}, \ldots, x^{(m-1)}\right) \quad \text{where} \quad x^{(i)} = A^i x^{(o)}$$

and let M be the smallest number for which $v \in l_M$, then the vector v can be expanded in the basis $x^{(o)}, x^{(n)}, \ldots, x^{(M-1)}$:

$$v = \sum_{i=0}^{M-1} \beta_i \, x^{(i)}$$

For any n we can easily obtain:

$$\lambda^n v = \sum_{i=0}^{M-1} \beta_i \, x^{(n+i)}$$

$$\lambda^n \cdot = \sum_{i=0}^{M-1} \beta_i \, s_{n+i}$$

$$0 = \sum_{i=0}^{M-1} \beta_i \, c_{n+i}$$

where s_n, c_n have the same meaning as before.

Solving the above system of linear equations we get:

$$v = \frac{\operatorname{Det} \begin{vmatrix} c_n & c_{n+1} & \cdots & c_{n+M-1} \\ c_{n+1} & c_{n+2} & \cdots & c_{n+M} \\ \vdots & \vdots & & \\ c_{n+M-2} & c_{n+M-1} & \cdots & c_{n+2M-3} \\ x^{(n+M-1)} & x^{(n+M)} & \cdots & x^{(n+2M-2)} \end{vmatrix}}{\operatorname{Det} \begin{vmatrix} c_n & c_{n+1} & \cdots & c_{n+M-1} \\ c_{n+1} & c_{n+2} & \cdots & c_{n+M} \\ \vdots & & & \vdots \\ c_{n+M-2} & c_{n+M-1} & \cdots & c_{n+2M-3} \\ s_{n+M-1} & s_{n+M} & \cdots & s_{n+2M-2} \end{vmatrix}}$$

$$v = F_{M-1}\left[x^{(n)}\right] / S_{M-1}^{(n)} = V_{M-1}^{(n)}$$

This theorem shows that for a given eigenvalues λ the F-algorithm defined by $(2')$, $(2'')$ and $(2''')$ can be used as a direct method for computing the corresponding eigenvector.

Let $x^{(n)}$ and $y^{(n)}$ be two sequences of vectors, $x^{(n)} \approx y^{(n)}$ will be used to mean that

$$\lim_{n\to\infty} \frac{\left(x^{(n)}, y\right)}{\left(y^{(n)}, y\right)} = 1 \qquad \text{for any } y, \text{ if } \left(y^{(n)}, y\right) \neq 0 \ \forall n$$

The same notation will be used for real numbers:

$$a_n \approx b_n \qquad \text{iff} \qquad \lim_{n\to\infty} a_n / b_n = 1$$

Let $\sum_{i=0}^{M-1} \alpha_i V_i$ be the expansion of the vector $x^{(0)}$ in the basis of eigenvectors $\{v_i\}$, then the sequence $\{s_n\}$ can be expressed in the form:

$$s_n = \sum_{i=0}^{M-1} a_i \lambda_i^n \qquad \text{where} \qquad a_i = \alpha_i \cdot (v_i, y)$$

For the Hankel determinant:

$$H_m^{(n)} = \text{Det} \begin{vmatrix} s_n & s_{n+1} & \cdots & s_{n+m} \\ s_{n+1} & s_{n+2} & \cdots & s_{n+m+1} \\ \vdots & \vdots & & \\ s_{n+m} & s_{n+m+1} & \cdots & s_{n+2m} \end{vmatrix}$$

we have the following estimate

$$H_m^{(n)} \approx a \cdot \left(\lambda_0 \cdot \lambda_1 \cdots \lambda_m\right)^n \text{ where } a = a_0 \cdot a_1 \cdots a_m \cdot \prod_{i=1}^{m-1} \prod_{j=0}^{i-1} \left(\lambda_i - \lambda_j\right)^2$$

Having this estimate one can prove the following theorem:

<u>Theorem 2.</u> If $|\lambda_0| > |\lambda_1| \ldots > |\lambda_{M-1}|$ then for $m = 0,1, \ldots, M-1$ and any natural n the following relations are valid:

$$S_m^{(n)} = O\left(\lambda_0 \cdot \lambda_1 \cdots \lambda_m\right)^n \qquad (4')$$

$$C_m^{(n)} = O(\lambda_1 \cdot \lambda_2 \cdot \ldots \cdot \lambda_{m+1})^n \tag{4''}$$

$$\lambda_{m+1} \approx \frac{S_{m+1}^{(n)} \; S_m^{(n)}}{S_{m+1}^{(n-1)} \; S_m^{(n+1)}} \approx \frac{C_m^{(n)} \; C_{m-1}^{(n)}}{C_m^{(n-1)} C_{m-1}^{(n+1)}} \tag{4'''}$$

$$\lambda_o \approx \frac{S_m^{(n)} \; C_{m-1}^{(n-1)}}{S_m^{(n-1)} C_m^{(n+1)}} \tag{4^{IV}}$$

$$V_{m+1}^{(n-1)} \approx \frac{\lambda_o \; V_m^{(n+1)} - \lambda_{m+1} \; V_m^{(n)}}{\lambda_o - \lambda_{m+1}} \tag{4^{V}}$$

This theorem, together with the eqs. (2') and (2''), gives the possibility of computing all eigenvalues using the same procedure. The following theorem is the conclusion of the relation (4^{V}):

Theorem 3. If $|\lambda_o| > |\lambda_1| \; \ldots > |\lambda_{M-1}|$ then there exists a set of numbers $a_i^{(m)}$ for $i > m$ such that:

$$V_m^{(n)} - v_o \approx \sum_{i>m} a_i^{(m)} \left(\frac{\lambda_i}{\lambda_o}\right)^n \cdot (v_i - v_o)$$

Now we can formulate the next theorem:

Theorem 4. Let $l_M = l(x^{(o)}, x^{(1)}, \ldots, x^{(M-1)})$ be the linear subspace generated by a sequence of vectors $x^{(i)} = A^i x^{(o)}$, let $v_o, v_1, \ldots, v_{M-1} \in l_M$ be the eigenvectors of matrix A and $|\lambda_o| > |\lambda_1| > \ldots > |\lambda_{M-1}|$ be the corresponding eigenvalues, then the sequences of vectors $V_m^{(n)}$ defined by (2'), (2'') and (2''') tend to the eigenvector v_o and the following relation is valid:

$$\frac{\| V_m^{(n)} - v_o \|}{\| v^{(n)} - v_o \|} = O\left(\lambda_{m+1} / \lambda_o\right)^n$$

This theorem shows the rate of the convergence acceleration due to the F-algorithm.

3. Application of the F-algorithm

In practical calculations the eigenvalue λ_0 is unknown and the exact eigenvalue must be substituted by the computed value $\lambda_0^{(n)}$.
If we replace in definition of c_n the exact value λ_0 by its estimation $\lambda_0^{(n)}$ we get

$$\hat{c}_{n+i} = s_{n+1+i} - \lambda_0^{(n)} s_{n+i} = c_{n+i} + \varepsilon_n s_{n+i} \qquad \text{for } i = 0,1,\ldots,M-1$$

where $\varepsilon_n = \lambda_0 - \lambda_0^{(n)}$

One can see that the computed vectors $\hat{V}_m^{(n)}$ and the exact vectors $V_m^{(n)}$ satisfy the following relation:

$$\left\| \hat{V}_m^{(n)} - V_m^{(n)} \right\| = o(\varepsilon_n)$$

If $\varepsilon_n = o\left(\frac{\lambda_{M-1}}{\lambda_0}\right)^n$ then

$$\left\| \hat{V}_m^{(n)} - V_m^{(n)} \right\| = o\left(\frac{\lambda_m}{\lambda_0}\right)^n$$

and the following estimate is valid:

$$\left\| \hat{V}_m^{(n)} - v_0 \right\| = o\left(\frac{\lambda_m}{\lambda_0}\right)^n \qquad \text{for } m = 1,2,\ldots,M-1 \qquad (5)$$

C.Brezinski [1] shows that the well known scalar ε-algorithm applied to the sequence

$$\lambda^{(n)} = \frac{(x^{(n)}, y)}{(x^{(n-1)}, y)}$$

accelerates the convergence of $\lambda^{(n)}$ and the following relation is satisfied:

$$\varepsilon_{2m}^{(n)} \approx \lambda_0 + o\left(\frac{\lambda_{m+1}}{\lambda_0}\right) \qquad \text{for } m = 0,1,\ldots$$

The following application of the F-algorithm is proposed:
given $x^{(o)} \neq 0$

$$x^{(o)} = x^{(o)} / \sqrt{\left(x^{(o)}, x^{(o)}\right)}$$

$$A^{(n)} = A\, x^{(n-1)}$$

$$\lambda^{(n)} = \left(x^{(n)}, x^{(o)}\right) \qquad n = 1, 2, \ldots, 2M-1$$

$$x^{(n)} = x^{(n)} / \lambda^{(n)}$$

$$\hat{\lambda}_o = \varepsilon^{(1)}_{2M-2}\left(\lambda^{(1)}, \lambda^{(2)}, \ldots, \lambda^{(2M-1)}\right) : \quad \varepsilon\text{- algorithm}$$

$$x^{(o)} = \hat{v}^{(1)}_{M-1}\left(x^{(1)}, x^{(2)}, \ldots, x^{(2M-1)}\right) : \quad F\text{ - algorithm}$$

If the computed eigenvalue $\hat{\lambda}_o$ is close to the exact value then this iterative process satisfies the relation (5).

REFERENCES

1. C.Breziński, Computation of the Eigenelements of a Matrix by the ε - Algorithm, UER d'IEEA - Informatique,France

2. P.Wynn, On the Convergence and Stability of the Epsilon algorithm, SIAM J.Numer.Anal 3 (1966)

GENERALIZED ORDER STAR THEORY

Arieh Iserles
King's College
University of Cambridge
Cambridge CB2 1ST
England

Abstract: In this paper we generalize the theory of order stars of Wanner, Hairer and Nørsett [14]. We show that there is a geometric relation between the location of the zeros and the poles of a rational approximation to the exponential and the distribution of its interpolation points.

By applying this theory we find that the A-acceptability and the general form of the denominator impose bounds on the number and location of the interpolation points. These bounds are used to characterize the A-acceptability properties of various families of rational approximations to exp(x), in particular to verify a conjecture of Ehle [2] on the order-constrained Chebishev approximations.

Much of the paper is based upon a joint work of the author with M.J.D. Powell [8].

1. Introduction

The stability analysis of one-step methods for numerical solution of ordinary differential equations can be reduced to the investigation of rational approximations to the exponential function. Hence a large amount of research has been devoted during recent years to this form of approximation.

Throughout this paper $R = P/Q$ will denote a rational function, $P\epsilon\pi_m$, $Q\epsilon\pi_n$. We are concerned with the following properties of R:

(i) Order of approximation at the origin,

$$R(x) - e^x = O(|x|^{p+1}) .$$

(ii) Negative interpolation points, $R(z_i) - \exp(z_i) = 0$, $z_i<0$, $i = 1, \ldots, k$. These points correspond to exponential fitting of a numerical method [9]. They also appear in the analysis of Chebishev approximations [1, 2].

(iii) A-acceptability, namely $|R(z)|<1$ for every complex z such that $\mathrm{Re}z<0$. This notion is equivalent to the A-stability of a one-step numerical method.

(iv) The distribution of the zeros of Q. It is advantageous in applications to large sparse differential systems [10, 11] if Q has real zeros only, and even only a single real zero of multiplicity n.

For given n and m, and possibly also the distribution of the zeros of Q, one is ultimately interested in using all the available coefficients to satisfy interpolation conditions. This is important to obtain high order, high degree of exponential fitting or Chebishev approximation.

In this paper we address ourselves to this problem. We show that there are two kinds of barriers on the number of interpolation points. The first is imposed by the very fact that we approximate the exponential function, while the second, more strict, stems out of A-acceptability considerations.

Our technical tool throughout this paper is the order star theory. This theory, which was developed by Wanner, Hairer and Nørsett [14], was already successfully used to obtain bounds on the order for a given number of real zeros of Q (which is a barrier of the first kind) and to prove the conjecture of Ehle on the A-acceptability of Padé approximations (a barrier of the second kind).

To suit our purposes, we need to generalize this theory, and this is done in Section 2.

In Section 3 we prove the following theorems:

Theorem 1: The number of real interpolation points of R, counted with their multiplicity, cannot exceed n+m+1.

Theorem 2: If Q has only real zeros, of multiplicities $\eta_1, \eta_2, \ldots, \eta_s$, then the number of real interpolation points of R, counted with their multiplicity, cannot exceed $m+1+\sum_{i=1}^{s}\min\{\eta_i,2\}$.

Theorem 3: If Q has only positive zeros then the number of non-positive interpolation points of R, counted with their multiplicity, cannot exceed m+2.

Theorem 4: Let R be A-acceptable and let k be the sum of the multiplicities of the negative interpolation points. Then the order cannot exceed $m-k+\min\{n,m+2-k\}$.

Theorem 5: The sum of the multiplicities of the negative interpolation points of an A-acceptable approximation R cannot exceed m+1.

Finally, in Section 4 we apply our results to various families of rational approximations to the exponential, like the generalized Padé approximations [3,5], the Padé interpolations [6] and the Chebishev approximations over $(-\infty,0]$ [1,2]. In particular, we verify a conjecture of Ehle on the A-acceptability of order-restricted Chebishev approximations.

Parts of the paper are based upon a joint work of the author with M.J.D. Powell [8]. The author would like to express his gratitude to Professor Powell for his kind permission to publish here the joint results.

2. Order stars

The main idea behind the theory of order stars is to establish a connection between the A-acceptability of R, the location of its zeros and poles and the distribution of the interpolation points. The original theory of Wanner, Hairer and Nørsett [14] considered only the interpolation at the origin, i.e. the order. As we are interested in an arbitrary distribution of real interpolation points, we need to generalize their theory. It should be pointed out that several other generalizations of the theory of order stars are possible - to the approximation of the exponential by global analytic functions [14], to the examination of stability properties of methods for numerical solution of second-order differential equations [4] and to the analysis of upwind methods for hyperbolic conservation laws [7].

We assume that R has real coefficients and is not identically zero and let $S(z) = \exp(-z)R(z)$. We define the *order star* $A = \{z \in \mathbb{C} : |S(z)| > 1\}$ and we denote by D its complement $\mathbb{C}/A$. Both A and D are symmetric about the real axis. Figure 1 depicts some examples of order stars.

The following six propositions hold:

Proposition 1: R is A-acceptable if and only if

(i) A has no intersection with the imaginary axis; and

(ii) R has no poles in $\mathbb{C}^{(-)} = \{z \in \mathbb{C} : \mathrm{Re}\, z < 0\}$.

Proof: By using the maximum principle [14]. ■

Proposition 2: For any positive number τ let $B_\tau = \{t \in S^1 : \tau t \in A\}$, where $S^1 = \{z \in \mathbb{C} : |z| = 1\}$. Then there is a number τ_0 such that for $\tau \geqslant \tau_0$ B_τ is just a single arc of S^1, which for $\tau \to \infty$ tends to $\{\exp(i\theta) : \frac{\pi}{2} \leqslant \theta \leqslant \frac{3\pi}{2}\}$.

Proof: By considering the asymptotic behaviour of $|S(z)|$ for large $|z|$ [14]. ■

We say that $z_0 \in \mathbb{C}$ is an interpolation point of degree $q \geqslant 1$ if $R(z) - \exp(z) = a(z-z_0)^q + O(|z-z_0|^{q+1})$, $a \neq 0$.

Proposition 3: z_0 is an interpolation point of degree q if and only if for $z \to z_0$ A consists of q sectors of angle π/q, separated by q sectors of D, each of the same angle.

Proof: We allow R to have complex coefficients, in order to assume, without loss of generality, that $z_0 = 0$.

Our goal is to show that, if $r > 0$ is sufficiently small and if θ varies between 0 and 2π, then $|S(re^{i\theta})| - 1$ changes sign 2q times, at the points $\theta_k(r)$, $k = 1,\ldots,2q$, and $\{\lim_{r\to 0+} \theta_k(r)\}_{k=1}^{2q}$ form an equispaced mesh in $[0,2\pi)$ with the mesh size π/q.

$|S(z)| = 1$ is equivalent to $\mathrm{Re}\{(S(z)-1)\,(\overline{S(z)+1})\} = 0$. Hence, because of the analycity of S in the neighbourhood of the origin and the interpolation condition $S(z) = 1 + O(|z|^q)$, it is sufficient to prove that if h is an analytic function around the origin which is not identically zero and $h(z) = a_0 z^q + O(|z|^{q+1})$, $a_0 \neq 0$, then $r_0 > 0$ exists such that, for every $0<r \leqslant r_0$, $\mathrm{Re}\, h(re^{i\theta})$ has 2q simple zeros for $0 \leqslant \theta < 2\pi$, which satisfy the equispacing condition of the previous paragraph.

By the analycity of $h \not\equiv 0$ there exists $r_1 > 0$ such that for every r in $(0,r_1]$

$$\frac{\partial}{\partial\theta} \arg h(z) = \mathrm{Re}\left[\frac{zh'(z)}{h(z)}\right]$$

holds, where $z = re^{i\theta}$. Hence $\frac{\partial}{\partial\theta} \arg h(z) = q + O(|z|)$ and $r_0 \in (0,r_1]$

Figure 1: Examples of order stars (The dark-shaded regions denote A)

a. Padé [2/2] approximation

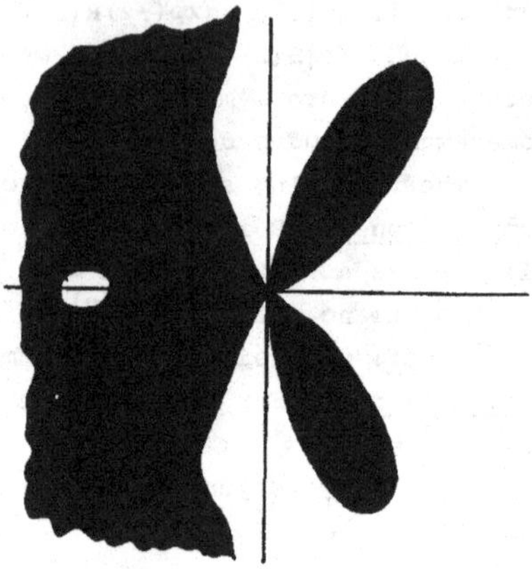

b. Ehle-Picel-Nørsett [2/2] approximation [3]

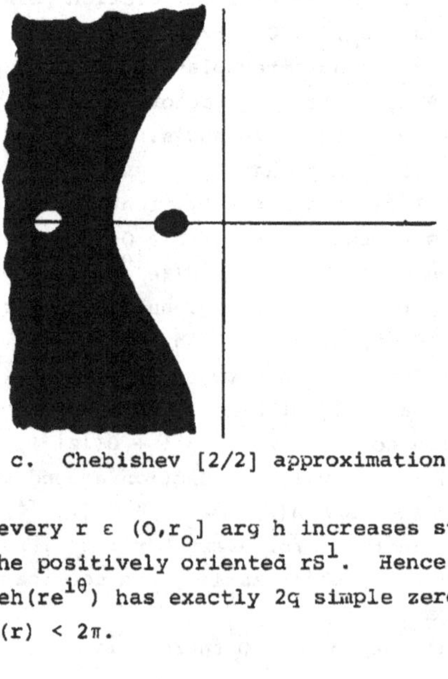

c. Chebishev [2/2] approximation [1].

exists such that for every $r \in (0, r_0]$ arg h increases strictly monotonically along the positively oriented rS^1. Hence, by the argument principle, $\mathrm{Re}\,h(re^{i\theta})$ has exactly 2q simple zeros $0 \leq \hat{\theta}_1(r) < \ldots < \hat{\theta}_{2q}(r) < 2\pi$.

Finally,

$$\frac{\partial}{\partial r} \arg h(z) = \mathrm{Im}\left[\frac{h'(z)}{h(z)} e^{i\theta}\right]$$

and so $\frac{\partial}{\partial r} \arg h(z)$ is uniformly bounded in the neighbourhood of the origin. Hence, by the implicit function theorem, $\lim_{r \to 0+} \hat{\theta}_k(r)$ exist for $k = 1,\ldots,2q$ and equal the $2q$ equispaced zeros of $\cos(q\theta + \arg a_0) = 0$. The proof follows [8]. ■

We call the connected components of A *A-regions*. By Proposition 3, at most q A-regions adjoin the point z_0. We say that an A-region is of *multiplicity* μ if its directed boundary passes through exactly μ interpolation points, which need not be either distinct or real. Proposition 2 implies that there is exactly one unbounded A-region. We denote it by A_∞. The analogous subsets for D we call *D-regions*, D-regions of *multiplicity* μ and D_∞.

Proposition 4: The argument of S decreases strictly monotonically along the positively oriented boundary of an A-region and increases strictly monotonically along the positively oriented boundary of a D-region.

Proof: By examining the normal derivative of S along the boundary of A [14]. ■

Proposition 5: Each bounded A-region of multiplicity μ contains exactly μ poles of R, counted with their multiplicity. Each bounded D-region of multiplicity μ contains exactly μ zeros of R. In both cases $\mu \geqslant 1$.

Proof: By Proposition 4 and the argument principle [8]. ■

Figure 2 illustrates the relation between the zeros, the poles and the interpolation points of a rational approximation to exp(x).

Figure 2:

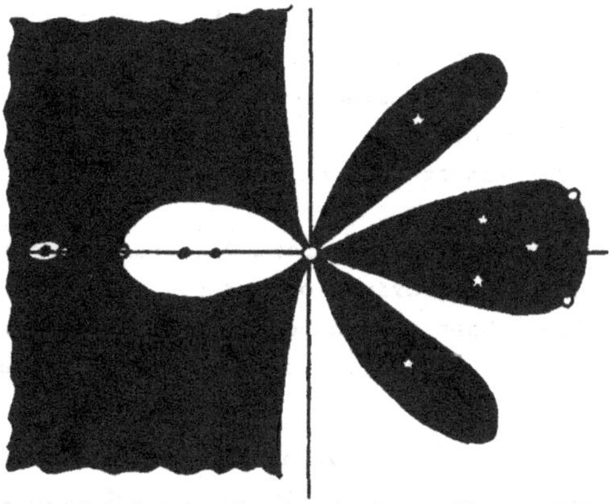

The positions of the interpolation points (denoted by 'o'), the zeros (denoted by '.') and the poles (denoted by '*') of a rational approximation to exp(x).

Proposition 6: There is at most one real interpolation point which adjoins both A_∞ and D_∞.

Proof: If there are two such points then D_∞ separates two unbounded A-regions, in contradiction to the uniqueness of A_∞ [8]. ∎

3. Barrier theorems:

By using the order stars we first prove three theorems which provide barriers on the number of real interpolation points without regard of A-acceptability. Theorem 1 was already proven in [5], using a different technique.

Theorem 1: The number of real interpolation points of R, counted with their multiplicity, cannot exceed $n + m + 1$.

Proof: Let x_i, $i = 1,\ldots,s$, be the real interpolation points with the interpolation degrees η_i, $i = 1,\ldots,s$, respectively. By Proposition 3 there are $2\eta_i$ sectors of A and D which adjoin x_i. Unless both A_∞ and D_∞ touch x_i, at least η_i of those belong either to bounded A-regions or to bounded D-regions.

Let γ be the portion of ∂A_∞ from $i\infty$ to the first intersection of ∂A_∞ with the real axis, together with its reflection in the lower half plane. If both A_∞ and D_∞ adjoin x_i then all the sectors of D which touch x_i to the left of γ and all the sectors of A which touch x_i to the right of γ belong to bounded regions. Hence, by Proposition 3, at least $\eta_i - 1$ sectors belong to bounded regions.

Therefore, by Proposition 6, the bounded A-regions and the bounded D-regions touch $x_1,\ldots,x_s$ at least $\Sigma_{i=1}^{s}\eta_i - 1$ times. By Proposition 5 the sum of the multiplicities of the bounded regions cannot exceed $n + m$, the total number of poles and zeros of R. Thus $\Sigma_{i=1}^{s} \eta_i \leq n+m+1$ and the proof follows [8]. ∎

Let $\alpha_1,\ldots,\alpha_r$ be distinct real numbers, $\xi_1,\ldots,\xi_r$ natural numbers such that $\Sigma_{i=1}^{r} \xi_i = n$ and $Q(x) = \Pi_{i=1}^{r} (1 - \alpha_i x)^{\xi_i}$. Following [12], we call approximations of this form *N-approximations*.

Theorem 2: Let R be an N-approximation and $\xi_1,\ldots,\xi_r$ the multiplicities of its poles. Then the number of real interpolation points of R, counted with their multiplicity, cannot exceed $m+1 + \Sigma_{i=1}^{r} \min\{\xi_i,2\}$.

Proof: Let d_o denote the sum of multiplicities of the bounded D-regions. By Proposition 5 $d_o \leq m$. The number of real interpolation points is maximal if as many as possible are adjoined only by D_∞. It is easy to see by Propositions 3 and 5 that this happens only if (i) every pole is contained in a separate, bounded A-region, (ii) for $\xi_i = 1$ the corresponding A-region adjoins one real interpolation point, (iii) for $\xi_i \geq 2$ the corresponding A-region adjoins two real interpolation points.

By Propositions 5 and 6 and by counting the multiplicities of the bounded D-regions we obtain that the number of real interpolation points cannot exceed $d_o + 1 + \Sigma_{i=1}^{r} \min\{\xi_i, 2\}$. This gives the desired result. ■

<u>Theorem 3</u>: If R is an N-approximation with positive poles then the number of non-positive interpolation points, counted with their multiplicity, cannot exceed $m+2$.

<u>Proof</u>: Once again we count the multiplicities of the bounded D-regions. By Propositions 3 and 5 the greatest number of non-positive interpolation points is attained only if the rightmost non-positive interpolation point is adjoined by the leftmost bounded A-region. In this case only two sectors of D which adjoin the non-positive interpolation points belong to D_∞. The statement of the theorem follows by Proposition 5. ■

More severe bounds on the number of negative interpolation points and on the order at the origin are attained if we demand the A-acceptability of R. The fundamental result in this respect is that the A-acceptability implies $\mathbb{C}^{(-)} \cap A = A_\infty$: No bounded A-region F such that $\mathbb{C}^{(-)} \cap F \neq \emptyset$ may exist, because if $F \subset \mathbb{C}^{(-)}$ then, by Proposition 5, R has a pole in $\mathbb{C}^{(-)}$, while if $F \cap i\mathbb{R} \neq \emptyset$ then $A \cap i\mathbb{R} \neq \emptyset$ and R cannot be A-acceptable by Proposition 1. Finally, by Proposition 2 $\mathbb{C}^{(-)} \cap A_\infty \neq \emptyset$ and, if $A_\infty \cap i\mathbb{R} \neq \emptyset$ then $A \cap i\mathbb{R} \neq \emptyset$ and Proposition 1 gives a contradiction.

<u>Theorem 4</u>: Let R be A-acceptable and let k be the sum of the interpolation degrees of the negative interpolation points. Then the order cannot exceed $m - k + \min\{n, m+2-k\}$.

<u>Proof</u>: Let p denote the order and let $x_q < x_{q-1} < \ldots < x_1$ be the negative interpolation points. By Theorem 1 $-1 \leq p \leq n+m-k$, where by $p = -1$ we mean that there is no interpolation at the origin at all.

If $-1 \leq p \leq 0$ then $\mathbb{C}^{(-)} \cap A = A_\infty$ and symmetry about the real axis imply that D_∞ cannot touch $x_2, \ldots, x_q$ and at most one sector of D_∞ touches x_1. Hence, by counting the multiplicities of bounded D-regions which adjoin $x_1, \ldots, x_q$, Propositions 3 and 5 and $n \geq m$, we obtain $k \leq m+1$ and, consequently, $k+p \leq m + \min\{n, m+2-k\}$.

Let $1 \leq p \leq n+m$. By Proposition 3 and $\mathbb{C}^{(-)} \cap A = A_\infty$, A_∞

adjoins the origin and all the sectors of D which touch $x_1,\dots,x_q$ belong to bounded D-regions. By considering the pattern of approach to the origin of sectors of D to the left of the imaginary axis, we find that at least $[(p+1)/2] - 1$ of them belong to bounded D-regions. Therefore, by Proposition 5, $k + [(p+1)/2] - 1 \leqslant m$. This, together with $n \geqslant m$ and $p+k \leqslant n+m$, gives the required result. ∎

By considering the proof of the last theorem, we can derive an extra result, which deserves to be stated on its own:

Theorem 5: The sum of the interpolation degrees of the negative interpolation points of an A-acceptable approximation R cannot exceed $m+1$.

Proof: By $\mathbb{C}^{(-)} \cap A = A_\infty$ and by counting the multiplicities of bounded D-regions. ∎

4. Applications:

Many approximations to the exponential function share the property that all their coefficients are used to obtain interpolation conditions in $(-\infty,0]$. Theorems 1 and 4 provide us with an A-acceptability characterization of these approximations.

(a) Generalized Padé approximations:

We say that R is a generalized Padé approximation to exp(x) [5] if $R(x) - \exp(x) = O(|x|^{n+m+1-k})$ and negative distinct $x_1,\dots,x_k$ exist such that $R(x_i) = \exp(x_i)$, $i = 1,\dots,k$. By [3] and Theorem 4 the only A-acceptable approximations of this kind are obtained for $n = m$, $0 \leqslant k \leqslant 2$; for $n = m+1$, $0 \leqslant k \leqslant 1$ and for $n = m+2$, $k = 0$. The case $k = 0$ corresponds to Padé approximations [14].

(b) Padé interpolations:

According to [6], R is Padé interpolation if $R(-j\alpha) = \exp(-j\alpha)$, $j = 0,1,\dots,n+m$, for some $\alpha > 0$. By Theorem 4 only the cases $(n,m) = (0,1)$ and $(n,m) = (1,1)$ give rise to A-acceptable approximations.

(c) Order-constrained Chebishev approximations:

In [2] Ehle considers the order-constrained Chebishev approximations to exp(x): $R(x) - \exp(x) = O(|x|^{n+m+1-k})$, where $1 \leqslant k \leqslant n+m+1$, and where the remaining degrees of freedom are used to minimize $\max|R(x) - \exp(x)|$ over $(-\infty,0]$. He shows that if $n = m$ and $1 \leqslant k \leqslant 2$ and if $n = m+1$ and $k = 1$ then R is A-acceptable and he conjectures that those are the only A-acceptable approximations of this kind. This conjecture is verified at once by Theorem 4.

In particular, the Chebishev approximations to exp(x), without any order constraints [1], are not A-acceptable.

Our theorems on N-approximations can be used to derive a result on Chebishev approximations of this kind.

(d) Chebishev approximations with a single real pole:

Chebishev approximation of exp(x) over $(-\infty,0]$ by means of N-approximations with a single pole was investigated in [13]. It is easy to show by standard means and Theorem 3 that if R is an approximation of this kind then distinct $x_1,\ldots,x_{m+2} < 0$ exist such that $R(x_i) = \exp(x_i)$, $i = 1,\ldots,m+2$. Hence, because of Theorem 5, these approximations cannot be A-acceptable.

References:

[1] W.J. Cody, G. Meinardus, R.S. Varga, Chebishev rational approximations to e^{-x} in $[0,+\infty)$ and applications to heat-conduction problems, *J. Approx. Th.* 2 (1969) 50-65.

[2] B.L. Ehle, On certain order constrained Chebishev rational approximations, *J. Approx. Th.* 17 (1976) 297-306.

[3] B.L. Ehle, Z. Picel, Two-parameter, arbitrary order, exponential approximations for stiff equations, *Math. of Comp.* 29 (1975) 501-511.

[4] E. Hairer, Unconditionally stable methods for second order differential equations, *Num. Math.* 32 (1979) 373-379.

[5] A. Iserles, On the generalized Padé approximations to the exponential function, *SIAM J. Num. Anal.* 16 (1979) 631-636.

[6] A. Iserles, Rational interpolation to exp(-x) with application to certain stiff systems, to appear in *SIAM J. Num. Anal.*

[7] A. Iserles, Order stars and a saturation theorem for conservation laws, in preparation.

[8] A. Iserles, M.J.D. Powell, On the A-acceptability of rational approximations to the exponential, DAMTP NA/3, Univ. of Cambridge (1980).

[9] W. Liniger, R.A. Willoughby, Efficient integration methods for stiff systems of ordinary differential equations, *SIAM J. Num. Anal.* 7 (1970) 47-66.

[10] S.P. Nørsett, Restricted Padé approximations to the exponential function, *SIAM J. Num. Anal.* 15 (1978) 1008-1029.

[11] S.P. Nørsett, G. Wanner, The real-pole sandwich for rational approximations and oscillation equations, *BIT* 19 (1979) 79-94.

[12] S.P. Nørsett, A. Wolfbrandt, Attainable order of rational approximations to the exponential function with only real zeros, *BIT* 17 (1977) 200-208.

[13] E.B. Saff, A. Schönhage, R.S. Varga, Geometric convergence to e^{-z} by rational functions with real poles, *Num. Math.* 25 (1976) 307-322.

[14] G. Wanner, E. Hairer, S.P. Nørsett, Order stars and stability theorems, *BIT* 18 (1978) 475-489.

SINGULARITIES OF FUNCTIONS DETERMINED BY THE POLES OF PADÉ APPROXIMANTS

J. Karlsson
Department of Mathematics
University of Umeå
S-901 87 Umeå
Sweden

E. B. Saff †
Center for Mathematical Services
University of South Florida
Tampa, Florida 33620
USA

1. Introduction.

Given a (formal) power series $f(z) = \sum a_n z^n$, the classical Padé approximant of type $[n/\nu]$ is a rational function $P_{n,\nu}/Q_{n,\nu}$, where $P_{n,\nu}, Q_{n,\nu} (\not\equiv 0)$ are polynomials of respective degrees at most n and ν, and such that the power series of $Q_{n,\nu}f - P_{n,\nu}$ starts with terms of degree $\geq n+\nu+1$. When qualitative properties of $f(z)$ are known (such as the existence and number of poles, branch points, etc.), a fundamental question in the study of Padé approximants is whether the poles of these approximants tend to the singularities of $f(z)$. Some classical results in this direction include the theorem of Montessus de Ballore on meromorphic functions [9](cf. [13]) and results for Stieltjes series [11], [7].

Much less is known, however, concerning the <u>inverse problem</u>. Here the essential question is the following. Suppose that $f(z)$ is a formal power series and that the poles of some sequence of its Padé approximants converge to a set L. Does it follow that f (or some continuation of f) is singular on L? A related problem is whether the function f is actually analytic off L.

The purpose of this paper is to survey known results concerning the inverse problem and to present some proofs of theorems previously announced [6] by the authors. In sections 2 and 3 we consider rational interpolants with fixed denominator degree. For the special case of Padé approximants of type $[n/1]$ our result in §2 concerns the validity of what physicists call the Domb-Sykes method [3]. In §3 we study rational interpolation in more general triangular schemes, and in §4 we discuss diagonal sequences of Padé approximants.

2. Padé Approximants with Fixed Denominator Degree

The earliest result in the inverse direction is the following theorem due to Fabry [2, p. 377], which can be regarded as a refinement of the ratio test.

<u>Theorem</u> 2.1. (Fabry) <u>Suppose that</u> $f(z) = \sum_0^\infty a_n z^n$ <u>is a (formal) power series for which</u>

$$\lim_{n\to\infty} \frac{a_n}{a_{n+1}} = \alpha .$$

†The research of this author was supported, in part, by the National Science Foundation.

Then $f(z)$ is analytic in the disk $|z| < |\alpha|$ and α is a singularity of $f(z)$.

Of course the conclusion regarding the radius of convergence is evident, so the significant part is that α is actually a singularity of $f(z)$. Now it is easy to see, directly from the definition, that a_n/a_{n+1} is the pole of the Padé approximant of type $[n/1]$ to $f(z)$. Thus Fabry's theorem can be reformulated as a result concerning the second row of the Padé table: *If the poles of the* $[n/1]$ *Padé approximants for* f *converge to* α, *then* α *is a singularity of* f.

A substantial generalization of Theorem 2.1 to other rows of the Padé table was recently proved by Vavilov, Lopez, and Prohorov [12]. They established

Theorem 2.2. Suppose f is analytic at the origin and that for fixed $\nu>0$ and all sufficiently large n, the Padé approximant of type $[n/\nu]$ to f has exactly ν finite poles. Assume further that the Padé denominators $Q_{n,\nu}$ (suitably normalized) converge to a polynomial Q of degree ν with $Q(0)\neq 0$. If Q has a single zero α of largest modulus and this zero is simple, then f is meromorphic with precisely $\nu-1$ poles (at the smaller zeros of Q) in the disk $|z| < |\alpha|$. Moreover, f has a singularity at α.

If in Fabry's theorem (or Theorem 2.2), additional information is given regarding the degree of convergence of the poles of Padé approximants, can we then describe the precise nature of the singularity at α? The following fundamental theorem in this regard is implicitly contained in the work of Hadamard and is a special case of a result due to Kovačeva [8] concerning the geometric convergence of the poles.

Theorem 2.3. Suppose f is analytic at the origin and that for fixed $\nu>0$ and all sufficiently large n, the Padé approximant of type $[n/\nu]$ to f has exactly ν finite poles. Assume further that there exists a polynomial Q of degree ν, with $Q(0)\neq 0$, such that the denominators $Q_{n,\nu}$ (suitably normalized) satisfy $\| Q_{n,\nu}-Q \| = O(R^{-n})$, $R>1$, $n=0,1,\ldots$, where $\|\cdot\|$ represents the norm on the space of polynomial coefficients. Then f is meromorphic with precisely ν poles (at the zeros of Q) in the disk $|z|<R|\alpha_\nu|$, where $|\alpha_\nu|$ is the maximum modulus of the zeros of Q.

While Kovačeva's result [8] applies to Newton-type interpolation schemes, no extensions of Theorem 2.3 have thus far appeared for rational interpolation in general triangular schemes. Section 3 of this paper is devoted to establishing such an extension.

Regarding slower than geometric convergence, there is a technique related to algebraic singularities which is known as the Domb-Sykes method [3]. This method is based on the observation (letting the singular point $\alpha=1$ for convenience) that if $f(z) = \sum_0^\infty a_n z^n$ is of the form

(2.1) $f(z) = C(1-z)^{\lambda} + h(z),\ \lambda \notin \mathbb{N}$, or $f(z) = C(1-z)^{\lambda}\log(1-z) + h(z),\ \lambda \in \mathbb{N}$,†

where $h(z)$ is analytic on $|z| \leq R$, $R>1$, then

$$a_{n+1}/a_n = 1 - (1+\lambda)/(n+1) + O(R^{-n}) \ . \tag{2.2}$$

Graphically, (2.2) implies that when the successive ratios (the reciprocals of the poles of the Padé approximants of type [n/1]) are plotted against $1/(n+1)$, then they asymptotically lie on a straight line. In the inverse direction, we can easily prove that *if* (2.2) *holds for a function* f, *then* f *must be of the form* (2.1), *where* $h(z)$ *is analytic in* $|z|<R$. This is a special case of our

Theorem 2.4. Suppose that f is analytic at the origin and $\nu>0$ is fixed. Let $\left\{\alpha^{(k)}_{n,\nu}\right\}^{\nu}_{k=1}$ denote the zeros of the denominator $Q_{n,\nu}$ of the Padé approximant of type $[n/\nu]$ for f, where

$$|\alpha^{(0)}_{n,\nu}| \leq |\alpha^{(1)}_{n,\nu}| \leq \cdots \leq |\alpha^{(\nu)}_{n,\nu}| \ .$$

Suppose there exist ν nonzero complex numbers $\left\{\alpha^{(k)}\right\}^{\nu}_{k=1}$ such that

$$\left|\alpha^{(\nu-1)}/\alpha^{(\nu)}\right| < 1/R, \quad R>1 \ , \tag{2.3}$$

$$\lim_{n\to\infty} \alpha^{(k)}_{n,\nu} = \alpha^{(k)}, \qquad k = 1,2,\ldots,\nu-1, \tag{2.4}$$

and

$$\frac{1}{\alpha^{(\nu)}_{n,\nu}} = \frac{1}{\alpha^{(\nu)}}\left[1 - \frac{1+\lambda}{n+\nu}\right] + O(R^{-n}) \ , \text{ as } n\to\infty. \tag{2.5}$$

Then, if $\lambda \notin \mathbb{N}$, the function f must be of the form

$$f(z) = \frac{C(z-\alpha^{(\nu)})^{\lambda} + h(z)}{\prod_{k=1}^{\nu-1} (z-\alpha^{(k)})^2} \ , \tag{2.6}$$

---

† $\mathbb{N}$ denotes the set of nonnegative integers.

where h *is analytic in* $|z| < R|\alpha^{(\nu)}|$ *and* $h(\alpha^{(k)}) = -C(\alpha^{(k)}-\alpha^{(\nu)})^{\lambda}$, *for* k=1, ..., ν-1. *If* $\lambda \in \mathbb{N}$, *then*

$$(2.7) \qquad f(z) = \frac{C(z-\alpha^{(\nu)})^{\lambda}\log(z-\alpha^{(\nu)}) + h(z)}{\prod_{k=1}^{\nu-1} (z-\alpha^{(k)})^2},$$

where h *is analytic in* $|z| < R|\alpha^{(\nu)}|$ *and* $h(\alpha^{(k)}) = -C(\alpha^{(k)}-\alpha^{(\nu)}) \log(\alpha^{(k)}-\alpha^{(\nu)})$, *for* k=1, ..., ν-1.†

We remark that when λ=-1, the convergence in (2.5) is geometric, and Theorem 2.4 reduces to a special case of Theorem 2.3. In fact, if λ is any negative integer, then Theorem 2.4 states that f is meromorphic in $|z| < R|\alpha^{(\nu)}|$, with poles at the $\alpha^{(k)}$, k=1, ..., ν, and this more general situation is not covered by Theorem 2.3.

Proof of Theorem 2.4. Normalizing the $Q_{n,\nu}(z)$ so that they are monic, it follows from (2.4) and (2.5) that

$$\lim_{n\to\infty} Q_{n,\nu}(z) = \prod_{k=1}^{\nu} (z-\alpha^{(k)}) \equiv Q(z), \quad \forall\ z.$$

Further, from (2.3), the polynomial Q has a single zero of largest modulus, namely at $\alpha^{(\nu)}$, and this zero is simple. Hence Theorem 2.2 implies that in the disk $|z| < |\alpha^{(\nu)}|$, the function f is meromorphic with precisely ν-1 poles (at the $\alpha^{(k)}$, k=1, ..., ν-1). Now set

$$\hat{Q}_{n,\nu}(z) \equiv \prod_{k=1}^{\nu-1} (z-\alpha^{(k)}_{n,\nu}), \qquad \hat{Q}(z) \equiv \prod_{k=1}^{\nu-1} (z-\alpha^{(k)}).$$

From the Padé conditions, we have

$$(z-\alpha^{(\nu)}_{n,\nu})\hat{Q}_{n,\nu}(z)\hat{Q}(z)f(z)-\hat{Q}(z)P_{n,\nu}(z) = O(z^{n+\nu+1}), \quad \text{as} \quad z\to 0,$$

and since $\hat{Q}f$ is analytic in $|z| < |\alpha^{(\nu)}|$, Hermite's formula implies that for any $0 < \sigma < |\alpha^{(\nu)}|$

$$(2.8) \qquad \limsup_{n\to\infty} \left[\max_{|z|\le\sigma} \left|(z-\alpha^{(\nu)}_{n,\nu})\hat{Q}_{n,\nu}\hat{Q}f - \hat{Q}P_{n,\nu}\right|\right]^{1/n} \le \sigma/|\alpha^{(\nu)}|.$$

†More precisely, for any λ, the conditions on h are meant to indicate that f has poles in $\alpha^{(1)},\ldots,\alpha^{(\nu-1)}$ with corresponding multiplicity (not twice the multiplicity).

Since $(\hat{Q}f)(\alpha^{(k)})\neq 0$ and $(\hat{Q}P_{n,\nu})(\alpha^{(k)})=0$ for $k<\nu$, it easily follows from (2.8) and (2.3) that, on any compact set $K\subset \mathbb{C}$,

$$\max_{z\in K} |\hat{Q}_{n,\nu}(z) - \hat{Q}(z)| = O(R^{-n}), \quad \text{as } n\to\infty. \tag{2.9}$$

Next, for notational convenience, if g is analytic at $z = 0$, we let $I_n(g)$ be the coefficient of z^n in the Maclaurin expansion for g. Further, we set $\alpha\equiv\alpha^{(\nu)}$. Then from the Padé conditions and the fact that the degree of $\hat{Q}P_{n,\nu}$ is at most $n+\nu-1$, we have

$$I_{n+\nu}\left[\left(\frac{z}{\alpha}-1\right)\hat{Q}^2 f\right] = I_{n+\nu}\left[\left\{\left(\frac{z}{\alpha}-1\right)\hat{Q} - \left(\frac{z}{\alpha^{(\nu)}_{n,\nu}}-1\right)\hat{Q}_{n,\nu}\right\}\hat{Q}f\right].$$

Using (2.5), this last equation can be written in the form.

$$I_{n+\nu}\left[\left(\frac{z}{\alpha}-1\right)\hat{Q}^2 f\right] = I_{n+\nu}\left[\left(\frac{z}{\alpha}-1\right)\left(\hat{Q}-\hat{Q}_{n,\nu}\right)\hat{Q}f\right] - I_{n+\nu}\left[zO(R^{-n})\hat{Q}_{n,\nu}\hat{Q}f\right] + \tag{2.10}$$

$$\frac{(1+\lambda)}{(n+\nu)\alpha} I_{n+\nu}\left[z\left(\hat{Q}_{n,\nu}-\hat{Q}\right)\hat{Q}f\right] + \frac{(1+\lambda)}{(n+\nu)\alpha} I_{n+\nu}\left[z\hat{Q}^2 f\right].$$

Now the first, second, and third terms on the right-hand side of (2.10) are each $O(\rho^{-n})$ for every $1<\rho<R|\alpha|$. Hence on multiplying (2.10) by $z^{n+\nu}$ and summing, we find that

$$F(z) \equiv \int_0^z \hat{Q}^2(t)f(t)dt$$

satisfies the following differential equation

$$(z-\alpha)F'(z) = (1+\lambda)F(z) + G(z), \tag{2.11}$$

where G is analytic in $|z| < R|\alpha|$. Solving (2.11) by the usual method gives

$$F(z) = (z-\alpha)^{1+\lambda}\int_0^z \frac{G(t)dt}{(t-\alpha)^{2+\lambda}}.$$

Thus F is analytic in $|z| < R|\alpha|$ except for a branch cut emanating from α, and so the same is true for

$$(\hat{Q}^2 f)(z) = \frac{d}{dz}\left[(z-\alpha)^{1+\lambda}\int_0^z \frac{G(t)dt}{(t-\alpha)^{2+\lambda}}\right] .$$

Next, let $G(z) = \sum_0^\infty c_k(z-\alpha)^k$ be the Taylor expansion for G about α (which converges for $|z-\alpha| < (R-1)|\alpha|$) and select a point z_0 so that $|z_0| < |\alpha|$ and $|z_0-\alpha| < (R-1)|\alpha|$. Then, on consistently choosing the same branch of the logarithm (say with a radial cut from α to infinity) we can integrate term-by-term to obtain, for z off the cut and $|z-\alpha| < (R-1)|\alpha|$,

$$(\hat{Q}^2 f)(z) = \frac{d}{dz}\left[C_1(z-\alpha)^{1+\lambda} + (z-\alpha)^{1+\lambda}\int_{z_0}^z \frac{G(t)dt}{(t-\alpha)^{2+\lambda}}\right]$$

$$= C_2(z-\alpha)^\lambda + \frac{d}{dz}\left[(z-\alpha)^{1+\lambda}\sum_{k=0}^\infty c_k\int_{z_0}^z (t-\alpha)^{k-\lambda-2}dt\right] .$$

If $\lambda \notin \mathbb{N}$, this gives

$$(\hat{Q}^2 f)(z) = C_2(z-\alpha)^\lambda + \frac{d}{dz}\left[\sum_{k=0}^\infty c_k\frac{(z-\alpha)^k}{(k-\lambda-1)} - C_3(z-\alpha)^{1+\lambda}\right]$$

$$= C(z-\alpha)^\lambda + h(z),$$

where h is analytic in $|z-\alpha| < (R-1)|\alpha|$, and hence in $|z| < R|\alpha|$. If $\lambda \in \mathbb{N}$, then the integration gives rise to the logarithmic term of (2.7). Thus f must be of the form (2.6) or (2.7). □

We remark that Theorem 2.4 has a converse (a direct theorem) which is fairly straightforward to prove.

3. Generalized Taylor Series.

In this section we consider rational interpolation in general triangular schemes and deduce an inverse result related to Theorem 2.3. As in the setting of the second author's generalization [10] of the theorem of Montessus de Ballore, we let E be a closed bounded point set in the z-plane whose complement K (with respect to extended plane) is connected and regular in the sense that K possesses a Green's function $G(z)$ with pole at infinity. For $\sigma>1$, we let Γ_σ denote generically the level curve

$$\Gamma_\sigma : G(z) = \log \sigma,$$

and we denote by E_σ the interior of Γ_σ.

Next, we consider a triangular scheme of interpolation points

$$\begin{array}{l} \beta_0^{(0)} \\ \beta_0^{(1)}, \beta_1^{(1)} \\ \cdot\ \cdot\ \cdot\ \cdot\ \cdot\ \cdot \\ \beta_0^{(n)}, \beta_1^{(n)}, \ldots, \beta_n^{(n)} \\ \cdot\ \cdot\ \cdot\ \cdot\ \cdot\ \cdot\ \cdot\ \cdot\ \cdot\ \cdot\ \cdot\ \cdot \end{array} \tag{3.1}$$

(not necessarily distinct in any row) which lie on E.† Setting

$$w_n(z) \equiv \prod_{k=0}^{n} (z-\beta_k^{(n)}) \quad , \quad n = 0,1,2,\ldots, \tag{3.2}$$

we assume that

$$\lim_{n\to\infty} |w_n(z)|^{1/n} = \Delta \exp G(z) , \tag{3.3}$$

uniformly in z on each closed bounded subset of K, where Δ is the transfinite diameter [14, §4.4] of E. We remark that condition (3.3) is equivalent to

$$\limsup_{n\to\infty} \ [\max |w_n(z)| \ : \ z\in E]^{1/n} \leq \Delta \qquad (\text{cf. } [14, §7.4]). \tag{3.4}$$

While the assumption of (3.3) is sufficient for proving a generalization of the (direct) theorem of Montessus de Ballore, the study of the inverse problem for rational interpolation in the points (3.1) requires much more refined properties, which we now state.

For functions f analytic in the points (3.1), we let I_n denote the divided difference operator in the points $\beta_0^{(n)}, \beta_1^{(n)}, \ldots, \beta_n^{(n)}$, that is

$$I_n(f) = f[\beta_0^{(n)}, \ldots, \beta_n^{(n)}] = \frac{1}{2\pi i}\int_C \frac{f(z)}{w_n(z)}\, dz \ , \quad n = 0, 1, \ldots, \tag{3.5}$$

†With slight modifications in the subsequent discussion, it suffices to assume, more generally, that no limit points of (3.1) lie exterior to E.

where the contour C is suitably chosen so as to enclose all the points $\{\beta_k^{(n)}\}_{k=0}^n$. We remark that if $L_n(z)$ is the unique polynomial of degree at most n which interpolates f in the points $\{\beta_k^{(n)}\}_{k=0}^n$, then $I_n(f)$ is simply the coefficient of z^n in the expansion of $L_n(z)$.

Now for each $n = 0, 1, \ldots,$ it is easy to see that there exists a unique monic polynomial $P_n(z)$ of degree n, such that

(3.6) $$I_j(P_n) = \delta_{j,n}, \quad \text{for all } j = 0, 1, \ldots .$$

We shall refer to these polynomials $P_n(z)$ as <u>basis polynomials</u> for the scheme (3.1). They can be generated via the recurrence formula

$$P_n(z) = z^n - \sum_{k=0}^{n-1} I_k(z^n)P_k(z), \qquad P_0(z) \equiv 1 .$$

The divided difference operators together with the associated basis polynomials give rise to a <u>Generalized Taylor Series</u> (GTS) for f, namely

(3.7) $$f(z) \sim \sum_{n=0}^{\infty} I_n(f)P_n(z) .$$

A discussion of the algebraic properties of GTS can be found in Gelfond [4]. For our purposes we require that these series represent f on E as described in

<u>Definition</u> 3.1. The scheme (3.1) is said to have the <u>Walsh-Hadamard property</u> with respect to E if condition (3.3) holds and if, for every function f analytic on E, the series (3.7) converges to f uniformly on E.

The essential feature of such schemes is given in

<u>Lemma</u> 3.2. <u>If (3.1) has the Walsh-Hadamard property with respect to E and f is any function analytic on E, then</u>

(3.8) $$\limsup_{n\to\infty} |I_n(f)|^{1/n} = 1/\Delta\rho(f) ,$$

<u>where</u>

$$\rho(f) \equiv \sup\left\{\sigma\colon f \text{ is analytic in } E_\sigma\right\} .$$

<u>Moreover, the GTS for f converges to f uniformly on compact subsets of</u> $E_{\rho(f)}$.

Proof. We first show that the basis polynomials $P_n(z)$ satisfy

$$\lim_{n\to\infty} |P_n(z)|^{1/n} = \Delta \exp G(z) \quad , \tag{3.9}$$

uniformly on each closed bounded subset of K. For this purpose, let $\mu>1$ and select a point t on Γ_μ. Then for the function $g(z) = 1/(t-z)$, it is easy to verify that

$$I_n(g) = 1/w_n(t) \text{ , for all } n\geq 0 \ .$$

Hence, from the Walsh-Hadamard property, the series

$$\sum_0^\infty P_n(z)/w_n(t)$$

converges uniformly for z on E. Since the terms of this series are uniformly bounded on E, condition (3.3) implies that

$$\limsup_{n\to\infty} \left[\max |P_n(z)| \ : \ z\varepsilon E\right]^{1/n} \leq \lim_{n\to\infty} |w_n(t)|^{1/n} = \Delta\mu \ .$$

On letting μ tend to 1 in the last inequality, we have

$$\limsup_{n\to\infty} \left[\max |P_n(z)| \ : z\varepsilon E\right]^{1/n} \leq \Delta \quad ,$$

from which (3.9) follows.

Now let $f(z)$ be analytic on E. Then it follows immediately from (3.3) and (3.5) that

$$\limsup_{n\to\infty} |I_n(f)|^{1/n} \leq 1/\Delta\rho(f) \ . \tag{3.10}$$

Assume, to the contrary, that strict inequality holds in (3.10). Then it is easy to verify, from (3.9), that the GTS of f converges uniformly on compact subsets of E_λ, for some $\lambda > \rho(f)$. Since this series gives an analytic continuation of $f(z)$, we reach a contradiction to the definition of the number $\rho(f)$. □

We now consider rational interpolation in the scheme (3.1). For f analytic in these points and (n,ν) a given pair of nonnegative integers, we let $r_{n,\nu}$ be the unique rational function of the form

$$r_{n,\nu} = p_{n,\nu}/q_{n,\nu}, \quad \deg p_{n,\nu} \leq n, \quad \deg q_{n,\nu} \leq \nu, \quad q_{n,\nu} \not\equiv 0,$$

such that

(3.11) $$q_{n,\nu}(z)f(z) - p_{n,\nu}(z) = 0 \quad \text{for} \quad z = \beta_0^{(n+\nu)}, \beta_1^{(n+\nu)}, \ldots, \beta_{n+\nu}^{(n+\nu)} \ .$$

(In case of repeated points β, equation (3.11) is to be interpreted in the Hermite (derivative) sense.) Our main result is

Theorem 3.3. Suppose f is analytic on E and the scheme (3.1) satisfies the Walsh-Hadamard property with respect to E. Let $\nu>0$ be fixed and suppose there exists a polynomial q of the form

(3.12) $$q(z) = \prod_{k=1}^{\nu} (z-\alpha_k), \qquad \alpha_k \notin E \qquad \text{for} \quad 1 \le k \le \nu \ ,$$

such that the (denominator) polynomials $q_{n,\nu}$ (suitably normalized) of (3.11) satisfy

(3.13) $$||q_{n,\nu} - q|| = O(R^{-n}) \ , \qquad R>1 \ , \qquad n = 0, 1, \ldots,$$

where $\|\cdot\|$ represents any of the equivalent norms on the $(\nu+1)$-dimensional space of polynomial coefficients. Then, either

(*i*) f is meromorphic with at most $\nu-1$ poles in the whole plane, with the poles of f in zeros of q, or

(*ii*) f is meromorphic with precisely ν poles in $E_{R\sigma^*}$, where

(3.14) $$\sigma^* \equiv \max_{k=1}^{\nu} \ \{\sigma_k : \alpha_k \in \Gamma_{\sigma_k}\} \ ,$$

and these ν poles of f are the zeros, α_k, of q.

Proof. Suppose first that f is not meromorphic with at most $\nu-1$ poles in the whole plane. With the notation of Lemma 3.2 , we then need to show that

(3.15) $$\rho(qf) \ge R\sigma^* \ ,$$

and that f actually has poles in the ν zeros of q (with corresponding multiplicities).

Let $\tilde{q}(z)$ be the monic polynomial whose zeros are the poles of f in $E_{\rho(qf)}$ (if no such poles exist, we set $\tilde{q}(z) \equiv 1$). Then $\tilde{q}$ divides q, and

(3.16) $$\rho(qf) = \rho(\tilde{q}f) = \rho(\tilde{q}qf) \ .$$

We claim that $\tilde{q}(z) \equiv q(z)$. Suppose this is not the case. Then $\deg \tilde{q} \leq \nu-1$, and the interpolation conditions (3.11) imply that $I_{n+\nu}(\tilde{q}q_{n,\nu}f) = 0$. Hence we have

$$(3.17) \qquad I_{n+\nu}(q\tilde{q}f) = I_{n+\nu}[(q-q_{n,\nu})\tilde{q}f], \quad n = 0, 1, \ldots .$$

Next, select any point $z_0 \varepsilon E$ and write

$$q(z)-q_{n,\nu}(z) = \sum_{k=o}^{\nu} b_k^{(n)}(z-z_0)^k.$$

From (3.13) we have $|b_k^{(n)}| = O(R^{-n})$ as $n \to \infty$, and furthermore, by Lemma 3.2,

$$\limsup_{n\to\infty} |I_{n+\nu}[(z-z_0)^k\tilde{q}f]|^{1/n} = \limsup_{n\to\infty} |I_{n+\nu}(\tilde{q}f)|^{1/n} > 0, \qquad k=0,\ldots,\nu ,$$

where the positivity assertion follows from our assumption that f is not meromorphic with at most $\nu-1$ poles in the whole plane. Therefore, from (3.17), there holds

$$(3.18) \qquad \limsup_{n\to\infty} |I_{n+\nu}(q\tilde{q}f)|^{1/n} \leq \frac{1}{R} \limsup_{n\to\infty} |I_{n+\nu}(\tilde{q}f)|^{1/n} < \limsup_{n\to\infty} |I_{n+\nu}(\tilde{q}f)|^{1/n} ,$$

and so $\rho(q\tilde{q}f) > \rho(\tilde{q}f)$, which contradicts (3.16). Thus, as claimed, $\tilde{q}(z) \equiv q(z)$, which means that all the zeros of q lie in $E_{\rho(qf)}$ and that f has actual poles in these ν zeros. To establish (3.15), we simply repeat the above argument with $\tilde{q}$ replaced by $\hat{q}$, where $\hat{q}$ is the monic polynomial whose zeros are the poles of f in E_{σ^*} (compare (3.18)).

Finally, if f is meromorphic with at most $\nu-1$ poles in the whole plane, then as is easily seen from the proof of the Montessus de Ballore theorem, these poles must be limit points of zeros of the $q_{n,\nu}$, and hence must lie in the zeros of q. □

We now mention some examples of schemes (3.1) which satisfy the Walsh-Hadamard (WH) property.

Example 1. Let E be the disk $|z| \leq \tau$, $\tau>0$, and let all the $\beta_k^{(n)}$ be zero. Then $I_n(f) = f^{(n)}(0)/n!$, $P_n(z) = z^n$, and the GTS is simply the Maclaurin series for f. Hence, for a function f analytic at $z=0$, Theorem 3.3 applies to the Padé approximants of f. Notice, however, that Theorem 2.3 is somewhat stronger in this case since for possibility (i) it implies that f is rational with at most $\nu-1$ poles.

Example 2. Suppose that the scheme (3.1) is independent of n, that is, $\beta_k^{(n)} = \beta_k$ for all n. Then we have

$$P_n(z) = w_{n-1}(z) = \prod_{k=0}^{n-1} (z-\beta_k) ,$$

and the GTS is just the Newton series for f. Hence, if condition (3.3) holds, the Newton scheme has the WH property. In this case, as in the first example, the result of Kovačeva [8] is somewhat stronger than Theorem 3.3. However, her results do not apply to the examples which follow.

Example 3. Let E be the unit disk $|z| \leq 1$, and let the scheme (3.1) consist of the roots of unity, i.e., $w_n(z) = z^{n+1} - 1$, $n=0,1, \ldots$. If $f(z) = \sum_0^{\infty} a_k z^k$ is analytic on E, then it is easy to verify that

$$I_n(f) = \frac{1}{n+1} \sum_{i=0}^{n} \lambda_{n+1}^{i} f(\lambda_{n+1}^{i}) = a_n + a_{2n+1} + a_{3n+2} + a_{4n+3} + \cdots ,$$

where λ_{n+1} is a primitive $(n+1)$st root of unity. As shown by Ching and Chui [1] , the basis polynomials $P_n(z)$ are given by

$$P_n(z) = \sum_{k|n} \mu(n/k) z^k, \qquad n = 0, 1, \ldots,$$

where $k|n$ means k is a factor of n, and μ is the Möbius function defined by

$$\mu(j) = \begin{cases} 1 , & \text{if } j = 1 \\ (-1)^k , & \text{if } j = \Pi \text{ (} k \text{ distinct primes)} \\ 0 , & \text{if } p^2 | j \text{ for some } p>1 . \end{cases}$$

It is readily shown (cf. [1]) that the GTS gives maximal convergence to f, so this interpolation scheme has the WH property.

Example 4. Suppose the points β are the same in each fixed row of (3.1), that is, $w_n(z) = (z-\beta^{(n)})^{n+1}$, $n=0, 1, \ldots$. In this case, $I_n(f) = f^{(n)}(\beta^{(n)})/n!$, and the basis polynomials $P_n(z)$ are the Gontcharoff polynomials (cf. [4]). As discussed in the next example, if $E : |z| \leq \tau$, $\tau>0$, and the points $\beta^{(n)}$ tend to zero sufficiently fast, then these points have the WH property.

Example 5. For the scheme (3.1), suppose that the points $\beta_k^{(n)}$ tend to zero as $n \to \infty$ $(0 \leq k \leq n)$. Let Δ_n be the smallest convex polygon containing the points $\beta_0^{(n)}, \ldots, \beta_n^{(n)}$, and let d_n be the diameter of the smallest circle which encloses the polygons Δ_{n-1} and Δ_n. If $\Sigma d_n < \infty$, then, as shown by Gelfond [4, p. 47], the scheme (3.1) has the WH property with respect to any disk $E : |z| \leq \tau$, $\tau>0$.

4. Inverse Theorems for Diagonal Approximants.

With the notation of the previous section, we shall prove

Theorem 4.1. *Let E be a compact point set (not a single point) whose complement K is simply connected, and suppose the interpolation scheme (3.1) satisfies the WH property with respect to E. Let f be analytic on E and for $n=1,2,\ldots$, let $q_{n,n-1}$ and $q_{n,n}$ denote, respectively, the denominator polynomials (defined in (3.11)) for the rational interpolants $r_{n,n-1}$ and $r_{n,n}$. If all the zeros of these polynomials tend to infinity with n, i.e.,*

(4.1) $$\min\{|\zeta| : q_{n,n-1}(\zeta) = 0 \quad \text{or} \quad q_{n,n}(\zeta) = 0\} \to \infty \quad \text{as} \quad n\to\infty,$$

then f must be entire.

Proof. Since f is analytic on the closed set E, there exists a $\lambda>1$ such that f is analytic on $\overline{E}_\lambda$. Thus for $z\in\Gamma_\sigma$, $1<\sigma<\lambda$, Hermite's formula gives

(4.2) $$f(z) - \frac{p_{n,n}(z)}{q_{n,n}(z)} = \frac{1}{2\pi i}\int_{\Gamma_\lambda} \frac{w_{2n}(z)q_{n,n}(t)f(t)dt}{w_{2n}(t)q_{n,n}(z)(t-z)} .$$

Now note that (4.1) implies

$$\limsup_{n\to\infty} [\max\{|q_{n,n}(t)/q_{n,n}(z)| \; ; \; t\in\Gamma_\lambda, \; z\in\Gamma_\sigma\}]^{1/n} \leq 1 ,$$

and so, on estimating the integral in (4.2) and using (3.3), we obtain

(4.3) $$\limsup_{n\to\infty} \left[\max_{z\in\Gamma_\sigma} \left|f(z) - \frac{p_{n,n}(z)}{q_{n,n}(z)}\right|\right]^{1/n} \leq \left(\frac{\sigma}{\lambda}\right)^2 < 1, \qquad 1<\sigma<\lambda .$$

Next, let $L_{2n}(z)$ be the unique polynomial of degree $\leq 2n$ which interpolates $f(z)$ in the points $\{\beta_k^{(2n)}\}_{k=0}^{2n}$, and let $L_{2n-1}(z)$ be the polynomial of degree $\leq 2n-1$ which interpolates $f(z)$ in the points $\{\beta_k^{(2n)}\}_{k=0}^{2n-1}$. Let $\rho>1$. Then, for n sufficiently large and $1<\tau<\rho$, we have

$$(4.4) \qquad \frac{p_{n,n}(z)}{q_{n,n}(z)} - L_{2n}(z) = \frac{1}{2\pi i} \int_{\Gamma_\rho} \frac{w_{2n}(z)}{w_{2n}(t)} \frac{p_{n,n}(t)}{q_{n,n}(t)} \frac{dt}{t-z} \quad , \quad z\varepsilon\Gamma_\tau \ .$$

Set

$$R_n \equiv \min\{\sigma : \zeta\varepsilon\Gamma_\sigma \quad \text{and} \quad q_{n,n}(\zeta) = 0\} \ ,$$

so that, from (4.1), $R_n \to \infty$ as $n \to \infty$. Since, from (4.3), the sequence $p_{n,n}/q_{n,n}$ is uniformly bounded on E, it follows from a lemma of Walsh [14 , p. 250] that for $R_n > \rho$

$$\left| \frac{p_{n,n}(t)}{q_{n,n}(t)} \right| \le A \left[\frac{R_n \rho - 1}{R_n - \rho} \right]^n \quad , \qquad \text{for all} \quad t\varepsilon\Gamma_\rho \ ,$$

where A is a constant independent of n. Using this estimate in (4.4) we obtain for $z\varepsilon\Gamma_\tau$

$$(4.5) \qquad \limsup_{n\to\infty} \left| \frac{p_{n,n}(z)}{q_{n,n}(z)} - L_{2n}(z) \right|^{1/n} \le \left(\frac{\tau^2}{\rho^2} \right) \rho = \frac{\tau^2}{\rho} \ .$$

Estimating the difference $p_{n,n}/q_{n,n} - L_{2n-1}$ in a similar way gives

$$(4.6) \qquad \limsup_{n\to\infty} \left| \frac{p_{n,n}(z)}{q_{n,n}(z)} - L_{2n-1}(z) \right|^{1/n} \le \frac{\tau^2}{\rho} \ , \qquad z\varepsilon\Gamma_\tau \ .$$

Thus, on combining (4.5) and (4.6), we have

$$\limsup_{n\to\infty} \left| I_{2n}(f) \right|^{1/n} = \limsup_{n\to\infty} \left| \frac{L_{2n}(z) - L_{2n-1}(z)}{\hat{w}_{2n}(z)} \right|^{1/n} \le \frac{\tau^2}{\rho(\Delta\tau)^2} = \frac{1}{\rho\Delta^2} \ ,$$

where $\hat{w}_{2n}(z) \equiv w_{2n}(z)/(z-\beta_{2n}^{(2n)})$. On letting $\rho\to\infty$, we therefore obtain

$$(4.7) \qquad \lim_{n\to\infty} \left| I_{2n}(f) \right|^{1/2n} = 0 \ .$$

Finally, if the above argument is repeated for the sequence $p_{n,n-1}/q_{n,n-1}$, we also get

$$(4.8) \qquad \lim_{n\to\infty} \left| I_{2n-1}(f) \right|^{1/(2n-1)} = 0 \ ,$$

and so, by the WH property, f must be entire. □

Remark 1. With the conditions of Theorem 4.1, it follows immediately from inequality (4.3), with $\lambda=\infty$, that the sequence of rational interpolants $\{r_{n,n}\}$ converges faster than geometrically to f on any compact set in the plane; the same being true for the sequence $\{r_{n,n-1}\}$.

Remark 2. Theorem 4.1 remains valid if the hypothesis concerning the two sequences $\{q_{n,n-1}\}$, $\{q_{n,n}\}$ is replaced by the same assumption for any two diagonal sequences $\{q_{n,n+k}\}$, $\{q_{n,n+j}\}$, where k and j have opposite parity.

Remark 3. There are several instances when it suffices to know the behavior of only one diagonal sequence in Theorem 4.1. This is especially the case when there is a relationship between the even and odd numbered rows of the scheme (3.1). For example, a slight modification in the proof of Theorem 4.1 shows that *if* $w_{2n-1}(z)$ *divides* $w_{2n}(z)$ *for all* n *large, then the assumption that the zeros of the sequence* $\{q_{n,n}\}$ *tend to infinity implies that* f *is entire.* This is certainly the case if the scheme (3.1) is Newton and, in particular, if the interpolants are Padé approximants. In the Padé case, even more can be said concerning the inverse problem. In this regard, we mention a recent result of Gončar and Lungu [5]:

Theorem 4.2. Let f be analytic at infinity and, for $n=0,1,\ldots$, let $\tilde{Q}_{n,n}$ be the denominator of the $[n/n]$ Padé approximant to f (interpolating at infinity), where $\tilde{Q}_{n,n}$ is assumed to be monic of degree n. Let L be the set of limit points of the zeros of the $\tilde{Q}_{n,n}$, and suppose that L is regular in the sense that its complement has a Green's function with pole at infinity. If

$$\lim_{n\to\infty} [\max |\tilde{Q}_{n,n}(z)| \; ; \; z\in L]^{1/n} = \Delta(L) ,$$

where $\Delta(L)$ is the transfinite diameter of L, then f is analytic off L.

References

1. C-H. Ching and C. K. Chui, Mean boundary value problems and Riemann series, *J. Approx. Theory* 10(1974), 324-336.

2. P. Dienes, The Taylor Series, Dover, New York, 1957.

3. D. S. Gaunt and A. J. Guttman, Asymptotic analysis of coefficients. In: Phase Transitions and Critical Phenomena III, (Domb, Green, eds.) Academic Press, New York, 1974, pp. 181-243.

4. A. O. Gelfond, Differenzenrechnung, VEB Deutscher Verlag der Wissenschaften, Berlin, 1957.

5. A. A. Gončar and K.N. Lungu, Poljuci diagonalnih approksimatsij Padé i analitičeskoe prodolženie funktsij, *Mat. Sb.* 111 (153)(1980), 279-292.

6. J. Karlsson and E. B. Saff, Singularities of analytic functions determined by Padé approximants, *Notices of Amer. Math. Soc.* 26(1979), p. A-584.

7. J. Karlsson and B. von Sydow, The convergence of Padé approximants to series of Stieltjes, *Ark. Mat.* 14(1976), 43-53.

8. R. K. Kovačeva, Obobščennie approksimatsii Padé i meromorfnoe prodolženie funktsij, *Mat. Sb.* 109 (151)(1979), 365-377.

9. R. de Montessus de Ballore, Sur les fractions continues algébriques, *Bull. Soc. Math. France* 301(1902), 26-32.

10. E. B. Saff, An extension of Montessus de Ballore's theorem on the convergence of interpolating rational functions, *J. Approx. Theory* 6(1972), 63-67.

11. T. J. Stieltjes, Recherches sur les fractions continues, *Ann. Fac. Sci. Univ. Toulouse* 8(1894), 1-122.

12. V. V. Vavilov, G. Lopez, and V. A. Prohorov, Ob odnoj obpatnoj zadače dlja strok tablitsi Padé, *Mat. Sb.* 110 (152)(1979), 117-127.

13. H. Wallin, Rational interpolation to meromorphic functions, *these Proceedings*.

14. J. L. Walsh, Interpolation and Approximation by Rational Functions in the Complex Domain, 5th ed., Colloq. Publ. Vol. 20, Amer. Math. Soc., Providence, 1969.

PADE APPROXIMANTS AND RELATED METHODS FOR COMPUTING BOUNDARY VALUES ON CUTS

S. Klarsfeld

Division de Physique Théorique
Institut de Physique Nucléaire
91406 Orsay Cedex / France

In many problems of current interest it is necessary to calculate the boundary value of an analytical multivalued function F(z) on a certain cut in the complex plane. More generally one might be interested in performing the analytic continuation of F(z) through the cut from one Riemann sheet to another. It has been shown recently by several authors that ordinary and generalized Padé approximants (PA) may help solving also such problems. The object of this talk is to present a short review of the various methods proposed so far.

1. Ordinary PA methods

Let us consider, for the sake of simplicity, a function F(z) with only two branch points located on the positive real axis at z=1 and $z=b>1$. We assume further that the complex z plane is cut along some curve C joining these points. A single-valued branch of F(z) is then completely defined by the Taylor expansion about the origin (provided the latter is not just another singular point, e.g. a pole)

$$F(z) = \sum_{n=0}^{\infty} a_n z^n \quad , \tag{1}$$

which converges at most in a disk $|z| < R \leq 1$. The problem at hand is now to calculate the boundary value F(x+i0) as z tends to a real point $1 < x < b$ from above (fig.1). The PA, which provide the analytic continuation of the power series (1), are unfortunately not adequate for this purpose since, as is well known [1], their poles and zeros usually cluster precisely along the segment [1,b]. Two different approaches have been used to overcome this difficulty.

1 a . Change of the expansion center

Chisholm et al. [2] suggested that a simple way to force the poles out of the real axis is to construct PA on starting not from Eq.(1), but from a Taylor expansion about a complex point z_0 :

$$F(z) = \sum_{n=0}^{\infty} \alpha_n (z-z_0)^n. \tag{2}$$

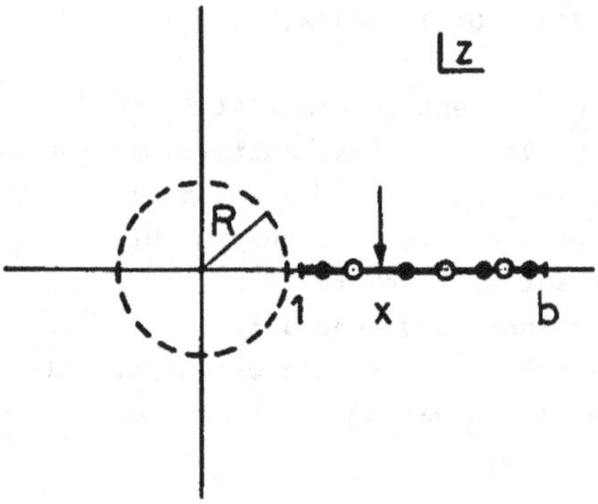

Fig. 1

In particular, if b=∞ and z_0 is chosen somewhere in the upper half-plane, one can expect the simulated cut to be rotated into the lower half-plane (fig.2), thus giving free access to points $x > 1$, and even to part of the second Riemann sheet. We shall denote the PA formed from Eq.(2) by $[M/N](z;z_0)$.

Of course, if Eq.(1) is the only information one has about F(z), the new expansion center should lie not too far from the origin, which restricts the efficiency of the method. However, in most applications some additional information (e.g. an integral representation) is available, allowing us to calculate F and its derivatives also outside the disk $|z| < R$.

Example : $F(z) = -\ell n(1-z)$. The expansion about the origin reads

$$F(z) = \sum_{n=1}^{\infty} z^n/n . \tag{3}$$

If we expand about $z_0 = 1+i$ then we get

$$F(z) = i\frac{\pi}{2} + \sum_{n=1}^{\infty} \frac{i^n}{n} (z-1-i)^n \tag{4}$$

It may be easily seen that the denominators of the PA to Eq.(4) are obtained from those associated with Eq.(3) by the substitution $z \to \zeta = i(z-1-i)$. Therefore to each pole z=p of the original PA corresponds a pole at $\zeta = p$, i.e. $z = 1-i(p-1)$, of the new PA. This is nothing but a clockwise rotation of angle $\pi/2$ around the branch point z=1.

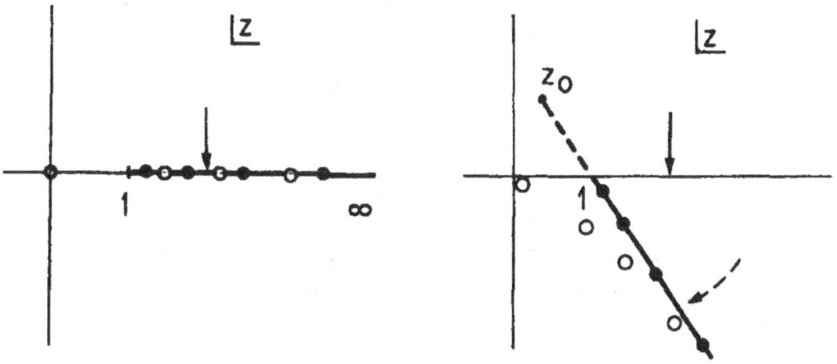

Fig. 2

Notice, however, that the zeros do not behave in the same way, because of the first term in the expansion (4). For instance, from Eq.(3) one gets

$$[2/2](z;0) = \frac{z-z^2/2}{1-z+z^2/6} \quad ; \tag{5}$$

the zeros are located at z=0, z=2, and the poles at $z=3 \pm \sqrt{3}$. On the other side, starting from Eq.(4) yields

$$[2/2]\ (z;1+i) = \frac{i\pi/2+(1-i\pi/2)\zeta - (1-i\pi/6)\zeta^2/2}{1-\zeta+\zeta^2/6}\ , \tag{6}$$

and the zeros are now at $z = 0.005 - 0.1i$ and $z = 0.351 - 0.761i$. The first one is near the zero of the function at z=0, while the second clearly does not lie on the rotated cut. Of course, this does not prevent us from using the new PA on the real axis. For instance, at z=2 one has $\zeta = 1+i$ and Eq.(6) gives $[2/2] \simeq 3.07i$, which is not bad, compared to the exact boundary value $F(2+i0) = i\pi$.

1b . Confórmal mapping

Instead of changing the expansion center one can change the variable through the simple non-linear transformation [3]

$$z = w(2-w) \quad , \tag{7}$$

which maps conformally the half-plane Re $w<1$ onto the z-plane cut along $[1,\infty)$. Substituting into Eq.(1) we get a new Taylor expansion

$$\tilde{F}(w) = F(z(w)) = \sum_{n=0}^{\infty} b_n w^n \quad , \tag{8}$$

where

$$b_n = 2^n \sum_{k=0}^{[n/2]} \binom{n-k}{k} (-1/4)^k a_{n-k} \quad . \tag{9}$$

It is obvious that the b_n's (unlike the α_n's above) are real if the a_n's were so, and that the first p coefficients in Eq.(1) determine the same number of coefficients in Eq.(8).

The variable w incorporates a square-root structure, as seen by inverting Eq.(7) :

$$w(z) = 1-(1-z)^{1/2} \quad , \qquad w(0) = 0 \quad , \tag{10}$$

and its net effect is to unfold the second Riemann sheet of $F(z)$.

If $F(z)$ has two branch points at $z=1$ and $z=b$, $\tilde{F}(w)$ will generally have three branch points, located at $w=1$ and $w=1\pm i\sqrt{b-1}$ (for $b=\infty$ the latter coalesce into $w=\infty$). In special cases, some of these points may be poles or even regular points. Such situations occur in the following examples :

F	$\tilde{F}$
$(1-z)^{\alpha}$	$(1-w)^{2\alpha}$
$\ell n(1-z)$	$2\,\ell n(1-w)$
$\left(\frac{1-z}{1-z/2}\right)^{1/2}$	$\frac{1-w}{(1-w+w^2/2)^{1/2}}$

In view of the way the new branch points are disposed it may be expected that the poles and zeros of the diagonal PA to $\tilde{F}(w)$, which we denote by $[N/N](w)$, will lie close to some lines completely contained in the right half-plane Re $w>1$. A typical situation is shown in fig.3 (for $b=\infty$ the new simulated cut would be, like the old one, along the positive real axis).

On the other hand, and this is the essential point of the method, the original cut $[1,b]$ is projected by the conformal mapping onto the vertical line $\mathrm{Re}\, w = 1$, with the upper lip corresponding to

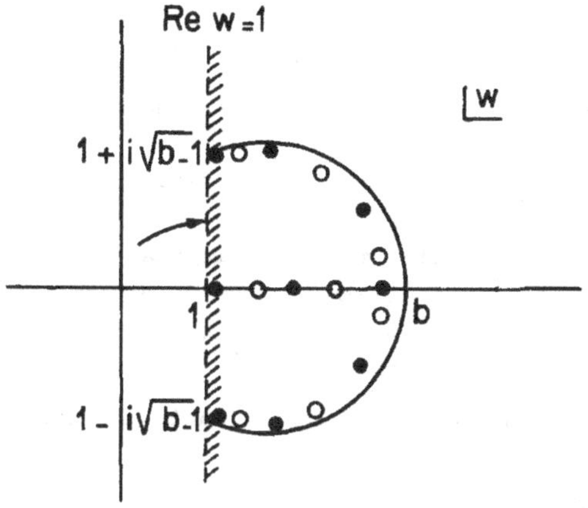

Fig. 3

$0<\mathrm{Im}\, w<\sqrt{b-1}$, and the lower lip to $0>\mathrm{Im}\, w>-\sqrt{b-1}$. In order to approximate the boundary values $F(x\pm i0)$ for $1<x<b$, or the discontinuity of F on the cut, we must therefore calculate $[N/N](w)$ at the two symmetrical points $w=1\pm i\sqrt{x-1}$ (notice that the two values so obtained are automatically complex conjugated if the series (1) has real coefficients). It is clear that the new simulated cut will do no harm in the process of analytically continuing $\tilde{F}$ on the line $\mathrm{Re}\, w=1$, and even beyond, thus allowing us also to explore to some extent the second Riemann sheet of $F(z)$.

2. Rational (Padé-type) approximants

Another interesting approach to the problem under discussion, proposed by Baumel et al.[4], consists of using rational approximants with preassigned poles in the z plane. For instance, if $F(z)$ is defined as before, with b finite, and we want to calculate $F(x+i0)$ for $1<x<b$, it seems advisable to seek approximants with poles disposed,

on (or close to) the semi-circle C connecting the two branch points through the lower half plane (fig.4). This can be achieved in many different manners, and fixes the denominator. Good convergence properties are obtained for Padé-type approximants (PTA), in which the denominators $Q_N(z)$ are given by

$$Q_N(z) = z^N q_N (1/z) \ , \tag{11}$$

where the q_N's form a set of polynomials orthogonal in the sense that

$$\int_\Gamma q_N(t)\ (t^k)^*\ |dt| = 0 \ , \qquad k=0,1,..,N-1. \tag{12}$$

Here Γ denotes the semi-circle image of C in the plane $t = 1/z$. The numerator $P_N(z)$ of the diagonal PTA is then determined from

$$P_N - Q_N F = O(z^{N+1}) \tag{13}$$

For brevity we shall use for these approximants the notation (N/N)(z) (cf. Brezinski [5]).

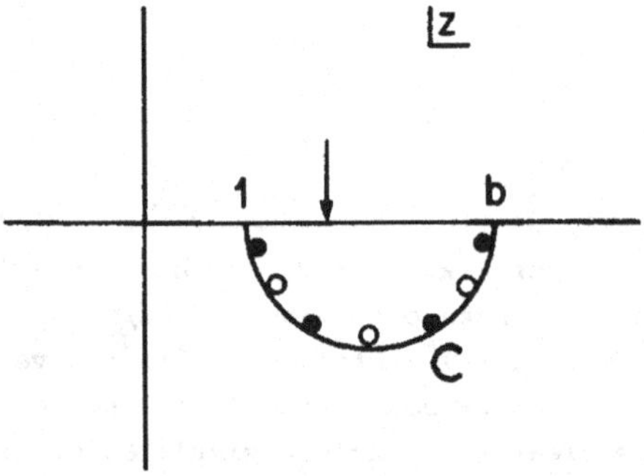

Fig. 4

It is easy to check that the orthogonality condition (12) forces the zeros of $q_N(t)$ to lie close to the arc Γ (and therefore the zeros of Q_N to lie close to C) when N is large. To this end we first replace Γ by the unit semi-circle γ with $|\arg t| \leq \pi/2$. A few of the correspon-ding orthogonal polynomials $\varphi_N(t)$ are given in Table I, together with their zeros. The latter clearly get closer and closer to γ as N increases.

Table I

Polynomials orthogonal on

the semi-circle $|t|=1$, $|\arg t|\leq\pi/2$

$\varphi_1(t) = t-\frac{2}{\pi}$	(0.637)
$\varphi_2(t) = t^2-\frac{2\pi}{\pi^2-4}t+\frac{4}{\pi^2-4}$	(0.535 ± i0.628)
$\varphi_3(t) = t^3-\frac{2}{3\pi}\frac{3\pi^2-16}{\pi^2-8}t^2+\frac{8}{3}\frac{1}{\pi^2-8}t-\frac{2}{3\pi}\frac{16-\pi^2}{\pi^2-8}$	(0.835, 0.355 ± i0.841)

$$\int_\gamma \varphi_N(t)(t^k)^*|dt| = 0\ ,\quad (k=0,1,\ldots,N-1)$$

Since $q_N(t)$ is obtained from $\varphi_N(t)$ by a linear transformation $t\to\alpha t+\beta$, the desired property of its zeros readily follows.

The convergence of the PTA is illustrated by a numerical example in Table II, where we compare them to the ordinary PA of Sec.1b. It may be noticed that $[N/N](w)$ is slightly more accurate than $(2N/2N)(z)$ (both require the same number of coefficients of the Taylor expansion).

Table II

Convergence of $(2N/2N)(z)$ PTA and of $[N/N](w)$

PA for $F(z)=\left(\frac{1-z}{1-z/2}\right)^{1/2}$ at $z=4/3+i0$ (exact value $F=-i$)

N	(2N/2N)(z)	[N/N](w)
3	-0.038005 - i0.915846	-0.042649 - i0.955974
4	0.036522 - i0.984012	0.011745 - i0.988140
5	0.007306 - i1.016619	0.003192 - i1.003181
6	-0.007601 - i1.003285	-0.000853 - i1.000852
7	-0.001537 - i0.996452	-0.000228 - i0.999772
8	0.001662 - i0.999288	0.000061 - i0.999939
9	0.000338 - i1.000787	0.000016 - i1.000016
10	-0.000374 - i1.000159	-0.000004 - i1.000004

3. Multivalued approximants

Non-rational approximants, which generalize the usual PA, have been considered by Padé himself in his celebrated Thesis, but did not receive much attention until recently [6]. The simplest of these are the so-called quadratic approximants (QA), defined by

$$[L/M/N](z) = \{-Q_M \pm (Q_M^2 - 4P_L R_N)^{1/2}\} / 2R_N, \tag{14}$$

where $P_L(z)$, $Q_M(z)$, $R_N(z)$, are polynomials of degrees L,M,N, respectively, determined from

$$P_L + Q_M F + R_N F^2 = O(z^{L+M+N+2}) , \tag{15}$$

and F(z) is given by Eq.(1). It is obvious that such approximants possess a built-in cut structure, so that they seem ideally suited for problems related to multivalued functions. Short [7] made extensive use of multivalued approximants to calculate boundary values of various Feynman parametric integrals, which define relativistic scattering amplitudes in the physical region. In doing this, he actually started from power expansions of the form (2), rather than from Eq.(1), and compared his results with those of Chisholm et al.[2]. He noticed that the QA are generally two orders of magnitude more accurate than the PA with the same number of coefficients. However, the conformal mapping method of Sec. 1b, which incorporates a cut structure in the new variable w, is capable of similar performances, while still working with ordinary PA. This was demonstrated on several numerical examples in ref. [3].

References

1. G.A. Baker, Jr. - Essentials of Padé approximants - Academic Press, New York, 1975.

2. J.S.R. Chisholm, A.C. Genz, and M. Pusterla - J. Comp. Appl. Math. 2 (1976) 73-76.

3. S. Klarsfeld - J. Phys. G : Nucl. Phys. 7 (1981) 1-7.

4. R.T. Baumel, J.L. Gammel, and J. Nuttall - preprint Univ. Western Ontario (1980), and private communication.

5. C. Brezinski - Padé-type approximation and general orthogonal polynomials - Birkhäuser Verlag, Basel, 1980.

6. R.E. Shafer - SIAM J. Numer. Anal. 11 (1974) 447-460.

7. L. Short - J. Phys. G : Nucl. Phys. 5 (1979) 167-198.

ACCELERATION DE LA CONVERGENCE

POUR CERTAINES SUITES

A CONVERGENCE LOGARITHMIQUE

Christine KOWALEWSKI

- * -

U.E.R I.E.E.A - Informatique

University of Lille

59655 Villeneuve d'Ascq Cédex

FRANCE

- * -

On considère une suite convergente de réels $(S_n) \to S^*$. On note $e_n = S_n - S^*$, $\Delta S_n = S_{n+1} - S_n$, $\Delta^2 S_n = \Delta S_{n+1} - \Delta S_n$. On dira qu'un ensemble de suites est accélérable s'il existe une transformation capable d'**accélérer** la convergence de toutes les suites de cet ensemble. Dans cet article, on cherche à déterminer des ensembles accélérables.

1. L'ENSEMBLE DES SUITES A CONVERGENCE LOGARITHMIQUE.

<u>*Définition.*</u> *(S_n) est à convergence logarithmique si* $\lim_{n\to\infty} \frac{\Delta S_n}{e_n} = 0$ (1) $\iff \lim_{n\to\infty} \frac{e_{n+1}}{e_n} = 1$ (2).
On appelle Log l'ensemble de toutes les suites à convergence logarithmique

$$\text{Log} : \frac{e_{n+1}}{e_n} = 1 - \lambda_n \left((\lambda_n) \in C_o \right) \qquad (3)$$

C_o *désignent l'ensemble des suites de limite nulle.*
(λ_n) est la suite dérivée première de (S_n). La connaissance de e_o et (λ_n) équivaut à celle de (e_n) tout entière.

<u>Premier résultat négatif</u>. Log n'est pas accélérable [R: 2] (4)

On va déterminer des sous-ensembles de Log en donnant des informations supplémentaires sur (λ_n).

2. DEFINITIONS DE SOUS-ENSEMBLES DE Log.

1°) LOG 1 : $\frac{e_{n+1}}{e_n} = 1 - \lambda_n \; (\lambda_n) \in C_o \; \exists N \in \mathbb{N}, \forall n \geq N, \lambda_n > 0$ (5)

Propriété 1.

LOG 1 est l'ensemble des suites à convergence logarithmique strictement monotones à partir d'un certain rang.

Démonstration. a) $(S_n) \in$ LOG 1 $\Rightarrow (\frac{e_{n+1}}{e_n}) \to 1 \Rightarrow e_n$ et e_{n+1} sont de même signe pour $n \geq N'$. De plus $|e_{n+1}| = |e_n||1-\lambda_n| < |e_n| \; \forall n \geq N \Rightarrow (S_n)$ monotone strictement à partir du rang Sup (N,N').

b) Soit (S_n) dans Log et strictement monotone à partir d'un certain rang. Si on avait $\forall N \; \exists n \geq N$ tq $\lambda_n \leq 0$, on aurait aussi $|e_{n+1}| = (1-\lambda_n)\,|e_n| \geq |e_n|$, ce qui contredit la stricte monotonie.

Deuxième résultat négatif. LOG 1 n'est pas accélérable [R: 2].

2°) De quel type est la convergence de (λ_n) vers 0 ?

Lemme 1.

Soit $(S_n) \in$ LOG 1 ; *alors*

. *ou bien* $(\frac{\lambda_{n+1}}{\lambda_n}) \xrightarrow[n\to+\infty]{} 1$

. *ou bien* $(\frac{\lambda_{n+1}}{\lambda_n})$ *n'a pas de limite.*

Démonstration. Soit $(\frac{\lambda_{n+1}}{\lambda_n}) \to K$

a) $\lambda_n > 0 \; \forall n \geq N \Rightarrow K \geq 0$

b) $(\lambda_n) \xrightarrow[n\to\infty]{} 0 \Rightarrow K \leq 1$

c) Si $K \in [0,1[$, $\exists B \in \mathbb{R}$ et $\exists N' \in \mathbb{N}$ telle que

$$\begin{cases} 0 \leq K < B < 1 \\ 0 < \lambda_{N'} < 1 \\ \forall n \geq N' \; \frac{\lambda_{n+1}}{\lambda_n} \in [0,B] \end{cases}$$

Donc, $\forall i > N'$, $0 < \lambda_i < B^{i-N'} \lambda_{N'}$.

En multipliant par $|e_{N'}|$ et en passant à la limite :

$$|e_{N'}| \prod_{i=N'}^{+\infty} (1-B^{i-N'}\lambda_{N'}) \leq \lim_{n\to\infty} |e_{n+1}| \leq 1.$$

Le produit infini est de même nature que $\sum_{j=0}^{+\infty} B^j$, donc convergent. La limite ℓ est strictement positive car $\lambda_N B^{i-N} \neq 1$ pour tout i. Donc $0 < |e_{N'}| \ell < \lim_{n\to+\infty}|e_n|$, ce qui est impossible

$$\text{LOGSF} : \frac{e_{n+1}}{e_n} = 1 - \lambda_n \quad (\lambda_n) \in C_o \quad (\lambda_n) \in \text{Log} \qquad (6)$$

C'est l'ensemble défini par Smith et Ford [R: 1]. LOGSF = $\{(S_n) \ / \ (\frac{e_{n+1}}{e_n}) \to 1$ et $(\frac{\Delta e_{n+1}}{\Delta e_n}) \to 1\}$.

Troisième résultat négatif. LOGSF n'est pas accélérable [R: 3] (7)

Toutes les suites de LOGSF ont une suite dérivée seconde $(\mu_n) \in C_o = \mu_n = 1- \frac{\lambda_{n+1}}{\lambda_n}$

3°) $$\text{LOG 2} : \frac{e_{n+1}}{e_n} = 1 - \lambda_n \quad (\lambda_n) \in C_o \quad (\lambda_n) \in \text{LOG 1} \qquad (8)$$

On ne sait pas si LOG 2 est accélérable.

4°) On peut donner une autre information sur (λ_n) : (λ_n) a une vitesse de convergence comparable à celle d'une suite dont on peut calculer le terme général

$$\text{L} : \frac{e_{n+1}}{e_n} = 1 - \lambda_n \quad (\lambda_n) \in C_o \quad (\lambda_n) \in \text{LOG 1} \quad \exists s \in \mathbb{R}, \ s \neq 0 \quad \left(\frac{\lambda_n}{1-\frac{\Delta S_{n+1}}{\Delta S_n}}\right) \xrightarrow[n\to+\infty]{} s \qquad (9)$$

On ne sait pas si L est accélérable.

Relations d'inclusion. L ⊂ LOG 2 ⊂ LOGSF ⊂ LOG 1 ⊂ Log (10).

Démonstration.

Soit $(S_n) \in$ LOGSF $(\frac{\lambda_{n+1}}{\lambda_n}) \xrightarrow[n\to+\infty]{} 1$, donc les λ_n sont tous de même signe à partir d'un certain rang N. S'ils étaient négatifs, le produit $e_{n+1} = e_N \prod_{i=N}^{n} (1-\lambda_i)$ ne pourrait diverger vers 0 lorsque $n \to +\infty$. Donc $(S_n) \in$ LOG 1. Les autres inclusions sont évidentes.

3. SOUS-ENSEMBLES DE Log CONDUISANT A DES ALGORITHMES D'ACCELERATION.

1°) $\widehat{LOG\ 1}$: $(S_n) \in \widehat{LOG\ 1}$ $\exists K \neq 0$, $K \in \mathbb{R}$ $\left(\frac{\Delta\lambda_n}{\Delta e_n}\right) \xrightarrow[n\to+\infty]{} K$ (11)

2°) $\widehat{LOG\ 2}$: $(S_n) \in \widehat{LOG\ 2}$ $\exists K \neq 0$, $K \in \mathbb{R}$ $\left(\frac{\Delta\mu_n}{\Delta\lambda_n}\right) \xrightarrow[n\to+\infty]{} K$ (12)

3°) LOG Raabe : (S_n) monotone strictement et $\exists\lambda > 1$, $\lambda \in \mathbb{R}$ $\left(\frac{1-\frac{\Delta S_{n+1}}{\Delta S_n}}{\frac{1}{n+1}}\right) \xrightarrow[n\to+\infty]{} \lambda$ (13)

C'est l'ensemble des suites Raabe convergentes (réalisent le critère de convergence de Raabe-Duhamel). Toute suite Raabe convergente est dans log.

4°) Pour $s \in \mathbb{R}$, $s \neq 0$ Ls : $(S_n) \in$ Log $\left(\frac{\lambda_n}{1 - \frac{\Delta S_{n+1}}{\Delta S_n}}\right) \xrightarrow[n\to+\infty]{} s$ (14)

Evidemment $L \subset \bigcup_{s\in\mathbb{R}} L_s$ (15)

Propriétés d'inclusion.

$\widehat{LOG\ 1} \subset LOG\ 2$ (16)

$\widehat{LOG\ 2} \subset LOG\ 2$ (17)

LOG Raabe $\subset$ LOGSF (18)

$\widehat{LOG\ 1} \subset L\ 1/2$ (19)

Démonstration. (16) : Si (S_n) $\widehat{LOG\ 1}$ d'après (11) et la propriété 1 $\left(\frac{\lambda_n}{e_n}\right) \xrightarrow[n\to+\infty]{} K$ (20)

Or $\frac{\Delta\lambda_n}{\Delta e_n} = \frac{\lambda_n}{e_n} \cdot \frac{\frac{\lambda_{n+1}}{\lambda_n} - 1}{\frac{e_{n+1}}{e_n} - 1} \to K$. Donc $\left(\frac{\lambda_{n+1}}{\lambda_n}\right) \xrightarrow[n\to+\infty]{} 1$ donc $(S_n) \in$ LOGSF

De plus $(\Delta\lambda_n)$ est de signe constant à partir d'un certain rang, donc $(S_n) \in$ LOG 2.

(18) : Si $(S_n) \in$ LOG Raabe, $\left(\frac{\Delta S_{n+1}}{\Delta S_n}\right) \xrightarrow[n\to+\infty]{} 1$ donc $(S_n) \in$ LOGSF.

(19) : Si $(S_n) \in \widehat{LOG\ 2}$, $\frac{\Delta\lambda_n}{\Delta e_n} = \frac{\lambda_n}{e_n} \cdot \frac{\mu_n}{\lambda_n}$. D'après (20), $\left(\frac{\mu_n}{\lambda_n}\right) \xrightarrow[n\to+\infty]{} 1$ (21).

Or $\frac{\lambda_n}{1-\frac{\Delta S_{n+1}}{\Delta S_n}} = \frac{\lambda_n}{1+\lambda_{n+1} - \frac{\lambda_{n+1}}{\lambda_n}} = \frac{1}{\frac{\lambda_{n+1}}{\lambda_n} + \frac{\mu_n}{\lambda_n}}$ (22) ; donc, ceci tend vers $\frac{1}{2}$.

Remarque : construction de suites de L_s :

Soient V un ouvert de $\mathbb{R}$ contenant S^* et $f : V \to \mathbb{R}$.

Si
(i) f est de classe C^{2p} sur V
(ii) $f(S^*) = S^*$
(iii) $f'(S^*) = 1$
(iv) $f''(S^*) = \ldots = f^{(p-1)}(S^*) = 0$
(v) $f^{(p)}(S^*) \neq 0$

alors la suite générée par $\begin{cases} S_o \in V \\ S_{n+1} = f(S_n) \end{cases}$ est dans L 1/p (23)

Démonstration. $e_{n+1} = e_n + C\, e_n^p + \phi(e_n)\, e_n^p$ avec $C = \dfrac{f^{(p)}(S^*)}{p!} \neq 0$ et

$$\phi(e_n) = \frac{f^{(p+1)}(S^*)}{(p+1)!} e_n + \ldots + \frac{f^{(2p)}(S^*+\theta_n)}{(2p)!} e_n^p \quad \theta_n \in]0,e_n[$$

$$\lambda_n = -e_n^{p-1} [C + \phi(e_n)]$$

$$\mu_n = 1 - [1+e_n^{p-1}(C+\phi(e_n))]^{p-1} \frac{C+\phi(e_{n+1})}{C+\phi(e_n)}$$

On calcule $\dfrac{\mu_n}{\lambda_n}$ en utilisant $(1+y)^{p-1} = 1+(p-1)y + y\varepsilon(y)$ avec $\varepsilon(y) \underset{y \to o}{\to} o$

$$\frac{\mu_n}{\lambda_n} = \frac{\phi(e_{n+1})-\phi(e_n)}{e_n^{p-1}[C+\phi(e_n)]^2} + \frac{C+\phi(e_{n+1})}{C+\phi(e_n)} [(p-1) + \varepsilon((C+ (e_n))e_n^{p-1}]$$

Le premier terme est égal à

$$\frac{\sum_{i=1}^{p-1} C_i(e_{n+1}^i - e_n^i)}{e_n^{p-1}(C+\theta(e_n))^2} + \frac{e_{n+1}^p f^{(2p)}(S^*+\theta_{n+1})}{(2p)!\, e_n^{p-1}(C+\phi(e_n))^2} - \frac{e_n^p f^{(2p)}(S^*+\theta_n)}{(2p)!\, e_n^{p-1}\, (C+\phi(e_n))^2}$$

où

$$C_i = \frac{f^{(p+i)}(S^*)}{(p+i)!} .$$

Donc, il tend vers 0.

Donc $\left(\dfrac{\mu_n}{\lambda_n}\right) \underset{n\to+\infty}{\longrightarrow} p-1$, et, d'après (22), $\left(\dfrac{\lambda_n}{1 - \frac{\Delta S_{n+1}}{\Delta S_n}}\right) \underset{n\to+\infty}{\longrightarrow} \dfrac{1}{p}$

Il est possible que l'hypothèse (i) puisse être affaiblie

4. ALGORITHMES D'ACCELERATION POUR LES SUITES L_s.

Remarque. L_o, ensemble des suites de Log telles que $\left(\dfrac{\lambda_n}{1 - \frac{\Delta S_{n+1}}{\Delta S_n}}\right) \underset{n\to+\infty}{\longrightarrow} 0$ est non vide.

On considerera toutefois toujours $s \neq 0$.

Propriété 2.

$$L_s \neq \emptyset \Longleftrightarrow s \in [0,1]$$

Démonstration. Soit $(S_n) \in L_s$. Si $s \notin [0,1]$, d'après (22), $(\frac{\lambda_{n+1}}{\lambda_n}) \xrightarrow[n\to+\infty]{} 1$, donc $(\frac{\mu_n}{\lambda_n}) \to \frac{1}{s} - 1 < 0$. D'après (10) (λ_n) est positive, donc (μ_n) négative à partir d'un rang N. Il y a contradiction avec la divergence vers 0 de $\lambda_{n+1} = \lambda_N \prod_{i=N}^{n} (1-\mu_i)$ lorsque $n\to+\infty$.
Réciproquement, pour tout $s \in [0,1]$, on peut construire une suite de L_s.
La propriété 2 et (15) entraînent que

$$L \subset \bigcup_{s\in]0,1]} L_s \qquad (24)$$

Propriété 3. *Si* $(S_n) \in L_s$ *avec* $s \neq 0$, *alors*, $(S_n) \in$ LOGSF.

Démonstration. C'est une conséquence de (22).

Propriété 4. *Si* $(S_n) \in L_s$ *avec* $s \neq 0$ *et* $s \neq 1$, *alors*, $(S_n) \in$ LOG 2 (25)

Démonstration. D'après (22) et les propriétés 2 et 3, $(\frac{\mu_n}{\lambda_n})$ tend vers une constante strictement positive, donc $\mu_n > 0$ à partir d'un certain rang.
De ce qui précéde on déduit que :

$$\bigcup_{s\in]0,1[} L_s \subset L \subset \bigcup_{s\in]0,1]} L_s \qquad (25)$$

Remarque. Une suite de L_1 est dans LOGSF mais pas forcément dans LOG 2.

Résultats d'accélération. Soit $s \in]0,1]$; toute suite de L_s est accélérée par chacune des deux transformations :

$$T_n = S_n - \frac{1}{s}\frac{(\Delta S_n)^2}{\Delta^2 S_n} \quad \text{et} \quad T'_n = S_n - \frac{\Delta S_n}{(\frac{\Delta S_{n+1}}{\Delta S_n})^s - 1}$$

(26) (27)

Remarque. Si s=1, on retrouve Δ^2 d'Aitken donc :

Propriété 5.

L_1 *est l'ensemble des suites de* Log *accélérées par* Δ^2 *d'Aitken. Les transformations (26) et (27) sont peu intéressantes en pratique car elles nécessitent la connaissance de* s *Des comparaisons numériques dans le cas où* $s = \frac{1}{2}$ *donnent un léger avantage à (27).*

5. ACCELERATION A PARTIR DE $\widehat{\text{LOG}}$ 1.

Une suite de $\widehat{\text{LOG}}$ 1 vérifie $\left(\frac{\lambda_n}{e_n}\right) \xrightarrow[n\to+\infty]{} K \neq 0$, donc $\left(\frac{\lambda_{n+1}}{\lambda_n} \cdot \frac{e_n}{e_{n+1}}\right) \xrightarrow[n\to+\infty]{} 1$.

Dans le sous-ensemble de $\widehat{\text{LOG}}$ 1 constitué des suites pour lesquelles $\frac{\lambda_{n+1}}{\lambda_n} \cdot \frac{e_n}{e_{n+1}} = 1$,

$S^* = S_n - \dfrac{\Delta S_n}{\sqrt{\dfrac{\Delta S_{n+1}}{\Delta S_n}} - 1}$. On considère la transformation

$$T_n \quad S_n - \frac{\Delta S_n}{\sqrt{\dfrac{\Delta S_{n+1}}{\Delta S_n}} - 1} \quad (28)$$

(28) accélère $L_{\frac{1}{2}}$, d'après (27). On avait $\widehat{\text{LOG}}$ 1 $\subset L_{\frac{1}{2}}$.

On applique cette "technique du sous-ensemble" à LOG 2, LOG Raabe et L qui sont défini chacun par un rapport tendent vers une constante non nulle.

6. ACCELERATION A PARTIR DE $\widehat{\text{LOG}}$ 2.

Une suite de $\widehat{\text{LOG}}$ 2 vérifie $\left(\frac{\Delta\mu_n}{\Delta\lambda_n}\right) \xrightarrow[n\to+\infty]{} K$, donc $\left(\frac{\mu_n}{\lambda_n}\right) \xrightarrow[n\to+\infty]{} K$, donc $\left(\frac{\mu_{n+1}}{\mu_n} \cdot \frac{\lambda_n}{\lambda_{n+1}}\right) \xrightarrow[n\to+\infty]{} 1$

En appliquant la technique du sous-ensemble, on trouve S^* comme racine d'un polynôme de degré 3.

Questions. Ce polynôme a-t-il plusieurs solutions réelles ? L'une des solutions conduit-elle à un procédé d'accélération ?

7. ACCELERATION A PARTIR DE LOG Raabe.

Procédé d'accélération. La transformation $T_n = S_n - \dfrac{\Delta S_n}{\dfrac{n+1}{n} \cdot \dfrac{\Delta S_{n+1}}{\Delta S_n} - 1}$ (29)

accélère la convergence de LOG Raabe $\cap$ LOG 2.

Remarque. (29) est analogue à la première colonne de la transformation u de Levin. La démonstration de (29) se fait à partir du

Lemme 2.

Soit (S_n) *strictement monotone à partir d'un certain rang et soit* (t_n) *vérifiant*

. $(t_n) \xrightarrow[n\to+\infty]{} S^*$

. $\left(\frac{\Delta t_n}{\Delta S_n}\right) \to C \neq 1,\ C \in \mathbb{R}$.

Alors $T_n = S_n - \dfrac{\Delta S_n}{\dfrac{S_{n+1} - t_{n+1}}{S_n - t_n} - 1}$ accélère la convergence de (S_n).

Démonstration du lemme. Les hypothèses du lemme entraînent que $\left(\frac{t_n - S^*}{S_n - S^*}\right) \xrightarrow[n\to+\infty]{} C$.

D'autre part, $\dfrac{T_n - S^*}{S_n - S^*} = 1 - \left(1 - \dfrac{t_n - S^*}{S_n - S^*}\right) \left(\dfrac{1}{1 - \dfrac{t_{n+1} - t_n}{S_{n+1} - S_n}}\right) \xrightarrow[n\to+\infty]{} 0$ puisque $C \neq 1$

Démonstration de (29). Prenons $t_n = S_n - n\Delta S_n$.

Pour toute suite de LOG 2, $\dfrac{\Delta S_{n+2} - \Delta S_{n+1}}{\Delta S_{n+1} - \Delta S_n} = (1-\lambda_n)(1-\mu_n)\ \dfrac{\lambda_{n+1}[\mu_{n+1}-1]-\mu_{n+1}}{\lambda_n[\mu_n - 1] - \mu_n} > 0$ d'après (8)

donc (ΔS_n) converge de manière monotone vers 0.

La convergence de (S_n) entraîne donc que $(n\Delta S_n) \xrightarrow[n\to+\infty]{} 0$. D'autre part $\Delta t_n = -(n+1)\Delta^2 S_n$, donc

$$\frac{\Delta t_n}{\Delta S_n} = \frac{1 - \dfrac{\Delta S_{n+1}}{\Delta S_n}}{\dfrac{1}{n+1}} \to \lambda > 1$$

Variante. (29) nécessite la connaissance du rang de chaque terme de (S_n) utilisé.
En pratique cette difficulté disparait car : soit k un entier fixé

$$T_n = S_n - \frac{\Delta S_n}{\dfrac{n+k+1}{n+k} \cdot \dfrac{\Delta S_{n+1}}{\Delta S_n} - 1} \qquad (30)$$

accélère la convergence de toute suite de LOG Raabe $\cap$ LOG 2.

8. ACCELERATION A PARTIR DE L.

Soit $L = \bigcup_{s \in]0,1[} L_s$ et soit (U_n) la suite transformée de (S_n) par Δ^2 d'Aitken.

Lemme 3.

L est l'ensemble des suites de LOG 2 vérifiant $\dfrac{U_n - S^*}{S_n - S^*} \to K \in]0,1[$.

Démonstration.

$$U_n = S_n - \frac{\Delta S_n}{\dfrac{\Delta S_{n+1}}{\Delta S_n} - 1} \ ; \text{ donc } \frac{U_n - S^*}{S_n - S^*} = 1 - \frac{\dfrac{e_{n+1}}{e_n} - 1}{\dfrac{\Delta e_{n+1}}{\Delta e_n} - 1} = 1 - \frac{\lambda_n}{1 - \dfrac{\Delta S_{n+1}}{\Delta S_n}}$$

Si on applique la technique du sous-ensemble au rapport $\dfrac{\lambda_n}{1 - \dfrac{\Delta S_{n+1}}{\Delta S_n}}$, on obtient la

transformation $T_n = S_n + \dfrac{S_n - U_n}{\dfrac{\Delta U_n}{\Delta S_n} - 1}$ qui est le θ_2-algorithme.

On a aussi $T_n = S_n - \dfrac{\Delta S_n}{\dfrac{S_{n+1} - U_{n+1}}{S_n - U_n} - 1}$ (31).

Propriété 6.

Le θ_2-algorithme accélère la convergence de $\widehat{LOG\ 2}$.

Démonstration. D'après les lemmes 2 et 3, il suffit de montrer que $(\frac{\Delta U_n}{\Delta S_n}) \to C \neq 1$.

Pour $(S_n) \in \widehat{LOG\ 2}$, on a $(\frac{\Delta\mu_n}{\Delta\lambda_n}) \to K > 0$ donc $(\frac{\mu_n}{\lambda_n}) \to K > 0$.

$$\frac{\Delta U_n}{\Delta S_n} = 1 - \frac{\dfrac{(\Delta e_{n+1})^2}{\Delta^2 e_{n+1}} - \dfrac{(\Delta e_n)^2}{\Delta^2 e_n}}{\Delta e_n} = \frac{\dfrac{\Delta e_{n+1}}{\Delta e_n}}{\dfrac{\Delta e_{n+1}}{\Delta e_n} - 1} - \frac{\dfrac{\Delta e_{n+1}}{\Delta e_n}}{\dfrac{\Delta e_{n+2}}{\Delta e_{n+1}} - 1} =$$

$$(1-\lambda_n)(1-\lambda_n)\ \frac{\dfrac{\lambda_{n+1}}{\lambda_n} \cdot \dfrac{\mu_{n+1}}{\mu_n} + \dfrac{\Delta\mu_n}{\Delta\lambda_n}}{(\mu_n - 1 - \dfrac{\mu_n}{\lambda_n})\ (\lambda_{n+1}\dfrac{\mu_{n+1}}{\mu_n} - \dfrac{\lambda_{n+1}}{\mu_n} - \dfrac{\mu_{n+1}}{\mu_n})}$$

Donc $$(\frac{\Delta U_n}{\Delta S_n}) \xrightarrow[n\to+\infty]{} \frac{K}{1+K}$$

REFERENCES.

[1] SMITH et FORD
Acceleration of linear and logarithmic convergence.
Siam Journal of Numerical Analysis (vol 16, n° 2, april 1979).

[2] DELAHAYE et GERMAIN-BONNE
Quelques résultats négatifs en accélération de convergence
A paraître dans Numerische Mathematik.

[3] DELAHAYE et GERMAIN-BONNE
Communication personnelle.

Difficulties of Convergence Acceleration

by

I.M. Longman
Department of Geophysics and Planetary Sciences
Tel-Aviv University
Ramat-Aviv, Israel

A discussion is given of various difficulties that arise in convergence acceleration problems. A distinction is drawn between monotonic and alternating series, and certain difficulties are shown to lie with the former. Particular attention is paid to one of Levin's transformations, which in the case of convergent monotonic series is difficult to use owing to near cancellation of large positive and negative terms. However it is shown that the use of a recursion formula can alleviate the difficulty to some extent.

When, however, the terms a_n of the series are given as <u>explicit</u> functions of n, rather than numerically, it is shown how transformations may readily be effected to yield a convergence rate as rapid as one pleases - without loss of accuracy.

As a step towards the treatment of the near cancellation of large positive and negative terms mentioned above, a method is suggested for the addition of large positive and negative terms in the computer without loss of accuracy - in certain cases. As a demonstration of this technique, the Maclaurin series for e^{-x} is accurately summed for (moderately) large positive x.

A discussion is given as to whether we must necessarily expect difficulties when accelerating the convergence of monotonic series whose terms a_n are given numerically but not as explicit functions of n.

This paper is essentially explorative, and definitive answers to questions are not always available at the time of writing.

1. Introduction

Let us suppose that we have obtained - perhaps as the solution of a physical problem - a monotonic series of positive terms

$$S = a_1 + a_2 + a_3 + \dots + a_n + \dots, \tag{1}$$

but that the convergence is too slow for convenient computation. It may happen that S is even divergent, but convergence will be assumed in this paper.

Various methods exist for convergence acceleration, a powerful one for many purposes being the so-called n transformation due to Levin [1]. This may be briefly derived as follows*:

Let us assume that a good approximation to S is given by

$$S_1 = a_1 + a_2 + \dots + a_n + a_n(\alpha_{-1}n + \alpha_0 + \alpha_1 n^{-1} + \dots + \alpha_k n^{-k}), \tag{2}$$

where the α_i $(i = -1,0,1,\dots,k)$ are constants (i.e. independent of n). This assumption appears to be reasonable for many series, but not for all. A similar assumption was made by Bickley and Miller [2].

We define the partial sums

$$A_n = a_1 + a_2 + \dots + a_n, \tag{3}$$

and write

$$S_1 = A_n + a_n(\alpha_{-1}n + \alpha_0 + \alpha_1 n^{-1} + \dots + \alpha_k n^{-k}). \tag{4}$$

We desire to eliminate the α_i and obtain S_1 which, in this approximation, is assumed to be independent of n. The elimination is readily carried out as follows:

We have

$$\frac{n^k S_1}{a_n} = \frac{n^k A_n}{a_n} + \alpha_{-1}n^{k+1} + \alpha_0 n^k + \alpha_1 n^{k-1} + \dots + \alpha_k,$$

and since the α_i are all included in a polynomial of degree $k+1$ (whose $k+2$ order difference must vanish), we find, assuming S_1 constant,

$$S_1 \Delta^{k+2}\left(\frac{n^k}{a_n}\right) = \Delta^{k+2}\left(\frac{n^k A_n}{a_n}\right)$$

*The author is indebted to Dr. D. Levin for this explanation.

or

$$S_1 = S_{kn} = \frac{\Delta^{k+2}\left(\frac{n^k A_n}{a_n}\right)}{\Delta^{k+2}\left(\frac{n^k}{a_n}\right)} \tag{5}$$

where Δ is the usual difference operator, and the differencing is done with respect to n. Of course it now appears that S_1 is <u>not</u> constant with respect to n, and so of course S_{kn} is only an approximation to S.

It is now convenient to replace $k+2$ by k and define

$$U_{kn} = \frac{\Delta^{k}\left(\frac{n^{k-2} A_n}{a_n}\right)}{\Delta^{k}\left(\frac{n^{k-2}}{a_n}\right)} = S_{k-2,n}, \qquad k = 1,2,3,\dots, \tag{6}$$

which is the u transform of Levin of order (k,n). It may be written in the form

$$U_{kn} = \frac{\sum_{r=0}^{k} (-1)^r \binom{k}{r} (n+r)^{k-2} \frac{A_{n+r}}{a_{n+2}}}{\sum_{r=0}^{k} (-1)^r \binom{k}{r} (n+r)^{k-2} \frac{1}{a_{n+r}}}, \quad k = 1,2,3,\dots, \tag{7}$$

and now the difficulty for convergent monotonic series is apparent. For example if $a_n = 1/n^2$, for which

$$S = \pi^2/6 = 1.644934067, \tag{8}$$

we find

$$U_{kn} = \frac{1}{k!} \sum_{r=0}^{k} (-1)^r \binom{k}{r} (n+r)^k A_{n+r}, \tag{9}$$

and we lose accuracy due to the near cancellation of large positive and negative terms. Taking $k = 4$, $n = 2$ we have

$$U_{42} = (16A_2 - 324A_3 + 1536A_4 - 2500A_5 + 1296A_6)/24,$$

and if we work to nine places of decimals we find

$$U_{42} = 1.644953667.$$

This result is, however, not correct to nine places of decimals, as explained above. The "exact" value is

$$U_{42} = 1.644953703,$$

and this has been obtained in a semi-analytical manner from (9). The loss of accuracy here is perhaps not so serious when we compare U_{42} with the exact sum (8), but when higher order U_{kn}'s are considered - as they must be if we are to obtain better approximations to S - the error becomes much worse, and limits very considerably the accuracy directly obtainable by this method.

In contradistinction it may be noted that if S is an alternating series whose terms decrease monotonically in absolute value, there is no problem in the accurate numerical evaluation of U_{kn}, and experience has shown that U_{kn} is often extremely useful in such cases.

Various other convergence acceleration methods - involving recursions or the solution of simultaneous linear equations - can be used, but in the author's experience always the above type of inaccuracy "creeps in" as we proceed to higher order approximations, when the a_n are given (or used) numerically, as distinct from analytically. This for convergent monotonic series, but not for alternating series with decreasing terms. This has led the author to pose the question whether, for monotonic series with numerically given terms, it is essentially difficult to accelerate the convergence.

In fact a number of basic questions seem to present themselves, and they are itemized in the following section.

2. The basic questions.

From the introduction it appears that certain basic questions arise here, and these are conveniently formulated as follows:

1. Is it necessary that we lose accuracy in this way for a monotonic series, i.e. is such accuracy loss intrinsic in the convergence acceleration problem for a monotonic series? Why is a convergent alternating series superior in this respect.

2. Is there a way of handling the addition of large positive and negative terms in the computer, so as to obviate loss of significant figures?

3. Is it a good idea to transform S into an alternating series, in order that convergence acceleration techniques may be more readily applied?

4. If the a_n are given analytically, rather than just numerically, can we obviate the loss of accuracy in convergence acceleration for a monotonic series?

5. Can we get over our difficulties by the proper use of recurrence relations? For example U_{kn} can be evaluated by the aid of recurrence relations.

It is the purpose of this paper to try to answer these questions, and to suggest lines for further investigation. The questions will be considered in turn, although completely definitive answers are not always forthcoming at this time of writing.

It is convenient to commence with question 5. If we use U_{kn} as our convergence acceleration method, we may seek to evaluate it more accurately by use of a recurrence relation. It is readily shown that both the numerators and denominators of U_{kn} satisfy the recurrence relation

$$f_{kn} = nf_{k-1,n} - (n+k)f_{k-1,n+1}. \tag{10}$$

This may be obtained by the use of Sister Celine's technique - see Rainville [3]. Using (10) we can build up a table

n=1	n=2	n=3	n=4	n=5	
f_{01}	f_{02}	f_{03}	f_{04}	f_{05}	k=0
f_{11}	f_{12}	f_{13}	f_{14}		k=1
f_{21}	f_{22}	f_{23}			k=2
f_{31}	f_{32}				k=3
f_{41}					k=4

etc., starting from the first row, and we have shown above all f's dependent on the first five elements in the first row.

For the first row, inspection of equation (7) shows that for the denominators

$$f_{on} = 1/(n^2 a_n),$$

while for the numerators

$$f_{on} = A_n/(n^2 a_n).$$

For example for the case $a_n = 1/n^2$ our table of numerators starts

n=1	n=2	n=3	n=4	n=5	n=6	
1.000	1.250	1.361	1.424	1.465	1.491	k=0
-1.500	-1.583	-1.611	-1.624	-1.634		k=1
3.250	3.278	3.285	3.287			k=2
-9.861	-9.865	-9.869				k=3
39.479	39.479					k=4
-197.394						k=5

the figures being quoted to three places of decimals, although they were actually calculated to nine places. For the table of denominators we have

n=1	n=2	n=3	n=4	n=5	n=6	
1	1	1	1	1	1	k=0
-1	-1	-1	-1	-1		k=1
2	2	2	2			k=2
-6	-6	-6				k=3
24	24					k=4
-120						k=5

etc. Here we find, for example, by simple division, that

$$U_{42} = 1.644953715$$

We have now obtained three values for U_{42} as follows:

		Δ
Recurrence method	1.644953715	
"Exact"	1.64495 3703	-12
Direct use of (9)	1.64495 3667	-36

A gain of accuracy has clearly been achieved over the use of (9), by means of the recurrence method. However if we proceed too far in the above tables for the numerators and denominators, we find a severe loss of accuracy due to error build-up. It is possible that by a different use of recurrence relation (i.e. not starting from the first row), and a subsequent normalization, we could get over the numerical instability. But this question has not been investigated to date.

The author has investigated other iterative schemes for convergence acceleration, but in every case where monotonic series, having numerically given terms, were treated, numerical instabilities were found. Once again, however, alternating series gave no problems by these methods.

Now let us consider question 4. Referring again to the series given in equation (1), we define S_n by

$$S_n = a_1 + a_2 + \ldots + a_n + u_n = A_n + u_n, \tag{11}$$

where u_n is a correction term designed to speed the convergence. Then clearly

$$S_{n+1} = a_1 + a_2 + \ldots + a_n + a_{n+1} + u_{n+1} \tag{12}$$

and if $S_{n+1} = S_n$ then

$$u_{n+1} = u_n - a_{n+1}, \tag{13}$$

and it is appropriate to try to solve this difference equation.

Equation (2) suggests that we seek a solution of (13) in the form

$$u_n = \alpha n a_n, \tag{14}$$

where α is a constant. If we succeed, and if S is a convergent monotonic series, then $na_n \to 0$ as $n \to \infty$ and so $u_n \to 0$ as $n \to \infty$, and $S = \lim_{n\to\infty} S_n$. But if we have satisfied (13), all the S_n's are equal, i.e.

$$S_1 = S_2 = \dots = S_n = \dots = \text{const.}, \tag{15}$$

and this constant is therefore the required sum S. Usually, however, it turns out that α, thus found, is a function of n, and then we may expect S_n to be a good approximation to S.

We substitute (14) into (13), assuming α to be independent of n, and find

$$(n+1)\alpha a_{n+1} = n\alpha a_n - a_{n+1}$$

so that

$$\alpha = \frac{a_{n+1}}{na_n-(n+1)a_{n+1}} . \tag{16}$$

We now denote this by $\alpha(n)$ to emphasize the fact that, after all, α is not in general constant.

Let us consider again our example $a_n = 1/n^2$, $S = \pi^2/6 = 1.644934067$. We find then

$$\alpha(n) = \frac{1/(n+1)^2}{1/n-1/(n+1)} = \frac{n}{n+1},$$

and it is important to notice that the difference of the nearly equal terms $1/n$ and $1/(n+1)$ in the denominator is obtained exactly by virtue of our analytic calculation. In general we would expect a loss of accuracy in calculating the denominator of (16) for a monotonic series whose terms a_n are given numerically, rather than analytically. We see that in this example $\alpha(n)$ turns out to be nearly constant, and so we may expect to have achieved a substantial improvement in convergence rate. We have

$$u_n = na_n\alpha(n) = \frac{1}{n+1}$$

and

$$S_n = \frac{1}{1^2} + \frac{1}{2^2} + \dots + \frac{1}{n^2} + \frac{1}{n+1},$$

and numerically we find (to three places of decimals)

n	A_n	S_n
1	1.000	1.5 00
2	1.250	1.583
3	1.361	1.611
4	1.424	1.624
5	1.464	1.630
6	1.491	1.634

and the improvement in convergence rate is manifest.

Returning to the general case, we have

$$S_n = a_1 + a_2 + \ldots + a_n + \frac{na_n a_{n+1}}{na_n - (n+1)a_{n+1}} . \tag{17}$$

It is expedient to express the S_n's as partial sums of a new series

$$b_1 + b_2 + \ldots + b_n + \ldots , \tag{18}$$

and to this end we define

$$b_{n+1} = S_{n+1} - S_n \tag{19}$$

so that

$$b_{n+1} = -\alpha(n)na_n + (n+1)\alpha(n+1)a_{n+1} + a_{n+1}, \tag{20}$$

$$n = 0,1,2,\ldots$$

We may note, in passing, that b_1 (case n=0) contains the undefined a_0, but this is multiplied by zero, and the intention is that

$$b_1 = a_1 + \alpha(1)a_1. \tag{21}$$

From (19), (21) we find that

$$S_n = b_1 + b_2 + \ldots + b_n \tag{22}$$

and $\sum_{n=1}^{\infty} b_n$ is our new series after transformation. We find

$$b_n = a_n\left[\frac{a_{n+1} - na_n}{(n+1)a_{n+1} - na_n} + \frac{(n-1)a_{n-1}}{na_n - (n-1)a_{n-1}}\right], \tag{23}$$

$$n = 2,3,4,\ldots .$$

but

$$b_1 = a_1\left[\frac{a_2-a_1}{2a_2-a_1}\right]. \tag{24}$$

The reason that b_1 is given separately from b_n is that we will see in the example which follows that when a_n is given analytically, na_n does not always tend to zero as n tends to zero, and then there is a more "natural" form for b_1. This will become clear in the example which follows.

As example we now consider again the series for which $a_n = 1/n^2$. From (23) we then find

$$b_n = \frac{1}{n^2(n+1)}, \quad n = 2,3,4,\ldots, \tag{25}$$

but from (24)

$$b_1 = 3/2, \tag{26}$$

and the reason for the separation of equations (23) and (24) now becomes clear. It is convenient to introduce a "regular" b_1, obtained by putting $n = 1$ in (25),

$$b_1(\text{reg.}) = \frac{1}{1^2 2}, \tag{27}$$

and we write our new series

$$S = \frac{3}{2} + \frac{1}{2^2 3} + \frac{1}{3^2 4} + \ldots + \frac{1}{n^2(n+1)} + \ldots \tag{28}$$

in the form

$$S - 1 = \frac{1}{1^2 2} + \frac{1}{2^2 3} + \ldots + \frac{1}{n^2(n+1)} + \ldots \tag{29}$$

The reason for making the first term "regular", is to facilitate further applications of this convergence acceleration method. The "regularity" of the first term on the right hand side of (29) helps to make the subsequent derived series better convergent.

We can apply our method to (29), for which we now take

$$a_n = \frac{1}{n^2(n+1)}, \tag{30}$$

and we find from (23)

$$b_n = \frac{1}{2n^2(n+1)^2}, \quad n = 2,3,4,\ldots, \tag{31}$$

but from (24)

$$b_1 = 5/8, \tag{32}$$

while from (31) we see that

$$b_1(\text{seg.}) = \frac{1}{2.1^2.2^2} = \frac{1}{8}. \tag{33}$$

Thus we now have

$$S - 1 - \frac{1}{2} = \frac{1}{2.1^2.2^2} + \frac{1}{2.2^2.3^2} + \ldots + \frac{1}{2n^2(n+1)^2} + \ldots \tag{34}$$

If we apply our method to the series on the right hand side of (34), we obtain an even more rapidly convergent series for $S = \pi^2/6$. We may summarize a few results as follows:

$$S = \sum_{n=1}^{\infty} \frac{1}{n^2}$$

$$S - 1 = \sum_{n=1}^{\infty} \frac{1}{n^2(n+1)}$$

$$S - \frac{3}{2} = \frac{1}{2} \sum_{n=1}^{\infty} \frac{1}{n^2(n+1)^2}$$

$$S - \frac{13}{8} = 2 \sum_{n=1}^{\infty} \frac{1}{n(n+1)^2(3n+1)(3n+4)}$$

$$S - \frac{355}{216} = 2 \sum_{n=1}^{\infty} \frac{57n^2 + 81n + 4}{(n+1)^2(3n+1)(3n+4)(12n^2+35n+27)(12n^2+11n+4)},$$

where each series after the first converges more rapidly than the preceding. It may be noted that in each of the first three derived series above, the number subtracted from S comes from the first term of the previous series. But this rule does not apply to the last series above. The reason is readily shown to be connected with whether or not na_n tends to infinity as $n \to 0$ for the series whose convergence is being accelerated. It is clear that we can, in principle, by proceeding far enough, achieve as fast a rate of convergence as we wish.

It is worthy of note that if we use the above iterative method numerically instead of analytically on the same series $\Sigma\, 1/n^2$, there is a gradual loss of accuracy due to error propagation. But this does not appear to happen when we treat an alternating series (whose terms decrease in absolute value) numerically in this way. Thus once again the problem of the monotonic series whose terms are given numerically asserts itself.

We may note that the above method of convergence acceleration may readily be applied to a power series by simply putting $a_n = c_n x^n$ in (1). After transformation we will then have a series of rational functions of x. Further transformations can also be effected (i.e. we may iterate), and we will obtain series of rather more complicated rational functions of x.

We now turn to question 3. First of all the point here is to try to transform a monotonic series into an alternating series, when the terms are given numerically. (If they are given analytically, it is often easy to transform into an oscillating series, but we have seen this is not then necessary. For example we can readily obtain the result

$$\sum_{n=1}^{\infty} \frac{1}{n^2} = 2 \sum_{n=1}^{\infty} \frac{(-1)^{n+1}}{n^2} .)$$

So let us start again with our series (1), and now we assume that the (positive) terms are decreasing, i.e. that

$$0 < a_n < a_{n-1},\ n = 2,3,4,\ldots$$

and

$$a_1 > 0.$$

We present a method for formally converting (1) into an alternating series. We commence by associating with (1) the power series

$$S(x) = a_1 x + a_2 x^2 + a_3 x^3 + \ldots + a_n x^n + \ldots ,$$

and consider the formal expansion

$$\begin{aligned} \frac{S(x)}{1+x} &= (a_1x+a_2x^2+\ldots+a_nx^n+\ldots)(1-x+x^2-x^3+\ldots) \\ &= a_1x + (a_2-a_1)x^2 + (a_3-a_2+a_1)x^3 + \ldots \\ &= b_1x + b_2x^2 + b_3x^3 + \ldots + b_nx^n + \ldots , \end{aligned}$$

where

$$
\begin{aligned}
b_1 &= a_1 > 0 \\
b_2 &= a_2 - a_1 < 0 \\
b_3 &= a_3 - a_2 + a_1 > 0 \\
&\cdots\cdots\cdots \\
b_n &= a_n - a_{n-1} + a_{n-2} - \ldots + (-1)^{n-1} a_1 \\
&\cdots\cdots\cdots\cdots\cdots
\end{aligned}
$$

and b_n has the sign $(-)^{n+1}$.

We also have the possibility of determining the b_n's by means of the recurrence relation

$$b_n = a_n - b_{n-1}\ ,\ n = 2,3,4,\ldots$$

with

$$b_1 = a_1.$$

Finally, putting $x = 1$ we have the formal expansion

$$\tfrac{1}{2}S = b_1 + b_2 + b_3 + \ldots + b_n + \ldots, \tag{35}$$

which is an alternating series.

Now one might think that this new series is more amenable to convergence acceleration, than the monotonic series from which it was derived. For example it might seem appropriate to apply Levin's u transformation to (35) instead of to (1). However a careful examination shows that this is not the case. The reason is that for the u transformation to be appropriate for convergence acceleration of series (1), it is necessary that the a_n's satisfy, at least approximately, a certain type of linear first order difference equation. It is not difficult to show that, starting from such a_n's, the b_n's will satisfy a linear difference equation of at least second order, and this means that the u transformation will not be a good convergence accelerator for the series (35). For a careful analysis of this general question, the reader is referred to the interesting paper by Levin and Sidi [4]. Thus once again our attempt to achieve an efficient convergence acceleration on a monotonic series is thwarted. In fact the more one works on this problem for a monotonic series with numerically given terms, the more one becomes convinced that there is an intrinsic difficulty here.

We consider now question 2, and propose to present a basic idea how to avoid loss of accuracy in the addition in the computer of large positive and negative terms.

The idea is best made clear by a simple example. Consider the exponential series

$$e^{-x} = 1 - x + \frac{x^2}{2!} - \frac{x^3}{3!} + \dots \quad . \tag{36}$$

If we wish to calculate e^{-x} directly from this series for large positive x, we have the difficulty that large positive and negative terms nearly cancel, and so in the computer we lose much if not all of the accuracy.

As an attempt to get over this difficulty, supposing first that x is a (positive) integer, let us replace each term $x^n/n!$ by

$$\frac{x^n - \left[\frac{x^n}{n!}\right]n!}{n!} , \tag{37}$$

where the square brackets are used to signify the integral part of the quantity enclosed. The point here is that although $n^n/n!$ may be a large number, we may nevertheless be able to truncate it in the computer to find the integer $\left[\frac{x^n}{n!}\right]$ exactly. Then in the numerator of (37), x^n and $\left[\frac{x^n}{n!}\right]n!$ are integers, which (if x and n are not too large) can be calculated in the computer exactly. And so the numerator of (37), which is their difference, can also be obtained in the computer exactly, despite being the difference of two large numbers. Finally the division by $n!$ yields (37) to full accuracy to which the computer is working. However (37) may be written in the form

$$\frac{x^n}{n!} - \left[\frac{x^n}{n!}\right] = \mathrm{frac}\left(\frac{x^n}{n!}\right),$$

the fractional part of $\frac{x^n}{n!}$.

We may note, in passing, that the above procedure is not the same as working in double precision, since we do not use more than single precision at any stage.

Now in summing the series (36) for large positive x, we know that we must expect a result less than unity (and such kind of qualitative knowledge may be assumed to be available in many problems - or even we could find this out by summing the series directly, since the loss in accuracy will not usually be so great that we do not obtain the correct order of mangitude of the answer), and this means that we can "throw away" from the outset the integral part of each term. (We could actually "throw away" much more than this in many cases, but the above suffices to explain the principle.) This is precisely what we do here when we use merely the fractional part of each term, but by our method the fractional part is calculated to the full accuracy of the computer, without loss of significant figures. For example, working to ten significant figures we find

$$\frac{10^4}{4!} = 416.6666667,$$

which, after subtraction of the integral part 416, leaves us with 0.666666700 to nine places of decimals. In contradistinction we have

$$\frac{10^4 - \left[\frac{10^4}{24}\right]24}{24} = \frac{10{,}000-416\times 24}{24} = \frac{16}{24} = 0.666666667$$

to nine places of decimals.

The following table shows results obtained in a calculator for the exponential series (36), working to nine places of decimals. The improvements in accuracy are not spectacular, but could presumably be enhanced considerably if instead of "throwing away" only the integral part of each term, we would also "throw away" the first few figures after the decimal point-for the larger values of x. We could then usefully employ the series for larger x.

Table I

Calculation of e^{-x} from the exponential series (36), working to nine places of decimals. The first column give exact values of e^{-x}, the second column gives results of calculations when the series is "treated" as above, while the third column gives results of straightforward addition of terms of (36).

x	e^{-x}	Treated series	Untreated series
2	0.135335283	0.135335283	0.135335283
3	0.049787068	0.049787068	0.049787069
4	0.018315639	0.018315639	0.018315638
5	0.006737947	0.006737947	0.006737955
6	0.002478752	0.002478751	0.002478755
7	0.000911882	0.000911882	0.000911970
8	0.000335463	0.000335466	0.000335525
9	0.000123410	0.000123434	0.000123449
10	0.000045400	0.000045335	0.000045477

We may note that for larger values of x we begin to lose accuracy due to the inexact evaluation of the integer

$$x^n - \left[\frac{x^n}{n!}\right]n! \tag{38}$$

in the computer, due to the largeness of the component parts of (38).

We see from the above that there still remain some difficulties in this particular case of summation of the exponential series (36). We may list them as follows:

(a) x may not be an integer.

(b) x(integer) and/or n may be too large for the exact evaluation of (38) in the computer.

With regard to (a) we may write

$$x = k + a$$

where $$k = [x], \quad 0 < a < 1.$$

Then $$\frac{x^n}{n!} = \frac{(k+a)^n}{n!} = \frac{k^n + \binom{n}{1}k^{n-1}a + \ldots + a^n}{n!}.$$

We have here all positive terms, and we can replace

$$\frac{k^n}{n!} \text{ by } \frac{k^n - \left[\frac{k^n}{n!}\right]n!}{n!} \text{ as before.}$$

These are still problems here, and further work seems called for.

With regard to (b), it is possible to "build up" (38) without loss of accuracy in some cases. As an example of "building up" in a computer that holds ten significant figures we have

$$11^{10} - \left[\frac{11^{10}}{10!}\right]10! = 11^{10} - 10!\times 7147$$

$$= 11 \times 11^9 - 90\times 7147\times 40320$$

$$= 10(11^9 - 9\times 7147\times 40320) + 11^9$$

$$= 2357947691 - 23555590 = 2391001,$$

the answer being exact. Here too further work needs to be done.

Of course we have only so far dealt with the exponential series, and one las to consider other cases that arise, e.g. in the numerator and denominator of U_{kn}.

Finally we should try to answer question 1, and can only say that all the evidence points to an intrinsic accuracy loss when we attempt to accelerate the convergence of a monotonic series, whose terms are given numerically, but that no such problem arises in the case of an alternating series whose terms decrease in absolute value. Here again definitive answers are not known to this author, and it seems that further research is desirable.

REFERENCES

[1] D. Levin, Development of non-linear transformations for improving convergence of sequences, Intern. J. Computer Math. B3 (1973), 371-388.

[2] W.G. Bickley and J.C.P. Miller, The numerical summation of slowly convergent series of positive terms, Phil. Mag. (Series 7) 22(1936), 754-767.

[3] E.D. Rainville, Special Functions, Macmillan, New York, 1967, Ch. 14.

[4] D. Levin and A. Sidi, Two new classes of non-linear transformations for accelerating the convergence of infinite integrals and series, J. Appl. Math. Comp., in press.

ON THE EVEN EXTENSION OF AN M FRACTION

by

John H. McCabe

Mathematical Institute

University of St Andrews

St Andrews, Fife, Scotland.

ABSTRACT: One result of the surge of interest in Padé approximations during the last two decades has been the study of two-point Padé approximations. In particular, rational functions which are derived from power series expansions about the origin and the point at infinity have found several applications and the theory associated with them has developed accordingly.

These particular two-point Padé approximations are convergents of continued fractions of the form

$$c_0 + c_1 z + c_2 z^2 + \dots + \frac{c_k z^k}{1 + d_1 z} + \frac{n_2 z}{1 + d_2 z} + \frac{n_3 z}{1 + d_3 z} + \dots, \qquad k \geq 0,$$

now generally known as M fractions or, alternatively, general T fractions. The coefficients of these continued fractions can be obtained by a variety of methods, including the well known q - d algorithm.

The purpose of this short talk is to discuss an even extension of the above continued fractions, that is a continued fraction whose even order convergents are the successive convergents of the above fraction. The extension is a continued fraction of a form not frequently met in the literature, but is of a simpler type than the M fraction. The same q - d algorithm, with two slight modifications, can be used to provide the coefficients of the even extension, and this will be described.

Finally, an example for which the extension will provide error bounds, whereas the M fraction will not, is considered.

1. INTRODUCTION

It is well known that the elements in the lower half of the (normal) Padé table for the series

$$c_0 + c_1 z + c_2 z^2 + \dots + c_r z^r + \dots . \tag{1}$$

are the convergents of the set of continued fractions

$$c_0 + c_1 z + \dots + c_{k-2} z^{k-2} + \frac{c_{k-1} z^{k-1}}{1} + \frac{e_1^k z}{1} + \frac{a_2^k z}{1} + \frac{e_2^k z}{1} + \frac{a_3^k z}{1} + \dots \tag{2}$$

for $k = 2,3, \dots$. In our notation, the convergents of the above fraction with k = 2 form the main staircase of the Padé table for the series (1).

The coefficients of the continued fractions (2) can be obtained from those of (1) by a variety of methods, and one possible technique is the q - d algorithm of Rutishauser. Setting $a_1^k = 0$ for all k and $e_1^k = -c_k/c_{k-1}$ we generate further coefficients by the rhombus rules

$$e_r^k + a_{r+1}^k = e_r^{k+1} + a_r^{k+1}$$

$$e_{r+1}^k * a_{r+1}^k = e_r^{k+1} * a_{r+1}^{k+1} .$$

The coefficients of the continued fractions (2) then lie on the *kth* row of the semi-infinite two-dimensional array shown in figure 1. The relationships between the coefficients are displayed by the rhombii in the figure.

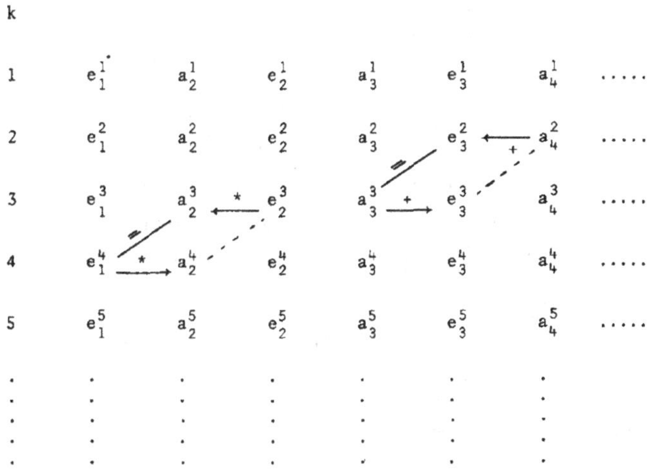

Figure 1

The notation used has been altered from that usually used in the q - d algorithm so that it merges better with what follows.

The recent interest in two-point Padé approximations has resulted in the natural extension of the above, in the sense that the elements of the two-point Padé table for the series (1) and the series

$$\frac{b_1}{z} + \frac{b_2}{z^2} + \frac{b_3}{z^3} + \ldots + \frac{b_r}{z^r} + \ldots \tag{3}$$

can be regarded as the convergents of continued fractions.

The continued fractions are of one of the forms

$$c_0 + c_1 z + \dots + \frac{c_k z^k}{1+d_1^k z} + \frac{n_2^k z}{1+d_2^k z} + \frac{n_3^k z}{1+d_3^k z} + \frac{n_4^k z}{1+d_4^k z} + \dots, \quad k \geq 0 \tag{4a}$$

or

$$\frac{b_1}{z} + \frac{b_2}{z^2} + \dots + \frac{b_k}{z^k} - \frac{b_k z^{-k}}{1+d_1^{-k} z} + \frac{n_2^{-k} z}{1+d_2^{-k} z} + \frac{n_3^{-k} z}{1+d_3^{-k} z} + \dots, \quad k > 0. \tag{4b}$$

These are M fractions or, alternatively, general T fractions, for the series (1) and (3). [See, for example, McCabe [1] or Jones and Magnus [2].] The coefficients can again be obtained by a quotient difference algorithm. Specifically,

$$n_{r+1}^k + d_r^k = n_r^{k+1} + d_r^{k+1} \tag{5a}$$

$$n_{r+1}^k * d_{r+1}^{k+1} = n_{r+1}^{k+1} * d_r^k \tag{5b}$$

where $n_1^k = 0$ for all k and $d_1^k = -c_k/c_{k-1}$, c_0/b or $-b_k/b_{k+1}$ according as k is greater than zero, equal to zero, or less than zero. The coefficients of the continued fractions (4) form the rows of the infinite two dimensional array shown in figure 2.

The elements related by the rules (5) again form rhombii as shown. The elements below the step line in figure 2 are related to those in figure 1 by

$$\left.\begin{aligned} n_r^{r+j} + d_r^{r+j} &= e_r^{j+1} \\ n_r^{r+j-1} &= a_r^{j+1} \end{aligned}\right\} r = 2,3,4, \dots, \; j = 0,1,2,3, \dots . \tag{6}$$

Thus the convergents of the continued fraction

$$\frac{c_0}{1} + \frac{(n_1^1+d_1^1)z}{1} + \frac{n_2^1 z}{1} + \frac{(n_2^2+d_2^2)z}{1} + \frac{n_3^2 z}{1} + \dots$$

are the elements on the main staircase of the Padé table for the series (1).

	d_1	n_2	d_2	n_3	d_3	n_4	d_4		
$\cdot$	$\cdot$	$\cdot$	$\cdot$	$\cdot$	$\cdot$	$\cdot$	$\cdot$	$\cdot$	$\cdot$
-4	d_1^{-4}	n_2^{-4}	d_2^{-4}	n_3^{-4}	d_3^{-4}	n_4^{-4}	d_4^{-4}	$\cdot$	$\cdot$
-3	d_1^{-3}	n_2^{-3}	d_2^{-3}	n_3^{-3}	d_3^{-3}	n_4^{-3}	d_4^{-3}	$\cdot$	$\cdot$
-2	d_1^{-2}	n_2^{-2}	d_2^{-2}	n_3^{-2}	d_3^{-2}	n_4^{-2}	d_4^{-2}	$\cdot$	$\cdot$
-1	d_1^{-1}	n_2^{-1}	d_2^{-1}	n_3^{-1}	d_3^{-1}	n_4^{-1}	d_4^{-1}		
0	d_1^{0}	n_2^{0}	d_2^{0}	n_3^{0}	d_3^{0}	n_4^{0}	d_4^{0}		
1	d_1^{1}	n_2^{1}	d_2^{1}	n_3^{1}	d_3^{1}	n_4^{1}	d_4^{1}	$\cdot$	$\cdot$
2	d_1^{2}	n_2^{2}	d_2^{2}	n_3^{2}	d_3^{2}	n_4^{2}	d_4^{2}	$\cdot$	$\cdot$
3	d_1^{3}	n_2^{3}	d_2^{3}	n_3^{3}	d_3^{3}	n_4^{3}	d_4^{3}	$\cdot$	$\cdot$
4	d_1^{4}	n_2^{4}	d_2^{4}	n_3^{4}	d_3^{4}	n_4^{4}	d_4^{4}	$\cdot$	$\cdot$
$\cdot$	$\cdot$	$\cdot$	$\cdot$	$\cdot$	$\cdot$	$\cdot$	$\cdot$		

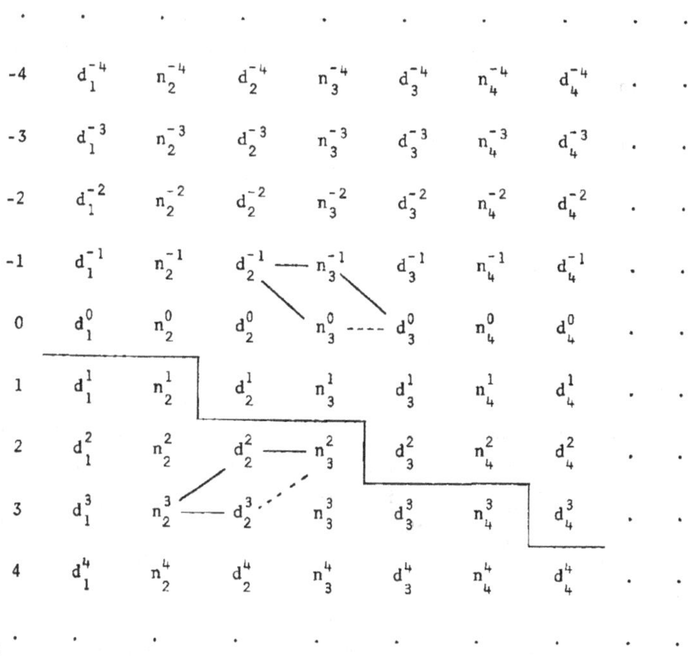

Figure 2

2. EVEN EXTENSIONS

The particular M fraction

$$\frac{c_0}{1+d_1^0 z} + \frac{n_2^0 z}{1+d_2^0 z} + \frac{n_3^0 z}{1+d_3^0 z} + \frac{n_4^0 z}{1+d_4^0 z} + \dots , \qquad (7)$$

obtained from (4a) by setting $k = 0$, can be regarded as the even contraction of the continued fraction

$$\frac{c_0}{1} + \frac{m_1^0 z}{1} + \frac{\ell_2^0}{1} + \frac{m_2^0 z}{1} + \frac{\ell_3^0}{1} + \frac{m_3^0 z}{1} + \frac{\ell_4^0}{1} + \frac{m_4^0 z}{1} + \dots . \qquad (8)$$

where $m_1^0 = d_1^0$ and

$$\ell_r^0 = -n_r^0/(n_r^0 + d_{r-1}^0)$$

$$m_r^0 = d_r^0 d_{r-1}^0/(n_r^0 + d_{r-1}^0) \qquad (9)$$

for r = 2,3,4, ... In addition of course, due to the correspondence of (7) with both of the series (1) and (2), the continued fraction (7) is also the even contraction of the continued fraction

$$\frac{b_1}{z} + \frac{\tilde{m}_1^0}{1} + \frac{\tilde{\ell}_2^0 z}{z} + \frac{\tilde{m}_2^0}{1} + \frac{\tilde{\ell}_3^0 z}{z} + \frac{\tilde{m}_3^0}{1} + \frac{\tilde{\ell}_4^0 z}{z} + \ldots . \qquad (10)$$

where $\tilde{m}_1^0 = 1/d_1^0$ and

$$\tilde{\ell}_r^0 = -n_r^0/(n_r^0 + d_r^0)$$

$$\tilde{m}_r^0 = 1/(n_r^0 + d_r^0)$$

for r = 2,3,4, ...

The continued fractions (8) and (10) are even extensions of the M fraction (7), and are of course, at least in form, equivalent to each other. It is easily verified that (8) and (10) will, on contraction, yield (4) by using the standard relations between continued fractions and their contractions. See, for example, Khovanskii [3,p.14]. Continued fractions of the forms (7) and (8) were introduced by Perron [4, section 31]. Since then the extended forms have appeared in the literature rather less frequently than those of the form (7), and when they do appear it is not always in connection with two-point Padé approximations. McCabe [5] mentioned them in this latter context and, more recently, they appear in the work of Sidi [6], as a result of some recurrence relations that he develops for two-point Padé approximants, and also in that of Thron [7].

The coefficients of the continued fraction (8) can of course be obtained by using the quotient-difference algorithm (5) and then the relations (9). However, they can themselves be generated directly by a rhombus algorithm similar to (5) but different in two respects.

Set $\ell_1^k = 0$ and $m_1^k = -c_k/c_{k-1}$, c_0/b_1 or $-b_k/b_{k+1}$ according as k is greater than, equal to, or less than zero. Subsequent coefficients are then generated by

$$\begin{aligned} \ell_r^{k-1} * m_r^k &= \ell_r^k * m_{r-1}^{k-1} \\ (1+\ell_r^k) * m_{r-1}^{k+1} &= m_{r-1}^k * (1+\ell_{r-1}^{k+1}) \end{aligned} \qquad (11)$$

for r = 2,3,4, ... and k = 0,±1,±2,±3, ...

In an array similar to that in figure 2, the elements related by the above rules form the rhombii

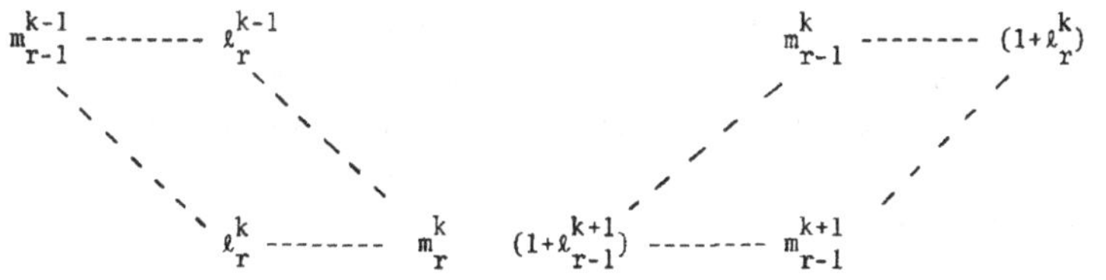

The rules have the same pattern as those in (5), extending upwards in the table for the m_r^k and downward for the ℓ_r^k. However, unlike (5), both rules are quotient rules and one, that which yields the ℓ_r^k, has unity added to each of the ℓ coefficients. The existence of (8) will of course depend on the existence of (7). See McCabe and Murphy [8] for existence conditions. Clearly, from (9), the algorithm will fail if n_r is equal to $-d_{r-1}$ for some value of r. This will indicate that c_r is zero, in which case the algorithm would fail anyway. See McCabe [10].

The even order convergents of (8) are of course those of (4), and agree with equal numbers of terms of (1) and (2) when expanded accordingly. The odd order convergents are, after the first, ratios of polynomials of degrees r-1 and r respectively, where 2r + 1 is the order of convergent, and the 'fit' r + 1 terms of (1) and r - 1 terms of (2) when expanded accordingly.

The coefficients ℓ_r^k and m_r^k for $k > 0$, $r = 1,2,3, \ldots$, are merely those of the even extension of the continued fraction (4a), without the initial terms $c_0 + c_1 z + \ldots$ $\ldots + c_k z^k$.

The n and d coefficients are related to the ℓ and m coefficients by the equations

$$n_j^k = -\ell_j^k\, m_{j-1}^k/(1+\ell_j^k).(1+\ell_{j-1}^k) \tag{12a}$$

$$d_j^k = m_j^k/(1+\ell_j^k). \tag{12b}$$

We can derive further equations, relating the a and e coefficients to the ℓ and m coefficients by making use of (5a), (6) and (12). We obtain

$$a^{k+1} = -\ell_j^{j+k-1}\, m_j^{j+k-1}/(1+\ell_j^{j+k-1}).(1+\ell_{j-1}^{k+k-1}) \tag{13a}$$

$$e_j^{k+1} = m_j^{j+k-1}/(1+\ell_j^{j+k-1}).(1+\ell_{j+1}^{j+k-1}) \tag{13b}$$

for $j = 2,3,4, \ldots$, and $k = 0,1,2, \ldots$.

In the table of ℓ and m coefficients, shown in figure 3, we need those coefficients between the two lines to construct the continued fraction

$$\frac{c_0}{1} + \frac{e_1^1 z}{1} + \frac{a_2^1 z}{1} + \frac{e_2^1 z}{1} + \ldots + \frac{a_j^1 z}{1} + \frac{e_j^1 z}{1} + \ldots , \tag{14}$$

corresponding to the series (1). In the case where only the series (1) is known we would not be able to calculate m_1^0 and ℓ_2^0 However, we would then give any non-zero value to m_2^1 and proceed with the calculation of the subsequent coefficients. These would of course have values that depend on m_2^1, but those in (14) would be the same whatever value m_2^1 has.

	.	.	.	.	.	.	.	.	.
-4	m_1^{-4}	ℓ_2^{-4}	m_2^{-4}	ℓ_3^{-4}	m_3^{-4}	ℓ_4^{-4}	m_4^{-4}	.	.
-3	m_1^{-3}	ℓ_2^{-3}	m_2^{-3}	ℓ_3^{-3}	m_3^{-3}	ℓ_4^{-3}	m_4^{-3}	.	.
-2	m_1^{-2}	ℓ_2^{-2}	m_2^{-2}	ℓ_3^{-2}	m_3^{-2}	ℓ_4^{-2}	m_4^{-2}	.	.
-1	m_1^{-1}	ℓ_2^{-1}	m_2^{-1}	ℓ_3^{-1}	m_3^{-1}	ℓ_4^{-1}	m_4^{-1}	.	.
0	m_1^{0}	ℓ_2^{0}	m_2^{0}	ℓ_3^{0}	m_3^{0}	ℓ_4^{0}	m_4^{0}	.	.
1	m_1^{1}	ℓ_2^{1}	m_2^{1}	ℓ_3^{1}	m_3^{1}	ℓ_4^{1}	m_4^{1}	.	.
2	m_1^{2}	ℓ_2^{2}	m_2^{2}	ℓ_3^{2}	m_3^{2}	ℓ_4^{2}	m_4^{2}	.	.
3	m_1^{3}	ℓ_2^{3}	m_2^{3}	ℓ_3^{3}	m_3^{3}	ℓ_4^{3}	m_4^{3}	.	.
4	m_1^{4}	ℓ_2^{4}	m_2^{4}	ℓ_3^{4}	m_3^{4}	ℓ_4^{4}	m_4^{4}	.	.
.	.	.	.	.	.	.	.	.	.

Figure 3

3. EXAMPLE

The initial elements of the table of ℓ and m coefficients for the two series

$$C(z) = 1 - \frac{z}{3} + \frac{z^2}{15} - \frac{z^3}{105} + \ldots + \frac{(-)^r z^r}{1.3.5....(2r+1)} + ... \tag{15}$$

and

$$B(z) = \frac{1}{z} + \frac{1}{z^2} + \frac{3}{z^3} + \frac{15}{z^4} + ... + \frac{1.3.5....(2r-3)}{z^r} + ... \tag{16}$$

are shown in figure 4.

k	m_1^k	ℓ_2^k	m_2^k	ℓ_3^k	m_3^k	ℓ_4^k	m_4^k	. . .
.	.	.	.	.	.	.	.	. . .
.	.	.	.	.	.	.	.	. . .
-5	-1/9	-2/9	-1/9	-4/9	-1/9	-6/9	-1/9	. . .
-4	-1/7	-2/7	-1/7	-4/7	-1/7	-6/9	-1/7	. . .
-3	-1/5	-2/5	-1/5	-4/5	-1/5	-6/5	-1/5	. . .
-2	-1/3	-2/3	-1/3	-4/3	-1/3	-6/3	-1/3	. . .
-1	-1	-2	-1	-4	-1	-6	-1	. . .
0	1	2	1	4	1	6	1	. . .
1	1/3	2/3	1/3	4/3	1/3	6/3	1/3	. . .
2	1/5	2/5	1/5	4/5	1/5	6/5	1/5	. . .
3	1/7	2/7	1/7	4/7	1/7	6/7	1/7	. . .
4	1/9	2/9	1/9	4/9	1/9	6/9	1/9	. . .
5	1/11	2/11	1/11	4/11	1/11	6/11	1/11	. . .
.	.	.	.	.	.	.	.	. . .

Figure 4

The continued fraction (8) for these series is thus

$$\frac{1}{1}+\frac{z}{1}+\frac{2}{1}+\frac{z}{1}+\frac{4}{1}+\frac{z}{1}+\frac{6}{1}+\frac{z}{1}+\frac{8}{1}+\cdots \tag{17}$$

which is an even extension of the M fraction

$$\frac{1}{1+z}-\frac{2z/3}{1+z/3}-\frac{4z/15}{1+z/5}-\ldots\ldots-\frac{2(r-1)z/(2r-1)(2r-3)}{1+z/(2r-1)}-\ldots\ldots \tag{18}$$

The continued fraction (2) is, with the aid of (13), easily seen to be

$$\frac{1}{1}+\frac{z/3}{1}+\frac{2z/15}{1}+\frac{3z/35}{1}+\frac{4z/63}{1}+\ldots\ldots, \tag{19}$$

corresponding to (15) only.

If z is real and positive the series $zC(2z^2)$ and $zB(2z^2)$ are, respectively, a convergent series and an asymptotic divergent series for Dawson's integral,

$$F(z) = e^{-z^2}\int_0^z e^{t^2}dt.$$

Since all the coefficients of (17) are then positive the convergents of the continued fraction

$$\frac{z}{1}+\frac{2z^2}{1}+\frac{2}{1}+\frac{2z^2}{1}+\frac{4}{1}+\frac{2z^2}{1}+\frac{6}{1}+\frac{2z^2}{1}+\dots \tag{20}$$

lie alternately on either side of F(z). Hence the error in approximating F(z) by the even convergent $A_{2k}(z)/(B_{2k}(z)$ is bounded by

$$\left|\frac{A_{2k}(z)}{B_{2k}(z)}-\frac{A_{2k-1}(z)}{B_{2k-1}(z)}\right| = \frac{1.2.4.6.\dots(2k-2)(2z^2)^k}{B_{2k}(z)B_{2k-1}(z)}$$

The denominators $B_{2k}(z)$ and $B_{2k-1}(z)$ can be expressed in terms of confluent hypergeometric functions and

$$B_{2k}(z) = 1.3.\dots(2k-1)\,{}_1F_1(-k;-k+1/2;z^2)$$

$$B_{2k-1}(z) = 1.3.\dots(2k-1)\,{}_1F_1(1-k;-k+1/2;z^2).$$

Hence, replacing the product $1.3.\dots(2k-1)$ by $2^k\Gamma(k+\frac{1}{2})/\sqrt{\pi}$ and using the relation that ${}_1F_1(a;b;z) = e^z\,{}_1F_1(b-a;b;-z)$ we find that the error of the $(2n)th$ convergent is bounded by

$$\frac{(\pi/2)n!(1+1/2n)e^{-z^2}z^{2n+1} \;/\; \Gamma(n+1/2).\Gamma(n+3/2)}{{}_1F_1(1/2;-n+1/2;-z^2)\;\;{}_1F_1(-1/2;-n+1/2;-z^2)}\,.$$

The error can be expressed exactly as

$$\frac{\pi n!e^{-2z^2}z^{2n+1}\,{}_1F_1(1/2;n+3/2;z^2)}{2\Gamma(n+3/2)\Gamma(n+1/2)\;\;{}_1F_1(1/2;-n+1/2;-z^2)}$$

obtained by applying a result of Laguerre to the differential equation satisfied by F(z). See McCabe [9] for details.

REFERENCES

[1] McCabe, J.H.: J.I.M.A., V.15, 1975 pp 363-372.

[2] Jones, W.B. and Magnus, A.: J. Comp. App. Maths., V.6, 1980, pp 105-120.

[3] Khovanskii, A.N.: 'The application of continued fractions and their generalisations to problems in approximation theory', translated by P. Wynn, 1963, Noordhoff, Groningen.

[4] Perron, O.: Die Lehre von den Kettenbrüchen, Band II, B. G. Teubner, Stuttgart, 1957.

[5] McCabe, J.H.: Ph.D. Thesis, Brunel University 1971.

[6] Sidi, A.: J. Comp. App. Maths., V.6, 1980, pp 9-17.

[7] Thron, W.J.: 'Padé and Rational Approximation', Academic Press, 1977.

[8] McCabe, J.H. and Murphy, J.A.: J.I.M.A., V.17, 1976, pp 233-247.

[9] McCabe, J.H.: Maths. Comp., V.28, 1974, pp 811-816.

[10] McCabe, J.H.: Maths. Comp., V.32, 1978, pp 1303-1305.

RATE OF CONVERGENCE OF SEQUENCES OF PADE-TYPE APPROXIMANTS AND POLE DETECTION IN THE COMPLEX PLANE

Alphonse MAGNUS

Summary. It is shown how to choose the poles of rational approximants of a function known by its Taylor coefficients at a point z_0 and a region of meromorphy, in order to optimize the approximation and the search for unknown poles.

1. Padé-type approximants of analytic functions.

The Padé-type (m,n,k) approximant of a function f is a rational function $\frac{N_m(z)}{Q_{n-k}(z)\, D_k(z)}$ where the numerator and the factors of the denominator are polynomials of degree less or equal than their subscripts. Complete freedom is left for the choice of Q_{n-k}, and the determination of a reasonable choice is the subject of this study. The remaining fraction $N_m(z)/D_k(z)$ is the Padé [m/k] approx--imant of $Q_{n-k}(z)\, f(z)$, which asks for the m+k+1 first Taylor coefficients of f at a point z_0 ; k is the expected number of poles in the region of investigation, the zeros of Q_{n-k} should be outside this region. The concept has been introduced by Brezinski ([3], [4]), in order to unify the study of rational approximation to [formal] power series, but the important influence of the determination of Q_{n-k} on the quality of approximation became immediately obvious ([3] § 3). The theoretical treatment is based on Walsh [17] and Gončar [7]. The main application will be the construction of cuts for functions with branch points, similar to what can sometimes be achieved with [n/n] Padé approximants, but with general assumptions for f, and where the cuts can be moved at will, respecting the position of the branch points (see [2], [10]). Other important problems which will not be investigated here include stable approximation to the exponential function ([3] § 3, [4] § 1.4, see remark e), and inversion of Laplace transforms ([3] § 3, [4] § 1.4).

2. Use of a region of holomorphy.

Let f be holomorphic in a region R (connected open set, in the terminology of Walsh [17] § 1.1), of boundary C. As k=0, the concept of Padé [m/o] approximation of Q_n f (Taylor partial sums) is easily extended to the polynomial interpolation at $z_0, \ldots, z_m \in R$: one has, using Cauchy integral formula ([17] § 2.6, [5] § 3.6)

$$Q_n(z)\ f(z) - N_m(z) = \frac{Z_{m+1}(z)}{2\pi i} \int_{C'} \frac{Q_n(t)\ f(t)dt}{Z_{m+1}(t)\ (t-z)}$$

with $Z_{m+1}(t) = (t-z_0) \ldots (t-z_m)$, and C' a contour in R close to C (if $z_i = \infty$, $t-z_i$ is replaced by 1 ; one must take $m \geq n$ if $\infty \in R$), or

$$f(z) - N_m(z)/Q_n(z) = \frac{1}{2\pi i} \int_{C'} \frac{S_n(t)\ f(t)dt}{S_n(z)\ (t-z)},\quad S_n(t) = \frac{Q_n(t)}{Z_{m+1}(t)}$$

hence the bound for the error :

$$|f(z) - N_m(z)/Q_n(z)| \leq |S_n(z)|^{-1} \max_{t\in C'} |S_n(t)|\ M(C'). \qquad (1)$$

We shall be interested in the asymptotic *geometric rate of convergence* of a sequence of Padé-type approximants indexed by n (i.e., m, Z_{m+1} and Q_n depend on n) :

$$\limsup_{n\to\infty} |f(z)-N_m(z)/Q_n(z)|^{\frac{1}{n}} \leq \rho(z) = \limsup_{n\to\infty} |S_n(z)|^{-\frac{1}{n}} \max_{t\in C} |S_n(t)|^{\frac{1}{n}}. \quad (2)$$

A *lower bound* for ρ(z) can be found if R is regular for the Dirichlet problem, which is the case if R is finitely many connected and if C is closed without isolated point (Lebesgue-Osgood conditions : see [17] § 4.1) ; this will also allow the move from C' in (1) to C in (2). Let then G(z;z') be the Green function of R singular at z', i.e. the harmonic function defined on $R \setminus \{z'\}$, vanishing on C and such that $G(z;z') + \ln|z-z'|$ is regular near z'(G(z;z') is therefore positive on R). Finally, let

$$G_n(z) = (G(z;z_0) + \ldots + G(z;z_m))/n \qquad G(z) = \liminf_{n\to\infty} G_n(z).$$

One has then $\rho(z) \geq \exp(-G(z))$.

Indeed, $\ln|S_n(z)| - n\,G_n(z)$ is harmonic and bounded from above in R, so that the maximum value is reached on C and is $\ln \max_{t\in C} |S_n(t)|$.

The next step is of course the designing of a sequence $\{Q_n\}$ such that $\rho(z) = \exp(-G(z))$. In some very special situations, the solution is easy : for instance, if R is the disk of center 0 and radius a, $G(z;z_k) = \ln|(a - \frac{\bar{z}_k z}{a})/(z-z_k)|$, so that one has just to take $a^2/\bar{z}_k$, $k=0,\dots,n-1$, as the zeros of Q_n, and $m \sim n$ (see [6] for similar examples) ; the case of the exterior of an arc of circle will be presented in § 6. The general problem has been studied extensively by Walsh ([17] chap. 4 and 7). Here is a satisfactory way (the solution is not unique) to dispose the zeros of Q_n on C :

THEOREM 1. *The best asymptotic rate of convergence for given sequences of interpolation points* $\{z_0, \dots, z_m\}$ *depending on* n, *i.e.* $\rho(z) = \exp(-G(z))$, *can be achieved when* $m \sim n$, *and if the zeros of* Q_n *are equidistant on* C *with respect to the measure* $d\mu(t) = \frac{\partial G_n(t)}{\partial \nu} \frac{|dt|}{2\pi}$, *where* ν *is the normal on* C *directed towards* R ([1] § 2.1, [17] § 4.2).

Indeed, from Gauss-Green formula,

$$G_n(z) + n^{-1} \ln|Z_{m+1}(z)| = \int_C \ln|t-z|\,d\mu(t) \quad (3)$$

$$\int_C d\mu(t) = (m+1)/n \quad (4)$$

which shows that $\ln|S_n(z)|^{1/n}$ will be close to $G_n(z)$, i.e. $\ln|Q_n(z)|^{1/n} = n^{-1} \sum_{r=1}^{n} \ln|q_r-z|$ will be close to the right-hand side of (3) if $\int_{q_r}^{q_{r+1}} d\mu(t) = n^{-1} \int_C d\mu(t) = (m+1)/n^2$: then,

$$\int_C \ln|t-z|\,d\mu(t) = \sum_{r=1}^{n} \int_{q_r}^{q_{r+1}} \ln|t-z|\,d\mu(t) \sim \sum_{r=1}^{n} \ln|q_r-z| \int_{q_r}^{q_{r+1}} d\mu(t)$$

$$\sim n^{-1} \sum_{r=1}^{n} \ln|q_r-z|.$$

3. Remarks.

a) If R is simply connected, the points q_r are such that $\Phi_n(q_r) = \exp(2\pi i r/n)$, $r=1,\dots,n$, where

$\Phi_n(z) = [\Phi(z;z_0) \ldots \Phi(z;z_m)]^{1/n}$, and $\Phi(z;z_k)$ maps conformally R on the exterior of the unit circle, with a pole at $z=z_k$ ($G(z;z_k) = \ln|\Phi(z;z_k)|$).

b) Practical use of theorem 1 should ask in general for numerical conformal mapping, or numerical solution of the integral equations (3) and (4) when $z \in C$ (if $\infty \in R$, $G(z)$ must be replaced by $G(z)-G(\infty)$). It is perhaps simpler to use orthogonal functions (in Szegö's sense : $\int_C S_m(t)\, \overline{S_n(t)}\, |dt| = 0$ if $m \neq n$ [15] chap. 16) as in [2], but then the zeros of Q_n are not always exactly on C. If a closed-form of $\Phi(z;z_k)$ is known, the present method is more convenient (see § 6).

c) If the interpolation points $z_0, \ldots, z_m$ may be chosen in a subset R' of R, these degrees of freedom may be used to optimize the rate of convergence at some point $z'' \notin R'$. The solution will then be a confluent interpolation (Taylor coefficients) at the point z_0 of R' which maximizes $G(z'';z_0)$. The problem of best approx--imation on a whole subset R" of R is more difficult, only the case R"=R' is classical (it uses harmonic functions constant on C and on the boundary of R' : [17] chap. 8, [7]).

d) More degrees of freedom can be used if C, or a part of C, joins branch points in some way. One can then study the influence of the shape of C on $\rho(z'')$ by Schiffer's variation formula ([1] chap. 7). The "Padé cut" corresponds to $z'' \sim z_0$ [12], [13].

e) An isolated essential singularity could be enclosed in a small contour C' which should be a part of C, but it seems better to make C' depend on n : $C' = C^{(n)}$, closer and closer to the singular point as n increases. Moreover, as f shows strong variations on $C^{(n)}$,

$$|f(z) - N_m(z)/Q_n(z)| \leq |S_n(z)|^{-1} \max_{t \in C^{(n)}} |S_n(t)\, f(t)|\, M(C^{(n)})$$

should be used instead of (1). For instance, with $f(z) = e^{-z}$, $C^{(n)} = n\, C^{(1)}$,

$$|e^{-z} - N_m(z)/Q_n(z)| \leq |V_n(z)|^{-1} \max_{t \in C^{(1)}} |V_n(t) e^{-nt}| \, M(C^{(n)}),$$

for z/n inside $C^{(1)}$, with $V_n(t) = S_n(nt)$, which allows the search of geometric rates of convergence for various approximants, including stable approximants (with poles with negative real parts).

4. Pole determination by Padé-type approximants.

We consider now the Padé-type approximant (m,n,k) of a meromorphic function f in R, in the sense of § 1 (Taylor coefficients at a fixed point z_0), although extension to rational interpolation is possible [16]. We have therefore, when $m \sim n$, $G(z) = G(z;z_0)$, $Z_{m+1}(z) = (z-z_0)^{m+1}$. R_M will denote the region defined by $G(z) > M$, C_M the boundary of R_M.

THEOREM 2. *If* f *has exactly* k *poles* $p_1, \ldots, p_k$ *in* R_M*, and if the polynomials* Q_n *are constructed in such a way that*

$$\limsup_{n \to \infty} \max_{C_M} \frac{|Q_n(t)|^{1/n}}{|t-z_0|} \Big/ \left(\frac{|Q_n(z)|^{1/n}}{|z-z_0|}\right) = \exp\,(M - G(z)) \text{ holds in } R_M$$

(as in theorem 1), *then, if* $m/n \to 1$ *when* $n \to \infty$,

$$\limsup_{n \to \infty} \left|f(z) - \frac{N_m(z)}{Q_{n-k}(z)\, D_k(z)}\right|^{\frac{1}{n}} \leq \exp(M-G(z)), \quad z \in R_M$$

$$z \neq p_r, \quad r=1,\ldots,k.$$

Moreover, the poles of f *are approximated by the zeros of* D_k *in such a way that* $\limsup_{n \to \infty} |D_k(p_r)|^{1/n} \leq \exp(M-G(p_r))$, $r=1,\ldots,k$ (*where* D_k *is monic or normalized by* $D_k(z_0) = 1$).

Indeed, residues must now be taken into account in the Cauchy integral formula of $Q_{n-k}\, f$:

$$Q_{n-k}(z)f(z) = \sum_{r=1}^{k} \frac{Q_{n-k}(p_r)\varphi_r}{z-p_r} + \frac{1}{2\pi i}\int_{C_M} \frac{Q_{n-k}(t)f(t)dt}{t-z}. \qquad (5)$$

The denominator $D_k(z)$ of the Padé $[m/k]$ approximant of $Q_{n-k}\, f$ uses a Toeplitz matrix of the Taylor coefficients of order $m+1-k,\ldots,$ $m+k$:

$$(Q_{n-k}\, f)_s = -\sum_{r=1}^{k} \frac{Q_{n-k}(p_r)\varphi_r}{(p_r-z_0)^{s+1}} + \frac{1}{2\pi i}\int_{C_M} \frac{Q_{n-k}(t)f(t)dt}{(p_r-z_0)^{s+1}} \qquad s=m+1-k,\ldots, m+k.$$

Solving $\sum_{j=0}^{k} (Q_{n-k}\, f)_{s-j}\, (D_k)_j = 0$, $s=m+1,\dots,m+k$ for

$D_k(p_r) = \sum_{j=0}^{k} (D_k)_j\, (p_r - z_0)^j$ shows that

$$D_k(p_r) \sim \max_{C_M} |S_n(t)|/|S_n(p_r)| \sim \exp n(M-G(p_r)).$$

There is no essential change if multiple poles are present ([8] § 3.1).

Finally, using (5) with $D_k\, Q_{n-k}\, f$ shows

$$D_k(z)\, f(z) - N_m(z)/Q_{n-k}(z) = O(\exp n(M-G(z))),\ z \in R_M.$$

5. Further remarks.

f) This extension of the Montessus de Ballore theorem allows the determination of poles which are the closest to z_0 in the Green function metric ; the speed of convergence can be visualized by the level lines of $G(z)$.

g) The theorem is still valid if f is meromorphic in a region $R_{M'}$, provided $M' < M$. At the limit $M' = M$, it is perhaps possible to discuss nonpolar singularities on C_M, as in [9].

6. Example.

The theorem 2 has been applied successfully to the Pindor function

$$f(z) = (z^2+4z+5)^{3/2}\,(z^2-6z+13)^{-1}\,\exp[0.3(z-\sqrt{5})/(z+\sqrt{5})]$$

for which poor performance of Padé approximation has been exposed [14]. The two branch points $-2\pm i$ have been joined by an arc of circle containing also the essential singular point $-\sqrt{5}$, which receives therefore no special treatment. In general, the Green function of the exterior of a circular arc of endpoints s_1 and s_2 is given by $G(z;z_0) = \ln|\Phi(z;z_0)|$, with the chain of conformal mappings $y = (z-s_1)/(z-s_2)$, $w = (y+y_0)/(y-y_0)$, $u = w+\sqrt{w^2-1}$, $\Phi = u\cos\alpha - i\sin\alpha$ (see, for instance, [11] § 5.7, Exercise 4), where the square root determination in u is such that $|\Phi| > 1$ outside C, or, equivalently, $|u|^2 > 1+2\,\mathrm{tg}\,\alpha\ \mathrm{Im}\,u$ (equality holds on C, so that α can be computed from a third point of C). If $\Psi(z;z_0)$ corresponds to the other determination of the square root, an acceptable $Q_n(z)$ is $Q_n(z) = (z-z_0)\ldots(z-z_m)\ [\Phi(z;z_0)\ldots\Phi(z;z_m) + \Psi(z;z_0)\ldots\Psi(z;z_m)]$.

If $z_0 = \ldots = z_m$, one has even recurrence relations

$$Q_0 = 1,\ Q_1 = (z-z_0)\ (w \cos\alpha - i \sin\alpha),$$

$Q_{n+1}(z) = 2(z-z_0)\ (w \cos\alpha - i \sin\alpha)\ Q_n(z) - (z-z_0)^2(\cos 2\alpha - iw \sin 2\alpha)\ Q_{n-1}(z)$, $n > 0$.

Acknowledgements.

I wish to thank prof. J. Karlsson for information about the reference [7].

REFERENCES
++++++++++

[1] L.V. AHLFORS, *Conformal Invariants Topics in Geometric Function Theory*. Mc Graw-Hill, N.Y. 1973.

[2] R.T. BAUMEL, J.L. GAMMEL, J. NUTTALL, Placement of cuts in Padé-like approximation. Preprint Univ. Western Ontario 1980.

[3] C. BREZINSKI, Rational approximation to formal power series. *J. Approx. Theory 25* (1979) 295-317.

[4] C. BREZINSKI, *Padé-Type Approximation and General Orthogonal Polynomials*, ISNM 50, Birkhäuser, Basel 1980.

[5] P.J. DAVIS, *Interpolation and Approximation*, Blaisdell, Waltham 1963 = Dover, N.Y. 1975.

[6] M. GOLOMB, Interpolation operators as optimal recovery schemes for classes of analytic functions, pp. 93-138 in C.A. MICCHELLI, T.J. RIVLIN, editors : *Optimal Estimation in Approximation Theory*, Plenum Press, N.Y. 1977.

[7] A.A. GONČAR, On the convergence of generalized Padé approximants of meromorphic functions, *Math. USSR Sb. 27* (1975) 503-514.

[8] A.S. HOUSEHOLDER, *The Numerical Treatment of a Single Nonlinear Equation*. Mc Graw-Hill, N.Y. 1970.

[9] J. KARLSSON, Singularities of functions determined by the poles of Padé approximants : these Proceedings.

[10] S. KLARSFELD, An improved Padé approximant method for calculating Feynman amplitudes : these Proceedings.

[11] Z. NEHARI, *Conformal Mapping*, Mc Graw-Hill, N.Y. 1952 = Dover, N.Y. 1975.

[12] J. NUTTALL, S.R. SINGH, Orthogonal polynomials and Padé approx--imants associated with a system of arcs, *J. Approx. Th. 21* (1977) 1-42.

[13] J. NUTTALL, The convergence of Padé approximants to functions with branch points, pp. 101-109 in E.B. SAFF, R.S. VARGA, editors : *Padé and Rational Approximation* A.P., N.Y. 1977.

[14] M. PINDOR, Padé approximants and rational functions as tools

for finding poles and zeros of analytical functions measured experimentally pp. 338-351 in L. WUYTACK, editor: *Padé Approximation and its Applications*, Lecture Notes Math. 765, Springer, Berlin 1979.

[15] G. SZEGÖ, *Orthogonal Polynomials*, AMS, Providence, 1939.

[16] H. WALLIN, Rational interpolation to meromorphic functions : these Proceedings.

[17] J.L. WALSH, *Interpolation and Approximation by Rational Functions in the Complex Domain*, AMS, Providence 1935.

Alphonse MAGNUS
Institut Mathématique U.C.L.
Chemin du Cyclotron 2
B-1348 Louvain-la-Neuve
Belgium.

RECURRENCE COEFFICIENTS IN CASE OF ANDERSON LOCALISATION

Alphonse MAGNUS

Summary. Questions related to Stieltjes transforms of jump functions with a dense set of jump points are presented.

1. The Anderson localisation.

Solid state systems can be studied through an element of the resolvent operator (Green function) of their Hamiltonians H :

$$G(z) = (F,(zI-H)^{-1} F), \quad z \notin \sigma, \text{ the spectrum of } H, \tag{1}$$

where F is in a Hilbert space, and is usually represented by a square summable sequence. From the spectral representation of the selfadjoint operator H ([11] chap. 6 and 10), one has

$$G(z) = \int_\sigma (z-x)^{-1}\, dW(x), \quad z \notin \sigma, \tag{2}$$

where W is an increasing function. The nature of W (jumpfunctions, absolutely continuous or singularly continuous) is of interest in problems of diffusion ([9] § 9, [10], [12]). Disordered systems (see [3] for a general presentation) were first studied from this point of view by Anderson [2], who found evidence for the presence of jump-functions only, with a dense set of jumppoints (Anderson localisation) in cases of strong disorder. His method consists in transformation of a divergent series of (1), and a rather free use of Stieltjes inversion formula ([18] § 65), so that his conclusions are closer to a suggestion for further research than a definite result... . Since then, many numerical simulations have been done (a review in [19]), using Hamiltonians represented by infinite symmetric matrices where some elements are randomly distributed. Everybody agrees that many questions remain open.

2. The recursion method.

All the information of W is present in the sequence of the coefficients of the Jacobi continued fraction of (2)

$$G(z) = b_0^2/(z-a_0-b_1^2/(z-a_1-b_2^2/\ldots)) \tag{3}$$

if the moment problem is determinate, which is always true if σ is bounded. Some cases of relations between features of continuous functions W and asymptotic behaviour of the sequences $\{a_n\}$ and $\{b_n\}$ have been discussed ([6], [8'], [13]), but the corresponding methods are unable to handle Anderson localisation in a satisfactory way. The first investigations for disordered systems [7] were concerned with the existence of limits of the sequences $\{a_n\}$ and $\{b_n\}$, and the interpretation of these limits. This line of research is based on perturbations of the constant coefficients case $a_n = a$, $b_n = b$ corresponding to the absolutely continuous spectrum $\sigma = [a-2b, a+2b]$, $dW(x) = c^t\,[(x-a+2b)(a+2b-x)]^{-1/2}\,dx$. The issue is still under discussion ([8], [16]).

3. Functional analysis approach.

From (3), $b_0^{-2}\,G(z)$ appears as the first diagonal element of the resolvent $(zI-T)^{-1}$, where T is an operator represented by a tridiagonal infinite symmetric matrix of diagonal elements a_0, a_1, ... and off-diagonal elements b_1, b_2, ... ([1] § 1.4 and § 4.2, [18] chap. 12). Functional analysis methods can explore some aspects of perturbation, for instance, the stability of the essential spectrum under compact perturbation ([11] chap. 4, theorem 5.35), i.e. the spectra of T' and T" differ only by isolated points if $a'_n - a''_n$ and $b'_n - b''_n \to 0$ when $n \to \infty$. However, the continuous or discontinuous nature of W is much more difficult to describe, if one considers the Weyl-von Neumann theorem ([11] chap. 10 § 2). Other possibilities include approximate eigenvectors and the behaviour of $\|(E+is-T)^{-1}v\|$ for small s ([15] § 11.5).

4. Behaviour of orthogonal polynomials.

A square-summable sequence $\{v_n\}$ is an eigenvector of T corresponding to the eigenvalue x if $(x-a_0)v_0-b_1v_1 = 0$, $-b_nv_{n-1}+(x-a_n)v_n-b_{n+1}v_{n+1} = 0$, $n=1,2,\ldots$, which means that the sequence of the values at x of the orthonormal polynomials $v_n = P_n(x)$ is square-summable. Probabilistic estimates of the behaviour of $P_n(x)$ when $n \to \infty$ are found in [9], [12'] but other approaches should be welcome. Much is done in [4], [5], [14], but does not seem to be easily related to Anderson localisation. Although the following theorem has almost no practical value, the method of proof could perhaps suggest how to use the orthogonal polynomials machinery in more relevant situations ... .

THEOREM. *If* W *is only made of an infinity of jump functions, with a set of jump points dense in some bounded interval, with jumps decreasing faster than the terms of a geometric progression of rate* $q < 1/16$*, at least one of the sequences* $\{a_n\}$*,* $\{b_n\}$ *must diverge.*

Indeed, if $x_1, x_2, \ldots$ are the jump points, ordered such that the corresponding jumps $w_1, w_2, \ldots$ decrease, one considers the monic orthogonal polynomials $\{B_n(x)\}$ satisfying

$$b_0^2 \ldots b_n^2 = \sum_{k=1}^{\infty} w_k (B_n(x_k))^2$$

and such that any other monic polynomial of degree n yields a larger right-hand side ([17] § 3.1) : for instance, $(x-x_1) \ldots (x-x_n)$, which will suppress the larger jumps (and is therefore a rough first estimate of $B_n(x)$ for rapidly decreasing sequences of jumps ...)

$$b_0^2 \ldots b_n^2 \leqslant \sum_{k=n+1}^{\infty} w_k (x_k - x_1)^2 \ldots (x_k - x_n)^2 .$$

From the hypothese on the jumps :

$$\limsup_{n \to \infty} (b_0 \ldots b_n)^{1/n} \leqslant q^{1/2} (\beta - \alpha),$$

where $[\alpha, \beta]$ is the smallest interval containing all the limit points of $\{x_k\}$. Now, if $\{a_n\}$ and $\{b_n\}$ have limits when $n \to \infty$, the essential spectrum must be $[a-2b, a+2b]$, densely filled with eigenvalues, so that $b = (\beta-\alpha)/4$, which is impossible if $q < 1/16$ ... .

Acknowledgements.

Thanks for expert advice and kind appreciation by J.P. Gaspard, R. Haydock, J. Heinrichs, C.H. Hodges and U. Krey are expressed here.

REFERENCES

++++++++++

[1] N.I. AKHIEZER, *The Classical Moment Problem*, Oliver & Boyd, Edinburgh 1965.

[2] P.W. ANDERSON, Absence of diffusion in certain random lattices, *Phys. Rev. 109* (1958) 1492-1505.

[3] R. BALIAN, R. MAYNARD, G. TOULOUSE, editors, *Les Houches Summer School Proceedings : Ill-Condensed Matter*, North-Holland, N.Y. 1979.

[4] T.S. CHIHARA, *An Introduction to Orthogonal Polynomials*, Gordon & Breach, N.Y. 1978.

[5] T.S. CHIHARA, Orthogonal polynomials whose distribution functions have finite point spectra. *SIAM J. Math. Anal. 11* (1980) 358-364.

[6] J.P. GASPARD, F. CYROT-LACKMANN, Density of states from moments. Application to the impurity band. *J. of Physics C : Solid State Phys. 6* (1973) 3077-3096.

[7] R. HAYDOCK, Study of a mobility edge by a new perturbation theory. *Phil. Mag. B 37* (1978) 97-109.

[7'] R. HAYDOCK, The recursive solution of the Schrödinger equation. Solid State Phys. 35 (1980) 215-294.

[8] C.H. HODGES, D. WEAIRE, N. PAPADOPOULOS, The recursion method and Anderson localisation. *J. Phys. C 13* (1980) 4311-4321.

[8'] C.H. HODGES, Van Hove singularities and continued fraction coefficients. *J. Phys. Lett. 38* (1977) L187-L189.

[9] K. ISHII, Localization of eigenstates and transport phenomena in the one-dimensional disordered system. *Suppl. Prog. Theor. Phys. 53* (1973) 77-138.

[10] R. JOHNSTON, Localisation and localisation edges - a precise characterisation. Preprint Blackett Laboratory, Imperial College London 1980.

[11] T. KATO, *Perturbation Theory for Linear Operators*, Springer, Berlin 1966.

[12] J.C. KIMBALL, Localisation and spectra in solid state systems. *J. Phys. C 11* (1978) 4347-4354.

[12'] J. KIMBALL, Two special cases of Anderson localisation. J. Phys.

C 13 (1980) 5701-5708.

[13] Al. MAGNUS, Recurrence coefficients for orthogonal polynomials on connected and non connected sets, pp. 150-171 in L. WUYTACK, editor : *Padé Approximation and its Applications*, Lecture Notes Math. 765, Springer, Berlin 1979.

[14] P.G. NEVAI, *Orthogonal Polynomials*, Memoirs AMS, Providence 1979.

[15] R.D. RICHTMYER, *Principles of Advanced Mathematical Physics*, Springer, N.Y. 1978.

[16] J. STEIN, U. KREY, Numerical studies on the Anderson localization problem. *Z. Physik B 34* (1979) 287-296 ; *37* (1980) 13-22.

[17] G. SZEGÖ, *Orthogonal Polynomials*. AMS Providence 1939.

[18] H.S. WALL, *Analytic Theory of Continued Fractions*, Van Nostrand, Princeton 1948.

[19] D. WEAIRE, B. KRAMER, Numerical methods in the study of the Anderson transition. *J. Non-Crystalline Solids 32* (1979) 131-140.

Alphonse MAGNUS
Institut Mathématique U.C.L.
Chemin du Cyclotron 2
B-1348 Louvain-la-Neuve,
Belgium.

ATOMIC RADIATIVE TRANSITIONS IN STRONG FIELDS VIA PADE APPROXIMANTS

A. MAQUET

Laboratoire de Chimie Physique*
Université Pierre et Marie Curie
11 rue Pierre et Marie Curie
F.75231 Paris Cedex 05 France

I. Introduction.

Most of the applications of Padé Approximants (PA) in theoretical physics are related to the study of the nature and convergence properties of the perturbative expansion. A good illustration of this tendency is the abundant literature devoted, for instance, to the connection between PA and the perturbation theory, applied to the anharmonic oscillator, [1,2] or to the Stark effect in hydrogen [3]. The usefulness of PA for describing the spin $\frac{1}{2}$ Ising model has been also thoroughly investigated [4]. Another area, which seems to have retained less attention but could reveal itself of great practical importance, concerns the evaluation of higher - order perturbative terms in the expression of atomic radiative transition amplitudes. The need for such calculations is a consequence of the advent of powerful laser sources allowing to observe multiphoton processes in the course of which several photons are simultaneously absorbed (or scattered) by an atomic system [5]. We illustrate in Sec II, the advantages of using PA for computing various second- and third-order radiative transition amplitudes in atomic Hydrogen [6].

Section III deals with another domain of interest in collision theory namely the summation of partial-wave expansions. By introducing the so-called Legendre-Padé Approximants [7], it is possible to sum such poorly convergent (or even divergent) series and at the same time to give a firm theoretical basis to the theory, at least in some elementary problems of potential scattering. (See also ref. (8)). In a more pedestrian way I wish to show here that, even when the analytical nature of the series is not known, which is precisely the case in most physical applications, useful numerical results can nevertheless be obtained by merely applying standard convergence acceleration techniques.

Finally I shall present in Sect.IV some results recently derived when considering the interaction of a multilevel atom with a very intense laser field, i.e.,beyond the range of validity of the usual perturbation theory [9]. The generalized transition amplitudes obtained in such a case exhibit a matrix-continued fraction structure, the convergence properties of which remain,to the best of our knowledge, an open problem.

II. 2-and 3-photon Processes in Atomic Hydrogen .

When irradiated by an intense laser light an atomic system may experience several multiphoton processes among which multiphoton ionization is often dominant [5]. Another interesting phenomenon observed in such experiments is the shift (usually referred as either the "light-shift" or the "ac-Stark shift") and broadening of atomic levels, both induced by the laser electromagnetic field [5]. Within the framework of the perturbation theory the description of such processes requires the evaluation of Nth- order transition amplitudes of the following general form :

$$T_{f/i} = \langle f | V \; G \; V \; G \; \ldots \; V \; | i \rangle \quad , \qquad (1)$$

where $|i\rangle$,$|f\rangle$ are respectively the initial and final atomic states; V is the usual reduced dipole interaction operator between the e.m. field and the atomic electrons; G is the resolvent operator :

$$G(E) = (E - H_0)^{-1} = \sum_{\nu} |\nu\rangle\langle\nu| \; (E - E_{\nu})^{-1} \; , \qquad (2)$$

associated to the atomic hamiltonian H_0: $H_0 |\nu\rangle = E_{\nu}|\nu\rangle$. The sum entering Eq.(2) runs over the entire atomic spectrum discrete + continuous : $\sum_{\nu} = \sum_{n} + \int dE \ldots$. G(E) is a meromorphic function of E in the whole complex plane, except for simple poles on the negative real axis and a cut on the positive real axis.

When specialized to the case of a non-relativistic hydrogenic atom undergoing 2- or 3-photon transitions the above amplitude $T_{f/i}$, Eq.(1), may be expressed, after separation of angular factors, in terms of reduced radial amplitudes :

$$\text{2-photon :} \quad T^{(2)}_{n'l',\,nl} = \langle n'l' | \, r \, G_{\lambda}(E) \, r \, | nl \rangle ; \; \lambda = l \pm 1 ; \quad l' = \lambda \pm 1 \qquad (3)$$

$$\text{3-photon :} \quad T^{(3)}_{n'l',\,nl} = \langle n'l' | \, r \, G_{\lambda_1}(E_2) \, r \, G_{\lambda_2}(E_1) \, r \, | nl \rangle ; \; \lambda_1 = l \pm 1 ; \; \lambda_2 = \lambda_1 \pm 1 \; ; \; l' = \lambda_2 \pm 1 . \qquad (3')$$

where $G_{\lambda}(E)$ represents the partial wave component, for angular momentum λ and energy E, of the Coulomb Green's function; $|n,l\rangle$ represents the initial hydrogenic bound state and $|n',l'\rangle$ the final state which may belong either to the discrete spectrum (bound-bound transitions) or to the continuous spectrum (multiphoton ionization). In the latter case the parameter n' becomes imaginary : $n' = (ik)^{-1}$, where $k = |\vec{k}|$ is the modulus of the wave vector of the photoelectron (atomic units are used).

Using the usual eigenfunction expansion, Eq. (2), of the Coulomb Green's function leads to very cumbersome computations owing to the presence of the integral over the continuous spectrum. Instead we found extremely convenient to use the so-called sturmian representation of G (E),[10,11] which enabled us to replace the sum + integral over the physical (discrete + continuous) spectrum by a sum over the (discrete) Sturmian spectrum of Hydrogen [12]. Eventually the amplitudes can be rewritten as series expansions :[6,11]

$$T^{(2)}_{n'l',\,nl} = \sum_{\mu} a_{\mu}\zeta^{\mu} \;;\; \zeta = \frac{1/n-x}{1/n+x}\,\frac{1/n'-x}{1/n'+x}\,,\quad x = \sqrt{-2E} \tag{4}$$

$$T^{(3)}_{n'l',\,nl} = \sum_{\mu_1}\sum_{\mu_2} a_{\mu_1\mu_2}\zeta_1^{\mu_1}\zeta_2^{\mu_2} \;;\; \zeta_1 = \frac{1/n-x_1}{1/n+x_2}\,\frac{x_2-x_1}{x_2+x_1}\,,\; x_1 = \sqrt{-2E_1}\,; \tag{4'}$$

$$\zeta_2 = \frac{x_1-x_2}{x_1+x_2}\,\frac{1/n'-x_2}{1/n'+x_2}\,,\quad x_2 = \sqrt{-2E_2}\,.$$

The coefficients a_{μ} and $a_{\mu_1\mu_2}$ (the explicit expressions of which are too long to be reproduced here) are given in terms of hypergeometric polynomials and thus can be computed within any chosen accuracy. The above simple or double series usually converge fairly well and the computation of approximate values of the sums does not give rise to any particular trouble. There are however important cases of physical interest [6,13] for which the series are divergent. This situation occurs if the energy argument E in the Coulomb Green's function is positive i.e. is located on the cut of G(E). In that case $|\zeta|>1$ and $|\zeta_2|>1$ in Eq. (4) and (4') and the expansions diverge. For extracting nevertheless the (finite!) numerical value of the physical amplitudes from those series we tentatively consider the expression (4) and (4') as representing Taylor expansions in terms of the variables ζ and ζ_2:

$$T^{(2)}_{n'l',\,nl}(\zeta) = \sum_{\mu} a_{\mu}\zeta^{\mu} \;; \tag{5}$$

$$T^{(3)}_{n'l',\,nl}(\zeta_2) = \sum_{\mu_2} b_{\mu_2}\zeta_2^{\mu_2}\,,\quad \text{with } b_{\mu_2} = \sum_{\mu_1} a_{\mu_1\mu_2}\zeta^{\mu_1} \tag{5'}$$

It is then an easy matter to use the partial sums of those divergent series as imput of an accelerating convergence process. We used principaly Wynn's ε- algorithm or its generalization i.e. the cross-rule connecting five adjacent entries in the Padé table [14,15]. Some numerical examples are presented in Table I. It is worth noting that usually the convergence was found along the main diagonal in the Padé table. Note also that in some case of very severe divergence of the series (typically $|u_{n+1}/u_n| \gtrsim 10$) we were compelled to resort to iteration techniques i.e. taking the diagonal sequence of the Padé table as imput of a new accelerating cycle.

Table 1. Typical convergence of the diagonal Padé sequence as compared to the partial sums of the expansion of $T^{(3)}_{n'l',\ nl}$, Eq.(5'); $|\zeta_2| \simeq 2.82$.

p	[p/0]	[p/p]
0	5.99(3) - i 2.05(3)	5.99(3) - i 2.05(3)
1	-2.88(4) + i 2.07(5)	-3.21(3) + i 1.29(3)
2	6.42(7) - i 3.70(7)	8.71(2) - i 4.70(2)
3	8.00(7) + i 1.09(9)	-1.28(2) + i 9.82(1)
.	.	.
.	.	.
.	.	.
10	1.44(15) + i 2.47(15)	8.13(-1) + i 1.10
11	-2.96(16) + i 2.37(15)	8.16(-1) + i 1.11
12	1.09(17) - i 2.81(17)	8.17(-1) + i 1.11
.	.	.
.	.	.
.	.	.
20	-5.11(25) - i 1.80(26)	
21	1.67(27) + i 2.99(26)	
22	8.99(27) + i 1.23(28)	
:		

In most cases the convergence of Padé sequences cannot be assessed on grounds of general theorems. However in the simple case of a two-photon bound-bound amplitude $T^{(2)}_{n'l',\ nl}$, as encountered in light-shift calculations, one can demonstrate the convergence of the diagonal sequences of PA. As a matter of fact the coefficients a_μ contain polynomials of order $p \leqslant (n-1)(n'-1')$ in the parameter μ and after a little algebra, $T^{(2)}$ may be rewritten as a linear combination of sums of the general form :

$$S(p,\ q;\ \alpha\ ,\zeta\ ) = \sum_{\mu=0}^{\infty} \mu^p \frac{(\mu+q)!}{(\mu+\alpha)} \frac{\zeta^\mu}{\mu!} \tag{6}$$

where p and q are positive integers and α and ζ are complex numbers. Such series obey the recurrence relation :

$$S(p,q;\alpha\ ,\zeta\ ) = R\ (p-1,\ q;\ \zeta\ ) - \alpha S(p-1,\ q;\ \alpha,\zeta\ )\ , \tag{7}$$

where $R(r,q;\ \zeta\ ) = \sum_n n^r (n+q)!\ \frac{\zeta^n}{n!}$. (8)

These latter series verify also the recurrence relation:

$$R(r,q;\ \zeta\ ) = R(r-1,q;\ \zeta\ ) - (q+1)\ R(r-1,\ q;\zeta\ )\ , \tag{9}$$

with $R(0,q;\ \zeta\ ) = q!(1-\zeta\ )^{-(q+1)}$. (10)

Thus the series $S(0,q;\alpha,\zeta\ )$ can always be recurrently expressed as a sum of algebraic terms of the form given in Eq.(10) plus one term proportional to :

$$S(0,q;\alpha,\zeta) = \sum_{\mu} \frac{(\mu+q)!}{(\mu+\alpha)} \frac{\zeta^{\mu}}{\mu!} = \frac{q!}{\alpha}\ {}_2F_1(\alpha,\ q+1;\ \alpha+1;\ \zeta)$$

$$= \frac{q!}{\alpha}\ (1-\zeta)^{-q}\ {}_2F_1\ (1,\alpha-q;\ \alpha+1;\ \zeta)\ . \qquad (11)$$

As a consequence the whole amplitude $T^{(2)}_{n'l',\ nl}$ may be reexpressed again as a sum of algebraic terms plus a term proportional to a Gauss hypergeometric function ${}_2F_1(1,\ b;\ c;\ \zeta)$which has a continued fraction expansion converging in the whole complex plane excepted on the cut $(1,\ +\infty)$ [16]. This result explains the observed good convergence of the diagonal sequences of PA. It should be kept in mind, however, that this proof holds only for second-order bound-bound transitions and that numerical convergence is the only test we have for assessing the validity of our procedure when applied to more sophisticated calculations. Nevertheless the excellent agreement between the PA results and those obtained by other methods, whenever available, lends further support to it.

II.Partial-wave expansions.

The use of PA, or their generalization the so-called Legendre PA, for summing poorly convergent or even divergent partial-wave expansions has recently received considerable attention [7,8]. Broadly speaking two classes of problems may be distinguished according to whether the analytical structure of the terms of the expansion is simple or not. For instance,those terms may be evaluated in closed form in many test cases of potential scattering. On the contrary, in most cases of realistic collisions problems, involving atomic or molecular targets, the coefficients of the expansion are determined only numerically. We present hereafter two numerical investigations corresponding respectively to these two classes.

II-A.Coulomb scattering.

The exact Coulomb scattering amplitude has been determined in closed form by Gordon [17]:

$$f_c(\theta,k) = -\gamma \exp\ (-2i\gamma\ \mathrm{Log}(\mathrm{Sin}\,\theta/2) + 2i\sigma_0)/(2k\ \sin^2(\theta/2)) \qquad (12)$$

where θ is the scattering angle; $\gamma = Z_1Z_2/k$; Z_1,Z_2 are the charges of the incident particle and of the target, k is the modulus of the wave vector of the scattered particle and σ_0 is the Coulomb phase shift: $\sigma_0 = \mathrm{Arg}\,\Gamma(1+i\gamma)$. Note the well known divergence of $f_c(\theta,k)$ as $\theta \rightarrow 0$, which is a direct consequence of the infinite range of the Coulomb potential.

The corresponding partial-wave expansion :[18]

$$f(\theta) = \sum_l (2l+1)\, f_l(\theta)\, P_l\,(\cos\theta) \qquad (13)$$

with : $f_l(\theta) = (\exp\,(2i\,\sigma_l)\ -1)/2\ ik$; $\exp\,(2i\,\sigma_l) = \Gamma(l+1+i\gamma)/\Gamma(l+1-i\gamma)$, has been thoroughly used, in particular in calculations related to the Coulomb excitation of nuclei. It should be stressed, however that Eqs. (12) and (13) are not equivalent since : i) the series (13) is divergent; ii) the function $f_c(\theta)$, Eq.(12), being singular at the origin cannot be expanded in a Legendre series. It has been shown, nevertheless that the equivalence of Eqs. (12) and (13) holds for distributions [19]. Moreover it has been demonstrated that the expansion (13) is Borel summable, [20] and Legendre-Padé summable [7]. In a more pedestrian approach we have shown that by merely accelerating the convergence of the series (13) on using Wynn's ε-algorithm , one can recover numerically Gordon's exact value. We present below (Fig.1) a geometrical comparison between the sequence of partial sums in Eq.(13) and the sequence $\varepsilon_{2n}^{(0)}$ in the corresponding ε-algorithm.

Fig.1. Complex plane representation of the convergence of Wynn's ε-algorithm as compared to the partial sums : $S_n = \sum_l^n (2l+1)\ f_l(\theta) P_l(\cos\theta)$ for Coulomb scattering. Dots correspond to $Re(S_n) + i\ Im(S_n)$. Circles correspond to $Re(\varepsilon_{2n}^{(0)}) + i\ Im(\varepsilon_{2n}^{(0)})$. The cross corresponds to Gordon's exact value. $\theta = 60°, \gamma = -1$, k = 1 (atomic units).

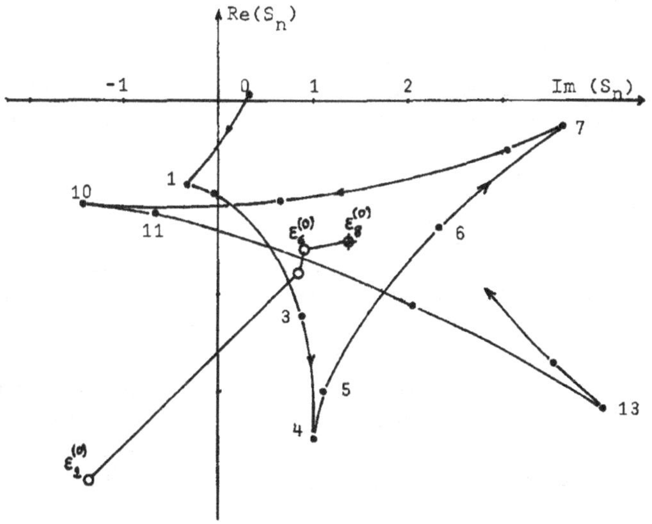

We are confident that the same procedure could be successfully applied to a very large class of potential scattering problems. See also Ref. (8).

II-B. Bremsstrahlung in a Coulomb field

We have been recently interested in interpreting experiments on multiphoton free-free transitions of a charged particle moving in the field of an ion, in the presence of a laser field [21]. As a first step towards achieving this purpose we have investigated the feasibility of expanding the transition amplitude in partial waves and we tested the method in the simpler but representative case of the bremsstrahlung in a Coulomb field.

Any charged particle scattered in a Coulomb field may simultaneously radiate or absorb a photon (Bremsstrahlung or free-free transitions). In the dipole approximation, the exact non-relativistic quantum mechanical calculation of the corresponding cross section for emission of a photon of energy within ω and $\omega+d\omega$, has been performed by Sommerfeld [22]. When specialized to the case of an electron in the field of a proton one has :

$$\frac{d\sigma}{d\omega} = \frac{32}{3}\,\alpha^3\,\frac{1}{k_1^3\;k_2\,\omega}\,b_o \tag{14}$$

where $\alpha = 1/137$ is the fine structure constant, k_1 and k_2 are the moduli of the wave vectors of respectively the incident and scattered electron and :

$$b_o = x_o\,\frac{d}{dx_o}\left[{}_2F_1\left(\frac{i}{k_1},\frac{i}{k_2};1;x_o\right)\right]^2 \;;\; x_o = -4\,\frac{k_1k_2}{(k_1-k_2)^2} \tag{15}$$

Independently, by using a partial-wave expansion of the Coulomb wave functions one can get another useful expression of b_o:

$$b_o = \sum\left\{l\left[T_{l,l-1}\right]^2 + (l+1)\left[T_{l,\,l+1}\right]^2\right\} \tag{16}$$

where $T_{l,\,l\pm1} = \int dr\, F_l\,(k_1r)\,F_{l\pm1}\,(k_2r)$,

and $F_l(kr)$ are radial wave functions for angular momentum l. The equivalence of formulae Eqs.(15) and (16) has been demonstrated explicitely by Biedenharn [23]. Nevertheless it is of interest to determine the convergence properties of the series (16) which provides useful informations for the bremsstrahlung and related problems. Although the matrix elements $T_{l,\,l\pm1}$, can be evaluated in closed form in terms of Gauss hypergeometric functions their structure is very intricate. Again the ε-algorithm proved very helpful for accelerating the convergence of this series giving us some confidence in the tractability of partial-wave expansions in the case of free-free transitions.

Table II. Convergence of the ε-algorithm as compared to the partial sums $S_n = \sum_1^n d\sigma_1/d\omega$ for Bremsstrahlung. Electron - proton collision; $k_1^2/2 = 4\text{eV}$; $\omega = 2\text{eV}$.

n	S_n	$\varepsilon_{2n}^{(0)}$
1	.606(-48)	.584(-47)
2	.131(-47)	.289(-47)
3	.193(-47)	.291(-47) exact.
.	.	.
.	.	.
.	.	
6	.275(-47)	
.	.	
.	.	
.	.	
9	.289(-47)	
.	.	
.	.	

IV Multilevel atoms in very intense fields

The calculation of generalized transition amplitudes for an atom irradiated by a very intense laser beam requires the knowledge of matrix elements of the resolvent $(E-H)^{-1}$ where the total Hamiltonian $H = H_{atom} + H_{field} + V$ and V is the interaction operator between the atom and the field. This resolvent is usually expanded in a Born series :

$$(E-H)^{-1} = G_0 + G_oVG_o + G_oVG_oVG_o + \ldots\ldots , \qquad (17)$$

where $G_0 = (E-H_{atom} - H_{rad})^{-1}$, but for higher intensities i.e., for higher values of the coupling constant included in V, the computation becomes rapidly unpractical, not to mention the fundamental question of the nature of the series so obtained. We present here an alternative matrix continued fraction expansion which should not suffer from such a drawback.

In the simplified case of an isolated atom interacting with an intense, coherent, single-mode field, the total Hamiltonian operator (E-H) may be straightforwardly represented by an infinite Hermitian matrix in the basis of the uncoupled atom-field states $|\{a\}, N\rangle = |\{a\}\rangle \otimes |N\rangle$. Here $\{a\}$ corresponds to the complete manifold of atomic states and N is the occupation number of the field-mode considered. Since the interaction operator V couples field states with occupation numbers differing only by one unit the matrix (E-H) exhibits the following tridiagonal block-structure :

$$
[E-H] = \begin{array}{c} \vdots \\ |\{a\},N-2\rangle \\ |\{a\},N-1\rangle \\ |\{a\},N\rangle \\ |\{a\},N+1\rangle \\ |\{a\},N+2\rangle \\ \vdots \end{array}
\overset{\cdots\ |\{a\},N-2\rangle\ |\{a\},N-1\rangle\ |\{a\},N\rangle\ |\{a\},N+1\rangle\ |\{a\},N+2\rangle \cdots}{\left[\begin{array}{ccccccc}
\ddots & \ddots & & & & & \\
\ddots & \boxed{N-2} & V_{-2,-1} & & & 0 & \\
 & V_{-1-2} & \boxed{N-1} & V_{-1,0} & & & \\
 & & V_{0,-1} & \boxed{N} & V_{0,1} & & \\
 & & & V_{1,0} & \boxed{N+1} & V_{1,2} & \\
 & 0 & & & V_{2,1} & \boxed{N+2} & \ddots \\
 & & & & & \ddots & \ddots
\end{array}\right]} \qquad (18)
$$

Here the diagonal blocks $\boxed{N+M}$ are diagonal matrices

$$
\boxed{N+M} = \begin{array}{c} |a_1,N+M\rangle \\ |a_2,N+M\rangle \\ |a_3,N+M\rangle \\ \vdots \end{array}
\overset{|a_1,N+M\rangle\ |a_2,N+M\rangle\ |a_3,N+M\rangle \cdots}{\left[\begin{array}{cccc}
E-E_1-(N+M)\omega & & 0 & \\
 & E-E_2-(N+M)\omega & & \\
0 & & E-E_3-(N+M)\omega & \\
 & & & \ddots
\end{array}\right]} , \qquad (19)
$$

where E_i are eigenenergies of the atomic state $|a_i\rangle$: $H_{atom}|a_i\rangle = E_i|a_i\rangle$ and ω is the photon energy. The elements $(V_{M,M\pm1})_{ij}$ of the blocks $\boxed{V_{M,M\pm1}}$ are assumed real and satisfy the symmetry relation :

$$
(V_{M,M\pm1})_{ij} = \langle a_i, N+M|V|a_j, N+M\pm1\rangle = (V_{M\pm1,M})_{j,i} \qquad (20)
$$

thus : $[V_{M\pm1,M}] = [V_{M,M\pm1}]^T$.

In an actual computation one has to determine only one element in the inverse matrix $[E-H]^{-1}$. The calculation can be performed by using standard numerical inversion of the truncated matrix, the convergence being checked with respect to the number of blocks retained in the truncation. Note however that as more blocks and more atomic states are included in the computation the size of the matrix increases very rapidly which makes the practical calculation very cumbersome. The method outlined below should not present such difficulties.

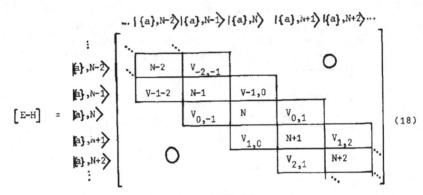

For the sake of illustration we present here the calculation for an elastic transition i.e. one involving the same initial and final state $|a,N\rangle$. The corresponding amplitude is of interest for describing forward- scattering processes and may be used in light - shift calculations. Generalization to inelastic multiphoton transitions is easily derived [9].

The first step consists of partitioning the Hamiltonian matrix $[E-H]$ as follows :

$$[E-H] = \begin{array}{|c|c|c|} \hline -1,-1 & -1,0 & 0 \\ \hline 0,-1 & N & 0,1 \\ \hline 0 & 1,0 & 1,1 \\ \hline \end{array} \quad (21)$$

where $[N]$ is the diagonal block Eq.(19) and the semi-infinite blocks $\boxed{1,1}$ and $\boxed{0,1}$ have the following form :

$$\boxed{1,1} = \begin{array}{|c|c|c|c} \hline N+1 & V_{I,2} & 0 & \\ \hline V_{2,1} & N+2 & V_{2,3} & \\ \hline 0 & V_{3,2} & N+3 & \ddots \\ \hline & & \ddots & \ddots \end{array} \quad ;$$

$$\boxed{0,1} = \begin{array}{|c|c|c|c} \hline V_{0,1} & 0 & 0 & \cdots \\ \hline \end{array} \quad . \quad (22)$$

The other blocks have symmetrical structures. The leading diagonal block $([E-H]^{-1})_{0,0}$ of the inverse matrix may be obtained straightforwardly and one has :

$$([E-H]^{-1})_{0,0} = \{N - [0,1][1,1]^{-1}[1,0] - [0,-1][-1,-1]^{-1}[-1,0]\}^{-1} \quad (23)$$

The next step is to determine the inverse $[1,1]^{-1}$ of the semi-infinite tridiagonal block matrix Eq.(22). Here also in the inverse matrix we only need to evaluate the leading block which can be expanded as a matrix continued (J-)Fraction. This may be demonstrated by partitioning again the matrix $[1,1]$ or more directly by generalizing to the matrix case the standard results on the inversion of J-matrices [16].

One gets finally :

$$[V_{0,1}][1,1]^{-1}[V_{1,0}] = [V_{0,1}] \cfrac{1}{[N+1] - [V_{1,2}] \cfrac{1}{[N+2] - [V_{2,3}] \cfrac{1}{[N+3] - \ldots} [V_{3,2}]} [V_{2,1}]} [V_{1,0}] \quad (24)$$

Such a continued fraction contains a lot of informations of physical interest. For instance the location of its poles in the complex plane furnishes useful data on the shift and broadening of the energy levels of the atomic system [9]. When specialized to the simplified case of a two-level atom, the matrices involved here reduce to scalar quantities and one recovers the usual continued fraction expansion of the amplitude which has been thoroughly investigated in the context of the dressed atom theory relevant,for instance,in laser theory [24]. In the case of a more realistic atomic model including an infinite discrete spectrum and one or several coupled continua (one has then to resort to L^2 discretization techniques) the structure of Eq.(23) becomes extremely intricated. It appears however than the Matrix Continued Fraction Eq.(23) truncated at a given stage furnishes a much better approximation than the corresponding perturbative expansion Eq.(17) truncated at the same order. As a matter of fact one can show that $([E-H]^{-1})_{0,0}$ Eq.(23), after inserting truncated continued fractions, represents the sum of an infinite number of terms in the perturbation expansion. The truncation corresponds then to a limitation on the maximum number of photons the atom can exchange with the field.

Aknowledgments - The results reported in Sec II have been obtained in collaboration with S.Klarsfeld and J. Bastian. The work described in Secs.III and IV has been done in collaboration with W.P. Reinhardt during a stay at the Joint Institute for Laboratory Astrophysics, University of Colorado and National Bureau of Standards, Boulder Co. 80309 , the support of which is gratefully aknowledged.

* Laboratoire Associé au C.N.R.S.

1 A.T. Amos, J. Phys.B 11, 2053 (1978) and references therein. A recent survey of Physical Applications may be found in "Padé and Rational Approximation", E.B. Saff and R.S. Varga, Editors, (Academic Press, New York, 1977).

2 B. Simon, Ann. Phys. (N.Y.) 58, 76 (1970).

3 H.J. Silverstone and P.M. Koch, J, Phys. B 12, L537 (1979).

4 D. Bessis, P. Moussa and G. Turchetti, J. Phys. A 13, 2763 (1980) and references therein.

5 P. Lambropoulos in "Advances in Atomic and Molecular Physics", edited by D.R. Bates and B. Bederson (Academic Press, New York, 1976) Vol. 12, p.87.

6 S. Klarsfeld and A. Maquet, Phys.Lett. 78A, 40 (1980) and refe-

rences therein.

7 A.K. Common and T. Stacey, J. Phys. A 11, 259 (1978); ibid. 275 (1978); A.K. Common, J. Phys. A 12, 1399 (1979); ibid. 2563 (1979). See also : J.Fleischer, Nucl. Phys. B 37, 59 (1972); J. Math. Phys. 14, 246 (1973).

8 C.R. Garibotti and F.F. Grinstein, J. Math. Phys. 19, 821 (1978); ibid. 2405 (1978); ibid. 20, 141 (1979).

9 A. Maquet, S.I. Shu and W.P. Reinhardt, submitted to Phys. Rev.A.

10 J. Schwinger, J. Math. Phys. 5, 1606 (1964); L. Hostler, J. Math. Phys. 11, 2966 (1970).

11 A. Maquet, Phys. Rev. A 15, 1088 (1977).

12 M. Rotenberg, Ann. Phys. (N.Y.) 19, 262 (1962); in "Advances in Atomic and Molecular Physics", Edited by D.R. Bates and I. Estermann (Academic Press, N.Y.1970) Vol. 6.

13 P. Agostini, F Fabre, G. Mainfray, G. Petite and N.K. Rahman, Phys. Rev. Lett. 42, 1127 (1979).

14 G.A. Baker Jr.,"Essentials of Padé Approximants" (Academic Press, New-York,1975).

15 C. Brezinski,"Accélération de la Convergence en Analyse Numérique" (Springer Verlag, Berlin, 1977).

16 H.S. Wall,"Analytic Theory of Continued Fractions" (Chelsea, Bronx N.Y., 1948).

17 W.Gordon, Z. Phys. 48, 180 (1928).

18 N.F. Mott and H.S.W. Massey,"The Theory of Atomic Collisions" (Oxford, London, 1965).

19 J.R.Taylor, Nuovo Cim. 23B, 313 (1974).

20 P. Hillion, J. Math. Phys. 16, 1920 (1975).

21 M. Gavrila and M. Van der Wiel, Comments on Atomic and Molecular Physics (1978).

22 A.J.F. Sommerfeld, "Atombau und Spektrallinien"(Ungar, New York, 1953). Vol.2.

23 L.C. Biedenharn, Phys. Rev. 102, 262, (1956).

24 S. Stenholm, Phys. Reports, 6, 1(1973); F.T. Hioe and E.W. Montroll, J. Math. Phys. 16, 1259 (1975).

On Two General Algorithms for Extrapolation with Applications to Numerical Differentiation and Integration

G. Mühlbach
University of Hannover
Hannover, Fed. Rep. of Germany

Dedicated to Prof. Dr. H. Tietz on the occasion of his 60th birthday

0. Summary

In this note we will discuss generalizations of the well known algorithms for extrapolation by algebraic polynomials due to NEVILLE-AITKEN and NEWTON. There are two steps of generalization. The first one applies to the problem of extrapolation of functions by linear combinations of functions forming a complete ČEBYŠEV-system, for instance consisting of rational functions with prescribed poles. Giving simple proofs by induction we will also show how one algorithm can be derived from the other. Nevertheless, numerically the algorithms are not equivalent. In a second step of generalization both algorithms will be extended to apply to the general problem of finite linear interpolation as well. In their most general form they may be used for solving recursively various problems of extra- or interpolation, in particular problems of accelerating convergence, problems of HERMITE-BIRKHOFF-interpolation, of generalized orthogonal polynomials and related problems. Also, the algorithms provide a convenient tool in establishing formulas for numerical differentiation and integration. In some cases of numerical differentiation the generalized NEVILLE-AITKEN algorithm is usefull in getting error estimates too. Some examples are given.

1. The Generalized NEVILLE-AITKEN Algorithm (GNA-algorithm)

We will assume throughout this paper that $\mathbb{K}$ is a commutative field of characteristic zero. By N we denote a nonnegative integer. Let G be an arbitrary set of cardinality N+1 at least. A (N+1)-tupel $(f_o,\dots,f_N)$ of functions $f_j: G \to \mathbb{K}$ will be called a ČEBYŠEV-system on G iff the generalized VAN-DER-MONDE determinant

$$(1.1)\quad V\binom{f_o,\dots,f_N}{x_o,\dots,x_N} := \det f_j(x_i) \neq 0$$

never vanishes whenever $x_o,\dots,x_N$ are pairwise distinct points of G. A ČEBYŠEV-system $(f_o,\dots,f_N)$ on G will be called complete iff for each $k = N(-1)0$ $(f_o,\dots,f_k)$ is a ČEBYŠEV-system on G.

Assume that $(f_o,\dots,f_N)$ is a complete ČEBYŠEV-system on G and that

$x, x_o, \ldots, x_N$ are pairwise distinct points of G. Given any function $f: G \rightsquigarrow \mathbb{K}$ we are looking for algorithms computing the "complete triangular scheme" of values at x

(1.2)
$$\begin{array}{llll} P_o^o & & & \\ & P_1^o & & \\ P_o^1 & & \ddots & \\ & P_1^1 & & P_N^o, \\ P_o^2 & \vdots & \cdot & \\ \vdots & P_1^{N-1} & & \\ P_o^N & & & \end{array} \qquad P_k^n = p_k^n[f](x) \quad (k=0(1)N), n=0(1)N-k)$$

where $p_k^n[f]$ is the unique linear combination of $f_o, \ldots, f_k$ that satisfies the interpolation conditions

(1.3) $p_k^n[f](x_i) = f(x_i) \quad (i=n(1)n+k)$.

We use the notations of [2] slightly modified which in fact are simpler then those originally introduced by the author [6].

Of course, one way of computing the P_k^n consists in solving first the systems of linear equations for the unknown coefficients associated with (1.3) and then computing P_k^n. An easy calculation shows that this requires $(1/60)N^5+O(N^4)$ point-operations. In contrast, computing the triangular field (1.2) by the GNA-algorithm (1.4) takes only $(2/3)N^3+O(N^2)$ such operations.

(1.4) <u>Theorem:</u> The algorithm

→compute for j=1(1)N

  →compute for n=0(1)N

$$F_{o,j}^n = f_j(x) - f_j(x_n) \cdot \frac{f_o(x)}{f_o(x_n)}$$

  →compute for k=1(1)j-1

    →compute for n=0(1)N-k

$$F_{k,j}^n = \frac{F_{k-1,k}^{n+1} \cdot F_{k-1,j}^n - F_{k-1,k}^n \cdot F_{k-1,j}^{n+1}}{F_{k-1,k}^{n+1} - F_{k-1,k}^n}$$

→compute for n=0(1)N

$$P_o^n = f(x_n) \cdot \frac{f_o(x)}{f_o(x_n)}$$

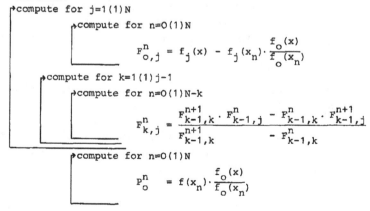

```
→compute for k=1(1)N
    →compute for n=0(1)N-k
```

$$p_k^n = \frac{F_{k-1,k}^{n+1}\cdot p_{k-1}^n - F_{k-1,k}^n\cdot p_{k-1}^{n+1}}{F_{k-1,k}^{n+1} - F_{k-1,k}^n}$$

computes the whole triangular field (1.2).

Proof by induction on N. For N=0 the GNA-algorithm gives $p_0^0 = f(x_0)\cdot f_0(x)/f_0(x_0)$ which evidently is the value at x of the unique multiple of f_0 that agrees with f at the point x_0. Assume now that the theorem is proved for all complete ČEBYŠEV-systems consisting of $M \leq N$ functions and all systems of M pairwise distinct points of G ("knots"), all $f:G \to \mathbb{K}$ and all $x \in G$ distinct from the knots. Then it remains to prove that for every complete ČEBYŠEV-system $(f_0,\ldots,f_N)$ on G consisting of N+1 functions and every system $(x_0,\ldots,x_N)$ of N+1 pairwise distinct points of G, every function $f:G \to \mathbb{K}$ and every $x \in G\setminus\{x_0,\ldots,x_N\}$ (actually, we can allow $x \in G\setminus\{x_1,\ldots,x_{N-1}\}$ assuming a fixed order of the knots or, equivalently, of the rows of the scheme (1.2))

(1.5) $\quad p_N^0 = p_N^0[f](x)$ where

$$p_N^0 = \frac{F_{N-1,N}^1\cdot p_{N-1}^0 - F_{N-1,N}^0\cdot p_{N-1}^1}{F_{N-1,N}^1 - F_{N-1,N}^0}$$

is the value computed by the algorithm, a weighted average of p_{N-1}^0 and p_{N-1}^1 with weights adding to one, and where as before $p_N^0[f]$ denotes the unique linear combination of $f_0,\ldots,f_N$ that agrees with f at $x_0,\ldots,x_N$. In fact, computing any other element of the triangular field (1.2) involves not more than N functions and interpolation points. Hence it is covered by the induction hypotheses. Before proving (1.5) we have to interpret the $F_{k,j}^n$. For $x \in G\setminus\{x_1,\ldots,x_{N-1}\}$, j=1(1)N, k=0(1)j-1 and n=0(1)N-k

(1.6) $\quad F_{k,j}^n = f_j(x) - p_k^n[f_j](x)$

is the extrapolation error at x when f_j is interpolated by that linear combination $p_k^n[f_j]$ of $f_0,\ldots,f_k$ which agrees with f_j at $x_n, x_{n+1},\ldots,x_{n+k}$. This is proved by induction on k. For k=0 (1.6) is immediate from the definition. Let $x \in G\setminus\{x_1,\ldots,x_{N-1}\}$ and $j \in \{1,\ldots,N\}$ be arbitrary but fixed. Assuming (1.6) to hold when k is replaced by k-1 then for the induction step we have to show that for n=0(1)N-k

$$(1.7)\quad \frac{F^{n+1}_{k-1,k}\cdot F^{n}_{k-1,j} - F^{n}_{k-1,k}\cdot F^{n+1}_{k-1,j}}{F^{n+1}_{k-1,k} - F^{n}_{k-1,k}} = f_j(x) - p^n_k[f_j](x).$$

To see this observe that $k < N$ and by induction hypotheses with respect to N, applied to $f = f_j$, we have

$$p^n_k[f_j](x) = \frac{F^{n+1}_{k-1,k}\cdot p^n_{k-1}[f_j](x) - F^{n}_{k-1,k}\cdot p^{n+1}_{k-1}[f_j](x)}{F^{n+1}_{k-1,k} - F^{n}_{k-1,k}}.$$

Subtracting this from $f_j(x)$ yields

$$f_j(x)-p^n_k[f_j](x) = \frac{F^{n+1}_{k-1,k}\cdot(f_j(x)-p^n_{k-1}[f_j](x)) - F^{n}_{k-1,k}\cdot(f_j(x)-p^{n+1}_{k-1}[f_j](x))}{F^{n+1}_{k-1,k} - F^{n}_{k-1,k}}$$

since the weights add to one. An application of the induction hypotheses with respect to k finally proves (1.7). Now we can prove (1.5). Define

$$q^o_N[f](x) = \frac{F^1_{N-1,N}\cdot P^o_{N-1} - F^o_{N-1,N}\cdot P^1_{N-1}}{F^1_{N-1,N} - F^o_{N-1,N}}$$

to be the left hand side of (1.5). We have to show first that $q^o_N[f](x)$ is well defined i.e. that the denominator does not vanish. When considered as a function of $x \in G$ according to (1.7) the denominator $F^1_{N-1,N} - F^o_{N-1,N} = p^o_{N-1}[f_N](x) - p^1_{N-1}[f_N](x)$ is a linear combination of $f_o,\ldots,f_{N-1}$ which has the zeros $x_1,\ldots,x_{N-1}$. Since $f_o,\ldots,f_{N-1}$ is a ČEBYŠEV-system on G it has no other zeros in G or is the zero function. But the latter would mean that f_N when reduced to $\{x_o,\ldots,x_N\}$ is a linear combination of $f_o,\ldots,f_{N-1}$ contradicting the fact that the ČEBYŠEV-system $(f_o,\ldots,f_N)$ must be linearly independent over any subset of G of cardinality N+1 at least. It remains to prove $q^o_N[f](x)=p^o_N[f](x)$. To do this note that $\mathbb{K}^G \ni f \mapsto Rf := q^o_N[f](x) - p^o_N[f](x)$ defines a linear functional R which is a linear combination of the evaluation functionals $L_i f = f(x_i)$ $(i=0(1)N)$. We claim that $Rf_j = 0$ $(j=0(1)N)$. Once this is proved from (1.1) we infer that R is the zero functional. Now, making use of the induction hypotheses with respect to N for $j=0(1)N-1$ $Rf_j = 0$ is obvious from

$$q^o_N[f_j](x) = \frac{F^1_{N-1,N}p^o_{N-1}[f_j](x) - F^o_{N-1,N}p^1_{N-1}[f_j](x)}{F^1_{N-1,N} - F^o_{N-1,N}} = f_j(x) = p^o_N[f_j](x).$$

Remembering that $F^1_{N-1,N}$ and $F^o_{N-1,N}$ are extrapolation errors we get

$$q_N^0[f_N](x)=\frac{(f_N(x)-p_{N-1}^1[f_N](x))\cdot p_{N-1}^0[f_N](x)-(f_N(x)-p_{N-1}^0[f_N](x))\cdot p_{N-1}^1[f_N](x)}{(f_N(x)-p_{N-1}^1[f_N](x))\qquad\qquad -(f_N(x)-p_{N-1}^0[f_N](x))}$$

for j=N. Subtracting this from $p_N^0[f_N](x)=f_N(x)$ finally gives $Rf_N=0$.

2. <u>The Generalized NEWTON Algorithm (GN-algorithm)</u> computes the complete triangular field (1.2) via the associated field of main coefficients $a_k^n[f]$ of $p_k^n[f]$ which usually are called divided differences. To be more specific assume that $(f_0,\ldots,f_N)$ is a complete ČEBYŠEV-system on G. Let $x_0,\ldots,x_N\in G$ be pairwise distinct points and let $f:G\to\mathbb{K}$ be a given function. Then the coefficient $a_k^n = a_k^n[f]$ before f_k in $p_k^n[f]$ will be called <u>divided difference</u> of f with respect to the functions $f_0,\ldots,f_k$ and the knots $x_n,\ldots,x_{n+k}$.

(2.1) Theorem: The algorithm

(i) → compute for j=1(1)N

  → compute for n=0(1)N

$$a_{0,j}^n = \frac{f_j(x_n)}{f_0(x_n)}$$

→ compute for j=2(1)N

  → compute for k=1(1)j-1

    → compute for n=0(1)N-k

$$a_{k,j}^n = \frac{a_{k-1,j}^n - a_{k-1,j}^{n+1}}{a_{k-1,k}^n - a_{k-1,k}^{n+1}}$$

→ compute for n=0(1)N

$$a_0^n = \frac{f(x_n)}{f_0(x_n)} .$$

→ compute for k=0(1)N

  → compute for n=0(1)N-k

$$a_k^n = \frac{a_{k-1}^n - a_{k-1}^{n+1}}{a_{k-1,k}^n - a_{k-1,k}^{n+1}}$$

computes the divided difference tables

$$a_k^n = a_k^n[f] \qquad (k=0(1)N;\quad n=0(1)N-k)$$

$$a_{k,j}^n = a_k^n[f_j] \qquad (j=1(1)N;\quad k=0(1)j-1;\quad n=0(1)N-k).$$

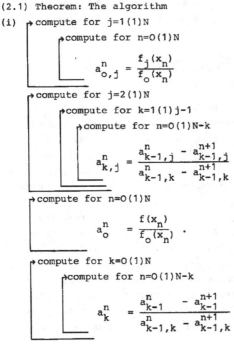

(ii) The algorithm

→ compute for n=0(1)N

$$r_0^n = f_0$$

$$p_0^n = a_0^n \cdot f_0$$

→ compute for j=1(1)N

→ compute for n=0(1)N-j

$$r_j^n = f_j - \sum_{k=0}^{j-1} a_{k,j}^n \cdot r_k^n$$

$$p_j^n = p_{j-1}^n + a_j^n \cdot r_j^n$$

computes the triangular fields

$$r_j^n = f_j - p_{j-1}^n[f_j] \qquad (j=0(1)N;\ n=0(1)N-j)$$

$$p_j^n = p_j^n[f]$$

of certain extrapolation error functions resp. of that linear combinations p_j^n of $f_0,\ldots,f_j$ which do agree with f at $x_n,\ldots,x_{n+j}$.

1. Proof by induction on N. For N=0 the assertion is trivially true. Assume now that the theorem is proved for all complete ČEBYŠEV-systems consisting of $M \leq N$ functions and all systems of M pairwise distinct points of G and all $f:G \longrightarrow \mathbb{K}$. It remains to prove that

$$p_N^0 := p_{N-1}^0 + a_N^0\, r_N^0 = p_{N-1}^0[f] + \frac{a_{N-1}^0[f] - a_{N-1}^1[f]}{a_{N-1}^0[f_N] - a_{N-1}^1[f_N]} \cdot (f_N - p_{N-1}^0[f_N])$$

produced by the algorithm is the unique linear combination $p_N^0[f]$ of $f_0,\ldots,f_N$ which agrees with f at $x_0,\ldots,x_N$. We remark that $a_{N-1}^0[f_N] - a_{N-1}^1[f_N] \neq 0$. For N=1 this follows from (1.1) and for $N \geq 2$ this is proved indirectly. If this difference would be zero then $p_{N-1}^0[f_N] - p_{N-1}^1[f_N] \in \operatorname{span}\{f_0,\ldots,f_{N-2}\}$ must be the zero function having N-1 different zeros $x_1,\ldots,x_{N-1}$. This would mean that f_N when restricted to $\{x_0,\ldots,x_N\}$ is a linear combination of $f_0,\ldots,f_{N-2}$, a contradiction.

Let $x \in G$ be arbitrary but fixed. Considering the linear functional

$$\mathbb{K}^G \ni f \overset{R}{\longmapsto} p_{N-1}^0[f](x) + \frac{a_{N-1}^0[f] - a_{N-1}^1[f]}{a_{N-1}^0[f_N] - a_{N-1}^1[f_N]} \cdot (f_N(x) - p_{N-1}^0[f_N](x)) - p_N^0[f](x)$$

it is easily seen that R is a linear combination of the evalution

functionals $L_i f = f(x_i)$ $(i=0(1)N)$ satisfying $Rf_j = 0$ $(j=0(1)N)$.
Consequently, R is the zero functional.

2. Proof of theorem (2.1): GNA $\Rightarrow$ GN. Let $x \in G\backslash\{x_1,\ldots,x_{N-1}\}$ be arbitrary. From theorem (1.4) we infer

$$p_N^0[f](x) = \frac{\begin{vmatrix} r_N^1(x) \cdot p_{N-1}^0[f](x) - r_N^0(x) \cdot p_{N-1}^1[f](x) \end{vmatrix}}{\begin{vmatrix} r_N^1(x) & -r_N^0(x) \end{vmatrix}} =$$

$$p_{N-1}^0[f](x) + r_N^0(x) \cdot \frac{p_{N-1}^0[f](x) - p_{N-1}^1[f](x)}{p_{N-1}^0[f_N](x) - p_{N-1}^1[f_N](x)} =: p_{N-1}^0[f](x) + h(x).$$

When considered as a function of $x \in G$ h must be a linear combination of $f_0,\ldots,f_N$ with main coefficient $a_N^0[f]$ vanishing at $x_0,\ldots,x_{N-1}$. Hence for all $x \in G$ $h(x) = a_N^0[f] \cdot r_{N-1}^0(x)$ and for $x \in G\backslash\{x_1,\ldots,x_{N-1}\}$

$$\frac{p_{N-1}^0[f](x) - p_{N-1}^1[f](x)}{p_{N-1}^0[f_N](x) - p_{N-1}^1[f_N](x)} = a_N^0[f] = \text{const.} \tag{2.2}$$

Comparing coefficients yields the recurrence formula

$$a_N^0[f] = \frac{a_{N-1}^0[f] - a_{N-1}^1[f]}{a_{N-1,N}^0 - a_{N-1,N}^1}.$$

Proof of GN $\Rightarrow$ GNA:
For all $x \in G\backslash\{x_1,\ldots,x_{N-1}\}$

$$p_N^0[f](x) = p_{N-1}^0[f](x) + a_N^0[f] \cdot r_N^0(x) = p_{N-1}^0[f](x) + a_N^0[f] \cdot F_{N-1,N}^0$$
$$= p_{N-1}^1[f](x) + a_N^0[f] \cdot r_N^1(x) = p_{N-1}^1[f](x) + a_N^0[f] \cdot F_{N-1,N}^1.$$

Subtracting we obtain (2.2) since as above the denominator $p_{N-1}^0[f_N](x) - p_{N-1}^1[f_N](x) = F_{N-1,N}^1 - F_{N-1,N}^0$ is proved to be nonzero. Introducting (2.2) into $p_N^0[f](x) = p_{N-1}^0[f](x) + a_N^0[f] \cdot F_{N-1,N}^0$ leads to

$$p_N^0[f](x) = \frac{\begin{vmatrix} F_{N-1,N}^1 \; p_{N-1}^0[f](x) - F_{N-1,N}^0 \; p_{N-1}^1[f](x) \end{vmatrix}}{\begin{vmatrix} F_{N-1,N}^1 & -F_{N-1,N}^0 \end{vmatrix}}.$$

Remarks:

(1) Asymptotically, the GN-algorithm needs only $(1/2)N^3+O(N^2)$ point operations for computing the whole field (1.2). When only the first row of this field consisting of elements p_k^0 $(k=0(1)N)$ is computed then

this number reduces to $(1/3)N^3+O(N^2)$.

(2) The GNA- and the GN-algorithm as formulated above compute the triangular field (1.2) columnwise. Of course, there are different ways for calculating the p_k^n, in particular when used for accelerating convergence computation row by row seems more natural.

(3) The GNA- and GN-algorithm have been considered earlier in [6],[7], [8],[9]. For the GN-algorithm without a recurrence relation for the divided differences see also [12]. More recently, BREZINSKI [2], [3] has given new proofs which are based upon SYLVESTER's identity for determinants. It is not easy to see how BREZINSKI's idea of proof can be modified to cover also the more general situation when $(f_o,\dots,f_N)$ is only assumed to be a ČEBYŠEV-system such that one of its proper subsystems again is a ČEBYŠEV-system on G. In contrast, the proofs given above generalize to that situation, see below and [8], [9], [11].

3. Algorithms for the General Problem of Finite Linear Interpolation

Let E be a linear space over $\mathbb{K}$ of dimension N+1 at least. By E^* we denote its algebraic dual. The general problem of finite linear interpolation [4] reads:

Suppose we are given N+1 elements $f_o,\dots,f_N$ of E, N+1 linear functionals $L_o,\dots,L_N \in E^*$ and N+1 elements $w_o,\dots,w_N \in \mathbb{K}$, can we find one and only one element $p_N^o \in \text{span}\{f_o,\dots,f_N\}$ such that

$$L_n\, p_N^o = w_n \qquad (n=0(1)N)?$$

Obviously, the answer is yes iff $\det(L_i f_j) \neq 0$. Assuming

$$(3.1)\quad \det(L_{n+s}\, f_j)_{\substack{s=0,\dots,k\\ j=0,\dots,k}} \neq 0 \qquad (k=0(1)N;\ n=0(1)N-k)$$

then p_N^o can be computed recursively by the GN-algorithm which moreover gives the complete triangular field of elements p_k^n $(k=0(1)N;n=0(1)N-k)$ where $p_k^n \in \text{span}\{f_o,\dots,f_k\}$ solves the interpolation problem $L_i\, p_k^n = w_i$ $(i=n(1)n+k)$. In theorem (2.1) one only has to make the replacements $f_j \to f_j$, $f_j(x_n) \to L_n f_j$, $f(x_n) \to w_n$. Moreover, condition (3.1) can be weakened allowing arbitrary gaps between the dimensions of nested subspaces [9], [11]. Applications of the GN-algorithm to HERMITE-BIRKHOFF interpolation with canonical complete ČEBYŠEV-systems as introduced by SCHUMAKER [13] can be found in [10].

Whereas the GN-algorithm is a method for computing certain elements of a linear space over $\mathbb{K}$ the GNA-algorithm is an algorithm running in the field $\mathbb{K}$ because division operations are basic. Extending the GNA-algorithm to the general problem of finite interpolation requires a modification of the problem. This was overlooked by BREZINSKI [2].

As before let E be a linear space over $\mathbb{K}$ of dimension N+1 at least. Let $f_o,\ldots,f_N \in E$ be linearly independent, let $w_o,\ldots,w_N \in \mathbb{K}$ and $L,L_o,L_1,\ldots,L_N \in E^*$ with $\det L_i f_j \neq 0$. The <u>modified problem of finite linear interpolation</u> is to compute

$$Lp_N^o \in \mathbb{K}$$

where p_N^o is the unique linear combination of $f_o,\ldots,f_N$ satisfying the conditions $L_i p_N^o = w_i \quad (i=0(1)N)$. Assuming that for k=0(1)N

(3.2) the restrictions of any k+1 of the functionals $L,L_o,\ldots,L_N$ to span $\{f_o,\ldots,f_k\}$ are linearly independent

then Lp_N^o can be computed recursively by the GNA-algorithm (1.4) which moreover gives the complete triangular field of values Lp_k^n (k=0(1)N; n=0(1)N-k) with $p_k^n \in \text{span}\{f_o,\ldots,f_k\}$ satisfying $L_i p_k^n = w_i \quad (i=n(1)n+k)$. In theorem (1.4) one only has to replace $f_j(x)$ by Lf_j, $f_j(x_n)$ by $L_n f_j$ and $f(x_n)$ by w_n. The proof consists in remarking that if $G=\{x,x_o,\ldots,x_N\}$ is an arbitrary set of cardinality N+2 and if we define for j=0(1)N $f_j(x):=Lf_j$, $f_j(x_n):=L_n f_j$ then according to (3.2) $(f_o,\ldots,f_N)$ is a complete ČEBYŠEV-system on G. Again, condition (3.2) can be considerably weakened. In what follows in addition to the notations introduced above for $f \in E$ $p_M^n[f]$ is defined to be that linear combination of $f_o,\ldots,f_M$ satisfying the interpolation conditions $L_i p_M^n[f] = L_i f \quad (i=n(1)n+M)$.

(3.3) <u>Theorem:</u> Let M and N be nonnegative integers such that $M < N$. Let E be a linear space over $\mathbb{K}$ of dimension N+1 at least. Let $f_o,\ldots,f_N \in E$ be linearly independent and let $L,L_o,\ldots,L_N \in E^*$. Suppose that

(i) for k=M,N and for i=0(1)N-k $\det (L_{i+p} f_j)_{\substack{p=o,\ldots,k \\ j=o,\ldots,k}} \neq 0$;

(ii) for n=0(1)N-M the restrictions of $L,L_n,L_{n+1},\ldots,L_{n+M-1}$ to $\{f_o,\ldots,f_M\}$ are linearly independent;

then there exist weights $\beta_n \in \mathbb{K}$ (n=0(1)N-M) adding to one such that for any $w_o,\ldots,w_N \in \mathbb{K}$ there holds

(3.4) $$Lp_N^o = \sum_{n=o}^{N-M} \beta_n \cdot Lp_M^n \quad ;$$

$\beta_o,\ldots,\beta_{N-M}$ are uniquely determined as solutions of the system of linear equations

$$\sum_{n=0}^{N-M} \beta_n = 1$$

$$\sum_{n=0}^{N-M} b_{s,n}\cdot\beta_n = 0 \qquad (s=1(1)N-M)$$

where $b_{s,n} := Lf_{M+s} - Lp_M^n[f_{M+s}]$.

Proof. Assuming that weights $\beta_n \in \mathbb{K}$ $(n=0(1)N-M)$ adding to one such that (3.4) holds do exist they are easily determined. In fact, by substituting $w_i = L_i\, f_{M+s}$ $(i=0(1)N)$ in (3.4) we are lead to (3.5). Let $b_{0,n} := 1$ $(n=0(1)N-M)$. It remains to show that the determinant

$$D := \det(b_{s,n})_{\substack{s=0,\dots,N-M\\ n=0,\dots,N-M}} = \det(b_{s,n}-b_{s,n-1})_{\substack{s=1,\dots,N-M\\ n=1,\dots,N-M}} \neq 0$$

is nonzero. To prove this consider

$$g_{s,n} := f_{M+s} - p_M^n[f_{M+s}] \quad (s=1(1)N-M;\ n=0(1)N-M).$$

Evidently, for $n=1(1)N-M$ every element

$$g_{s,n} - g_{s,n-1} \in \operatorname{span}\{f_0,\dots,f_M\} \quad (s=1(1)N-M)$$

lies in the null space of each functional $L_n, L_{n+1},\dots,L_{n+M-1}$. Consequently, this holds also true for any linear combination

$$g_n := \sum_{s=1}^{N-M} c_s\cdot(g_{s,n} - g_{s,n-1})$$

with arbitrary coefficients $c_s \in \mathbb{K}$. In view of the linearity of the interpolation operators we obtain with

$$g := \sum_{s=1}^{N-M} c_s\cdot f_{M+s}$$

that $g_n = p_M^{n-1}[g] - p_M^n[g]$ $(n=1(1)N-M)$. Now, $D = 0$ iff there do exist $c_s \in \mathbb{K}$ $(s=1(1)N-M)$ not all equal to zero such that $Lg_n = 0$ $(n=1(1)N-M)$. When $D = 0$ then g defined as above with such coefficients c_s is not the zero element of E. We are going to show that $g_n=0$ $(n=1(1)N-M)$. In fact, according to (ii) from $Lg_n = 0, L_n g_n = 0, L_{n+1}g_n=0, \dots, L_{n+M-1}g_n = 0$ we infer $g_n = 0$ $(n=1(1)N-M)$, i.e. $u := p_M^0[g] = p_M^1[g] = \dots = p_M^{N-M}[g]$ is a solution of the interpolation problem $L_i u = L_i g$ $(i=0(1)N)$. According to (i) from this we get $u=g$, a contraction because g cannot be a linear combination of $f_0,\dots,f_M$ since $f_0,\dots,f_N$ are assumed to be linearly independent.

The proof will be complete when we show that with $\beta_0,\dots,\beta_{N-M}$

defined to be the solutions of (3.5) for any $w_0,\dots,w_N \in \mathbb{K}$ equation (3.4) holds. Since according to (i) $\mathbb{K}^{N+1} \ni (w_0,\dots,w_N) \longmapsto p_N^0 \in \text{span}\{f_0,\dots,f_N\}$ is an isomorphism it will be sufficient to show that the linear functionals

$$f \longmapsto p_N^0[f] \quad \text{and} \quad f \longmapsto \sum_{n=0}^{N-M} \beta_n \cdot Lp_M^n[f]$$

do coincide for $f = f_j$ $(j=0(1)N)$. Because the weights add to one this is obvious for $j=0(1)M$. For $j=M+1(1)N$ this is a consequence of equations (3.5).

Hints for applications of the algorithms presented above to various problems of numerical analysis for instance to generalized orthogonal polynomials, Padé-type approximants for series of functions and rational interpolation can be found in [2]. Examples of applications to numerical differentiation and integration are discussed below.

Finally, it should be remarked that in computational practice the GNA-algorithm (3.3) and its NEWTONian analog can be used without any apriori knowledge about the existence of the various interpolants entering. The algorithm itself will check the relevant assumptions: it works iff all necessary existence conditions do hold. Here only some care with rounding off errors is necessary.

4. Applications to Numerical Differentiation and Integration; Examples

Obviously, deriving formulas for numerical differentiation or integration by differentiating or integrating (generalized) interpolation polynomials is subsumed under the modified problem of finite linear interpolation. Hence the GNA-algorithm (3.3) applies. It provides a convenient method for deriving formulas for numerical differentiation or integration recursively which often is simpler than computing, in a first step the interpolating polynomial p_N^0 and, in a second step, Lp_N^0. We will give three examples.

Example 1: SIMPSON's rule. Here $f_j(x) = x^j$, $L_j f = f(x_0+jh)$ $(j=0,1,2)$ for some $x_0 \in \mathbb{R}$, $h>0$, and

$Lf = \int_{x_0}^{x_0+2h} f(x)dx$. Applying the GNA-algorithm we compute the schemes

$$Lp_0^0 = 2h\cdot f(x_0) \qquad F_{01}^0 = 2h^2 \qquad F_{02}^0 = 4h^2\cdot(x_0+\tfrac{2}{3}h)$$

$$Lp_0^1 = 2h\cdot f(x_0+h) \qquad F_{01}^1 = 0 \qquad F_{02}^1 = \tfrac{2}{3}h^3$$

$$Lp_0^2 = 2h\cdot f(x_0+2h) \qquad F_{01}^2 = 2h^2 \qquad F_{02}^2 = -4h^2\cdot(x_0+\tfrac{4}{3}h).$$

Since in this case $F^1_{12} - F^0_{12} = 0$ the restrictions of L,L_0 or of L,L_1 to $\{f_0,f_1\}$ must be linearly dependent. It is easily seen that the latter holds true. Nevertheless, theorem (3.3) gives

$$Lp^0_2 = \beta_0 \cdot Lp^0_0 + \beta_1 \cdot Lp^1_0 + \beta_2 \cdot Lp^2_0 = \frac{h}{3}(f(x_0) + 4f(x_0+h) + f(x_0+2h))$$

where $\beta_0 = \frac{1}{6}$, $\beta_1 = \frac{4}{6}$, $\beta_2 = \frac{1}{6}$ are easily determined as solutions of

$$\beta_0 + \beta_1 \quad \beta_2 = 1$$

$$F^0_{01} \cdot \beta_0 + F^1_{01} \cdot \beta_1 + F^2_{01} \cdot \beta_2 = 0$$

$$F^0_{02} \cdot \beta_0 + F^1_{02} \cdot \beta_1 + F^2_{02} \cdot \beta_2 = 0.$$

Example 2: A formula for one sided differentiation based on interpolation by algebraic polynomials. Let $Lf = f'(x_0)$ (ordinary differentiation) and let f_j, L_j be as before. Since $Lf_0=0$ the GNA-algorithm starts with

$$Lp^0_1 = \frac{1}{h}(f(x_0+h) - f(x_0)) \qquad F^0_{12} = -h$$

$$Lp^1_1 = \frac{1}{h}(f(x_0+2h) - f(x_0+h)) \qquad F^1_{12} = -3h.$$

One "cross multiplication" yields immediately

$$Lp^0_2 = \frac{1}{2h}(-3f(x_0) + 4 \cdot f(x_0+h) - f(x_0+2h)),$$

cf. [15] p. 223 or [16], p.65. For an error estimate see example 3.

Our next example is more general in nature. Let on an interval $I \subset \mathbb{R}$ $(f_0,\dots,f_{N+1})$ be a real ECT-system in the sense of KARLIN and STUDDEN [5], i.e. there exist functions $w_k \in C^{(N+1-k)}(I)$ which are strictly positive and coefficients $c_{k,i} \in \mathbb{R}$ $(i=0(1)k-1;\ k=1(1)N+1)$ such that for all $x \in I$

$$f_0(x) = w_0(x)$$

(4.1)
$$f_k(x) = w_0(x) \cdot \int_a^x w_1(t_1) \cdot \int_a^{t_1} w_2(t_2) \dots \int_a^{t_{k-1}} w_k(t_k)\,dt_k \dots dt_1 + \sum_{i=0}^{k-1} c_{ki} \cdot f_i(x)$$

$$(a \in I,\ k=1(1)N+1).$$

For instance, with $I = [0,\infty)$ and $0<s_0<\dots<s_{N+1}$ the system of rational functions $f_j(x) = 1/(x+s_j)$ $(j=0(1)N+1)$ is an ECT-system on I with

$$w_0(x) = f_0(x),$$

(4.2)
$$w_k(x) = \left[\prod_{i=0}^{k-2} \frac{s_k - s_i}{s_{k-1} - s_i}\right] \cdot \frac{d}{dx}\left[\frac{s_{k-1}+x}{s_k+x}\right]^k > 0 \qquad (k=1(1)N+1).$$

The differential operators

$$(4.3)\quad L^j f = \left(\frac{1}{w_{j-1}}\left(\ldots\left(\frac{1}{w_1}\left(\frac{1}{w_0} f\right)'\right)'\ldots\right)'\right)' \qquad (j=1(1)N+1)$$

are called naturally associated with the system (4.1) because $f_0,\ldots,f_N$ is an integral basis of the homogeneous equation $L^{N+1}f = 0$ and f_{N+1} solves $(1/w_{N+1})L^{N+1}f = 1$.

Example 3: Numerical differentiation based on HERMITE-BIRKHOFF interpolation by ECT-systems. Let

$$E = (e_{ij})_{\substack{i=0,\ldots,k\\ j=0,\ldots,N+1}}$$

be a $(k+1)\times(N+2)$-incidence matrix satisfying the strong PÓLYA-condition and which is conservative [1],[10],[14], and let $e=\{(i,j):e_{i,j}=1\}$ be the index set corresponding to E. Suppose that $(s,t)\in e$ and that the matrix $\bar{E}$ obtained from E by deleting its last column and by replacing the element one in E with entry (s,t) by zero is a $(k+1)\times(N+1)$-incidence matrix with corresponding index set $\bar{e}$ which is poised with respect to a fixed system $x_0<\ldots<x_k$ of points $x_i\in I$. For instance, these assumptions are satisfied in any case of one sided differentiation (s=0 or s=k) when $\bar{E}$ is an HERMITEian incidence matrix (for all $j \geq 1$ $(i,j)\in\bar{e}$ implies $(i,j-1)\in\bar{e}$). Assume that from a function $f:I\rightarrow\mathbb{R}$ there are known

(i) for $(i,j)\in\bar{e}$ the N+1 values $L^j f(x_i)$ and that

(ii) f solves an inhomogeneous linear differential equation $(1/w_{N+1})L^{N+1}f = v$ where about the right hand side v only bounds $m \leq v(x) \leq M$ for $x_0 \leq x \leq x_k$ are assumed to be known.

The problem is to compute an approximate value of $L^t f(x_s)$ and to give an error estimate.

This is a typical situation where successfully both the GNA-algorithm (3.3) and a generalization of BIRKHOFF's remainder formula [2], [10] can be used. Let $(f_0,\ldots,f_N)$ be an integral basis of the homogeneous equation $L^{N+1}f = 0$ and let f_{N+1} be a solution of $(1/w_{n+1})\ L^{N+1}f = 1$. Then $L^t p[f](x_s)$ may be considered naturally as an approximation of $L^t f(x_s)$ where $p[f]\in \mathrm{span}\{f_0,\ldots,f_N\}$ solves the HERMITE-BIRKHOFF interpolation problem $L^j p[f](x_i) = L^j f(x_i)$, $(i,j)\in\bar{e}$. Applying the generalization of BIRKHOFF's remainder formula established in [10] we infer that there exists a $z\in[x_0,x_k]$ such that

$$(4.4)\quad L^t f(x_s) - L^t p[f](x_s) = v(z)\cdot(L^t f_{N+1}(x_s) - L^t p[f_{N+1}](x_s)).$$

Here both terms $L^t p[f](x_s)$ and $L^t p[f_{N+1}](x_s)$ can be computed exactly with the GNA-algorithm (3.3). Only the unknown $v(z)$ must be estimated. It should be remarked that even in the classical framework of interpolation by algebraic polynomials with simple knots (i.e. $w_o=1$, $w_k=k$ for $k=1(1)N+1$, $\frac{1}{w_k}L^k=\frac{1}{k!}(\frac{d}{dx})^k$ and $\bar{E}$ is an incidence matrix with entries 1 only in the first column) the remainder formula (4.4) is sharper than the usual estimates obtained from ROLLE's theorem more directly because besides the unknown factor $v(z)$ also the right hand side of (4.4) can be computed exactly and not only estimated as is suggested, for instance, in [16], p.66.

<u>Numerical example:</u> Consider the linear functionals $L_j f=f(j)$ $(j=0,1,2)$ and $Lf=f'(0)$. Assume that we know about $f:[0,2]\to\mathbb{R}$ $L_o f=0.1000$, $L_1 f=0.0368$, $L_2 f=0.0135$ and that $-\frac{1}{2}\leq\frac{1}{w_3(x)}L^3 f(x)\leq 1.5$ for $0\leq x\leq 2$ with $L^3 f$ defined under (4.3) and $w_o,\ldots,w_3$ by (4.2) with $s_j = j+1$ $(j=0(1)3)$. Based on these data give an approximation of $f'(0)$ and an error analysis. From (4.4) we know

$$(4.5)\quad L^1 f(0) - L^1 p[f](0) = \frac{1}{w_3(z)}\cdot L^3 f(z)\cdot(L^1 f_3(0) - L^1 p[f_3](0))$$

where $0\leq z\leq 2$ and $f_j(x) = \frac{1}{x+j+1}$ $(j=0(1)3)$ and $p[f]$ is the unique linear combination of f_o,f_1,f_2 that satisfies $L_j p[f]=L_j f$ $(j=0,1,2)$. From (4.5) we get

$$Lf - Lp[f] = \frac{1}{w_3(z)}\cdot L^3 f(z)\cdot(Lf_3 - Lp[f_3])$$

since $f(0) = p[f](0)$ and $w_o(0) = 1$. $Lp[f]$ will be computed by the GNA-algorithm:

$$Lp_o^o[f] = -f(0)$$

$$Lp_1^o[f] = 3f(1) - \tfrac{5}{2}f(0)$$

$$Lp_o^1[f] = -2f(1) \qquad\qquad Lp[f] = -5f(2)+8f(1) - \tfrac{10}{3}f(0)$$

$$Lp_1^1[f] = 15f(2) - 12f(1)$$

$$Lp_o^2[f] = -3f(2)$$

$$F_{o1}^o = Lf_1 - L_o f_1\cdot\frac{Lf_o}{L_o f_o} = \frac{1}{4} \qquad F_{o2}^o = \frac{2}{9}$$

$$F_{12}^o = -\frac{1}{36}$$

$$F_{o1}^1 = Lf_1 - L_1 f_1\cdot\frac{Lf_o}{L_1 f_o} = \frac{5}{12} \qquad F_{o2}^1 = \frac{7}{18}$$

$$F_{12}^1 = -\frac{1}{9}.$$

$$F_{o1}^2 = Lf_1 - L_2 f_1\cdot\frac{Lf_o}{L_2 f_o} = \frac{1}{2} \qquad F_{o2}^2 = \frac{22}{45}$$

It is easily computed $Lf_3 - Lp[f_3] = -\frac{1}{16}+\frac{1}{15} = \frac{1}{240}$ and $f'(0)\approx Lp[f]$

$= -0.1064 \pm 9\cdot 10^{-4}$ with error estimates

$$-0.0021 \leq -\frac{0.5}{240} \leq f'(0)-Lp[f] \leq \frac{1.5}{240} \leq 0.0063, \quad -0.1094 \leq f'(0) \leq -0.0992.$$

Actually, $f(x) = \frac{1}{10}e^{-x}$ has $f'(0) = -0.1000$.

References

[1] BIRKHOFF, G.D. General mean value and remainder theorems with applications to mechanical differentiations and integration, Trans. Amer. Math. Soc. 7 (1906), 107-136

[2] BREZINSKI, Cl. The Mühlbach-Neville-Aitken Algorithm and some Extensions, preprint, to appear in BIT

[3] BREZINSKI, Cl. A General Extrapolation Algorithm, Num.Math. 35 (1980), 175-187

[4] DAVIS, Ph. J. Interpolation and Approximation, Blaisdell Publishing Company, 1963

[5] KARLIN D.J., STUDDEN, W.J. Tchebycheff systems: with applications in analysis and statistics, Interscience Publ., New York 1966

[6] MÜHLBACH, G. Neville-Aitken algorithms for interpolation by functions of Čebyšev-systems in the sense of Newton and in a generalized sense of Hermite, In: Theory of approximation, with applications (A.G. Law, B.N. Sahney eds.) pp. 200-212, Academic Press, New York, 1976

[7] MÜHLBACH, G. Newton- und Hermite-Interpolation mit Čebyšev-Systemen, Z. Angew. Math. Mech. 54, 541-550 (1974)

[8] MÜHLBACH, G. The General Neville-Aitken-Algorithm and some Applications, Num. Math. 31 (1978), 97-110

[9] MÜHLBACH, G. The General Recurrence Relation for Divided Differences and the General Newton-Interpolation-Algorithm with Applications to Trigonometric Interpolation, Num. Math. 32 (1979), 393-408

[10] MÜHLBACH, G. An Algorithmic Approach to Hermite-Birkhoff-Interpolation, to appear

[11] MÜHLBACH, G. An Algorithmic Approach to Finite Linear Interpolation, to appear

[12] SALZER, H.E. Divided Differences for Functions of two Variables for Irregularly Spaced Points, Num. Math. 6 (1964), 68-77

[13] SCHUMAKER, L.L. On Tchebycheffian Spline Functions, J. Approx. Theory 18 (1976), 278-303

[14] SHARMA, A. Some Poised and Nonpoised Problems of Interpolation, SIAM Review vo. 14 no. 1, (1972), 129-151

[15] STIEFEL, E. Einführung in die numerische Mathematik, B.G. Teubner, 1970

[16] STUMMEL, F., HAINER, K. Praktische Mathematik, B.G. Teubner, 1971

FORMALLY BIORTHOGONAL POLYNOMIALS

H. VAN ROSSUM

1. INTRODUCTION

In 1967, J.D.E. Konhauser [10] considered biorthogonal polynomials $Y_n(x;k)$, $Z_n(x;k)$ of degree n in x and x^k respectively (x is a real variable, k a fixed integer ≥ 1) defined as follows:

$$(1.1)\qquad \int_0^\infty x^a e^{-x}\, Y_n^{(a)}(x;k)x^{km}dx \begin{cases} = 0, & m = 0,1,\dots,n-1,\\ \neq 0, & m = n.\end{cases} \qquad (a > -1)$$

$$(1.2)\qquad \int_0^\infty x^a e^{-x}\, Z_n^{(a)}(x;k)x^{m}dx \begin{cases} = 0, & m = 0,1,\dots,n-1,\\ \neq 0, & m = n.\end{cases} \qquad (a > -1)$$

The biorthogonality relations are

$$(1.3)\ \langle Y_m, Z_n \rangle := \int_0^\infty x^a e^{-x}\, Y_m(x;k)Z_n(x;k)dx \begin{cases} = 0, & m,n = 0,1,\dots,\ m \neq n,\\ \neq 0, & m = n;\ a > -1.\end{cases}$$

The moments c_n of the weight function $x^a e^{-x}$ are:

$$c_n = \int_0^\infty x^n x^a e^{-x}dx = \Gamma(a+n+1),\ (n = 0,1,\dots;\ a > -1).$$

$(c_n)_{n=0}^\infty$ is a Stieltjes sequence if $a > -1$. The inner product $\langle\cdot,\cdot\rangle$ above based on this sequence is positive definite.

In the present paper we use an inner product based on a (one-dimensional) moment sequence satisfying more modest requirements.

Polynomials in (an indeterminate) x, biorthogonal with respect to such inner product (usually indefinite), will be called formally biorthogonal polynomials, see DEFINITION 2.3.

2. We introduce several definitions.

DEFINITION 2.1. A sequence of complex numbers $(c_n)_{n=0}^\infty$ is called quasi normal if all determinants Δ_n where

$$\Delta_n := \begin{vmatrix} c_0 & c_1 & \cdots & c_n \\ c_1 & c_2 & \cdots & c_{n+1} \\ \vdots & \vdots & & \vdots \\ c_n & c_{n+1} & \cdots & c_{2n} \end{vmatrix} \quad (n = 0,1,\ldots),$$

differ from zero.

DEFINITION 2.2. A sequence $(c_n)_{n=0}^{\infty}$ is called k-quasi normal if for some integer $k \geq 1$ holds:
All determinants

$$\Delta_n^{(k)} := \begin{vmatrix} c_0 & c_1 & \cdots & c_n \\ c_k & c_{k+1} & \cdots & c_{k+n} \\ \vdots & \vdots & & \vdots \\ c_{nk} & c_{nk+1} & \cdots & c_{nk+n} \end{vmatrix} \quad (n = 0,1,\ldots),$$

differ from zero.

Remark. 1-quasi normal obviously means quasi normal.

Let P[x] denote the ring of polynomials in an indeterminate x over the real field. We introduce an inner product on P[x] (considered as a vector space in the usual way) as follows:

Let $(c_n)_{n=0}^{\infty}$ be quasi normal and $\Omega : P[x] \to \mathbb{R}$ given by $\Omega(x^n) = c_n$ $(n = 0,1,\ldots)$ then

(2.1) $\forall\ p,q \in P[x]:\ \langle p,q \rangle := \Omega(p(x)\cdot q(x))$,

where $p\cdot q$ is the ring product.

DEFINITION 2.3. The double sequence of polynomials $(Y_n(x;k),\ Z_n(x;k))_{n=0}^{\infty}$ where x is an indeterminate, k a fixed integer ≥ 1, Y_n and Z_n of degree n in x and x^k respectively is called formally biorthogonal with respect to the inner product in (2.1) if

$$\langle Y_n, Z_m \rangle \begin{cases} = 0, & m,n = 0,1,\ldots\ ; \quad (m \neq n), \\ \neq 0, & m = n. \end{cases}$$

This definition corresponds to Konhauser's definition in the case of a (positive) definite inner product [9], referred to as ("classical") biorthogonality in this paper.

THEOREM 2.1. *Let the real sequence* $(c_n)_{n=0}^{\infty}$ *be k-quasi normal, at least for* $k = 1$. *Define* $\Omega : P[x] \to R$ *by* $\Omega(x^n) = c_n$ $(n = 0,1,\ldots)$. *Then a system of polynomials, biorthogonal in the sense defined above, exists.*

PROOF. Put $Y_n(x;k) = x^n + a_{n-1}^{(n)} x^{n-1} + \ldots + a_1^{(n)} x + a_0^{(n)}$. Y_n is uniquely determined by the conditions

$$\Omega(x^{mk}\, Y_n(x;k)) \begin{cases} = 0, & m = 0,1,\ldots,n-1, \\ = N_n^{(k)} \neq 0, & m = n, \end{cases}$$

since the determinant of the system,

$$(2.2)\quad \begin{cases} c_0 a_0^{(n)} + c_1 a_1^{(n)} + \ldots + c_{n-1} a_{n-1}^{(n)} + c_n = 0, \\ c_k a_0^{(n)} + c_{k+1} a_1^{(n)} + \ldots + c_{k+n-1} a_{n-1}^{(n)} + c_{k+n} = 0, \\ \vdots \\ c_{nk-k} a_0^{(n)} + c_{nk-k+1} a_1^{(n)} + \ldots + c_{nk-k+n-1} a_{n-1}^{(n)} + c_{nk-k+n} = 0, \\ c_{nk} a_0^{(n)} + c_{nk+1} a_1^{(n)} + \ldots + c_{nk+n-1} a_{n-1}^{(n)} + c_{nk+n} = N_n^{(k)}, \end{cases}$$

differs from zero by assumption.

Let $\Delta_n^{(k)}(\underline{x})$ denote the determinant obtained from $\Delta_n^{(k)}$ upon replacing the last row by $1,x,\ldots,x^n$. Then we find:

$$Y_n(x;k) = \frac{\Delta_n^{(k)}(\underline{x})}{\Delta_{n-1}^{(k)}}, \quad N_n^{(k)} = \frac{\Delta_n^{(k)}}{\Delta_{n-1}^{(k)}}, \quad (n = 1,2,\ldots).$$

Now put $Z_n(x;k) = x^{kn} + b_{n-1}^{(n)} x^{kn-k} + \ldots + b_1^{(n)} x^k + b_0^{(n)}$,

and determine Z_n such that

$$(2.3)\qquad \Omega(x^m Z_n(x;k)) \begin{cases} = 0, & m = 0,1,\ldots,n-1, \\ = M_n^{(k)} \neq 0, & m = n. \end{cases}$$

The matrix of the system of equations in this case is the transpose of the matrix in (2.2). If we designate by $\Delta_n^{(k)}(x|)$ the determinant obtained from $\Delta_n^{(k)}$ by replacing the last column by $1,x,\ldots,x^n$, the solution of (2.3) can be written as

$$Z_n(x;k) = \frac{\Delta_n^{(k)}(x|)}{\Delta_{n-1}^{(k)}}, \quad M_n^{(k)} = \frac{\Delta_n^{(k)}}{\Delta_{n-1}^{(k)}}, \quad (n = 1,2,\ldots).$$

The biorthogonality of the system $(Y_n(x;k), Z_n(x;k))_{n=0}^{\infty}$ is shown as follows:

Let μ and ν denote non-negative integers with $\mu < \nu$. All terms in Y_μ have inner product with Z_ν equal to zero. If $\mu = \nu$ this holds for all terms in Y_μ except for the leading term, contributing $M_\mu^{(k)} \neq 0$ to the inner product. By symmetry, the assertion is now proven.□

For $k = 1$, the sets $(Y_n(x;k))_{n=0}^{\infty}$ and $(Z_n(x;k))_{n=0}^{\infty}$ coalesce to become an orthogonal system of polynomials.

If a quasi normal sequence is also k-quasi normal for some $k > 1$, then the corresponding biorthogonal system of polynomials is called the biorthogonal companion of order k of the orthogonal system of polynomials corresponding to $k = 1$.

In Konhauser's paper [10] it is shown that the system of Laguerre polynomials has biorthogonal companions of all orders in the classical sense.

We will show presently, that the same can be said about the Jacobi polynomials. Also the Bessel polynomials, the generalized Bessel polynomials and the totally positive polynomials have biorthogonal companions of all orders, in the formal sense.

To this end we need results on some determinants $\Delta_n^{(k)}$ of moment sequences (always real) associated with the systems of orthogonal polynomials mentioned above. We give them in the next section.

3. Theorems 3.1 and 3.2 are due to M.G. de Bruin (University of Amsterdam; private communication); as usual $(a)_0=1$, $(a)_n=a(a+1)\dots(a+n-1)$ for $n\in\mathbb{N}$ and $a\in\mathbb{C}$.

THEOREM 3.1. *Let for the moment sequence* $(c_m)_{m=0}^{\infty}$ *hold*

$$c_m = \frac{(a)_m}{(c)_m} \qquad (c \notin \mathbb{Z}\setminus\mathbb{N};\quad m \in \mathbb{N}_0).$$

Then we have for all integers $k \geq 1$:

$$(3.1)\qquad \Delta_n^{(k)} = (-1)^{\frac{1}{2}n(n+1)}\, k^{\frac{1}{2}n(n+1)} \prod_{j=0}^{n} \frac{(a)_{jk}}{(c)_{jk+n}} \cdot \prod_{j=1}^{n} \Psi_j(j) \cdot \prod_{j=1}^{n} j!$$

$$(3.2)\qquad \Psi_j(y) = \prod_{i=0}^{n-j} (a-c-n+i+y) \qquad (j = 1,2,\dots,n)$$

THEOREM 3.2. *Let* $c_m = \frac{1}{(c)_m}$ $(m = 0,1,\dots;\ c \notin \mathbb{Z}\setminus\mathbb{N})$.
For the determinants $\Delta_n^{(k)}$ *of the moment sequence* $(c_m)_{m=0}^{\infty}$ *holds:*

$$\Delta_n^{(k)} = \frac{(-1)^{\frac{1}{2}n(n+1)}\, k^{\frac{1}{2}n(n+1)} \prod_{j=1}^{n} j!}{\prod_{j=0}^{n} (c)_{jk+n}}, \qquad (n \in \mathbb{N}).$$

We postpone the proofs of Theorems 3.1 and 3.2 till section 4.

We have the following corollaries to Theorems 3.1 and 3.2.

Corollary 3.1. The moment sequence $((a)_m/(c)_m)_{m=0}^{\infty}$ $(c,a,c-a \notin \mathbb{Z}\setminus\mathbb{N})$, associated with the Jacobi polynomials $J_n(a,c;x)$, is k-quasi normal for $k = 1,2,\dots$.

Corollary 3.2. The moment sequence $(1/(c)_m)_{m=0}^{\infty}$ $(c \notin \mathbb{Z}\setminus\mathbb{N})$, associated with the generalized Bessel polynomials is k-quasi normal for $k = 1,2,\ldots$.

Our definition of Jacobi polynomials $J_n(a,c;x)$ is as follows (x is an indeterminate; $a,c \in \mathbb{R}$):

$$J_n(a,c;x) = x^n{}_2F_1(-n,-a-n;\ -c-2n+1;\ x^{-1})\quad (n = 0,1,\ldots).$$

$a,\ c,\ c-a \notin \mathbb{Z}\setminus\mathbb{N}\ ;\quad a,c \in \mathbb{R}.$

They are orthogonal with respect to an inner product based on the sequence

$$\frac{a}{c},\ \frac{a(a+1)}{c(c+1)},\ \frac{a(a+1)(a+2)}{c(c+1)(c+2)},\ \ldots$$

or, equivalently, on

$$1,\ \frac{a+1}{c+1},\ \frac{(a+1)(a+2)}{(c+1)(c+2)},\ \ldots\ .$$

If $c > a > -1$, we have the usual "classical" orthogonality of the $J_n(a,c;x)$ $(x \in \mathbb{R})$ on the real interval $[0,1]$ with respect to the weight function $x^a(1-x)^{c-a-1}$.

The connection of our $J_n(a,c;x)$ (for $c > a > -1$, $x \in \mathbb{R}$) and the Jacobi polynomial $P_n^{(\alpha,\beta)}(x)$ as defined in Szegö's well-known book is:

$$J_n(a,c;x) = \frac{(-1)^n n!}{(c+n)_n} P_n^{(a,c-a-1)}(1-2x),$$

$$P_n^{(\alpha,\beta)}(x) = (-1)^n\binom{2n+\alpha+\beta}{n} J_n(\alpha,\alpha+\beta+1;\ \tfrac{1-x}{2}).$$

THEOREM 3.3. *The system of Jacobi polynomials* $J_n(a,c;x)$ $(n = 0,1,\ldots)$, *where* x *is an indeterminate and* $c, a, c-a \notin \mathbb{Z}\setminus\mathbb{N}$ *has (formally) biorthogonal companions:* $(Y_n^{(a,c)}(x;k),\ Z_n^{(a,c)}(x;k))_{n=0}^{\infty}$ *for* $k = 2,3,\ldots$.

PROOF. Corollary 3.1 and Theorem 2.1. □

Remark. If $a,c,x \in \mathbb{R}$, $c > a > -1$, then the system of Jacobi polynomials $J_n(a,c;x)$ $(n = 0,1,\ldots)$ has biorthogonal companions of order $k = 2,3,\ldots$ in the classical sense. The biorthogonality relations are

$$\int_0^1 Y_n^{(a,c)}(x;k)\ Z_m^{(a,c)}(x;k)x^a(1-x)^{c-a-1}dx \quad \begin{cases} = 0, & m,n = 0,1,\ldots;\ m \neq n,\\ \neq 0, & m = n.\end{cases}$$

Next we consider the generalized Bessel polynomials,

$$B_n^{(c)}(x) = x^n {}_1F_1(-n;\ -c-2n+1;\ -x^{-1}) \quad (n = 0,1,\dots\ ;\quad c \notin Z\setminus N)$$

where x is an indeterminate.
These polynomials form an orthogonal system with respect to an (indefinite) inner product based on the sequence

$$\left(\frac{1}{(c)_m}\right)_{m=1}^{\infty} \quad \text{or} \quad \frac{1}{c},\ \frac{1}{c(c+1)},\ \frac{1}{c(c+1)(c+2)},\ \dots$$

or equivalently on

$$1,\ \frac{1}{c+1},\ \frac{1}{(c+1)(c+2)},\ \dots \qquad (c \notin Z\setminus N).$$

THEOREM 3.4. *The system of Bessel polynomials* $(B_n^{(c)}(x))_{n=0}^{\infty}$ *where* x *is an indeterminate and* $c \notin Z \setminus N$, *has biorthogonal companions of order* $k = 2,3,\dots$, *in the formal sense. Notation:* $(Y_n^{(c)}(x;k),\ Z_n^{(c)}(x;k))_{n=0}^{\infty}$.

PROOF. Corollary 3.2 and Theorem 2.1. □

The generalized Bessel polynomials $X^{(\alpha)}(z)$ $(z \in C)$ (in the notation of Al-Salam) introduced by Krall and Frink are related to the $B_n^{(c)}(z)$ as follows:

$$X^{(\alpha)}(z) = \sum_{m=0}^{n} \binom{n}{m} (n+\alpha+1)_m \left(\frac{z}{2}\right)^m = (-1)^n (n+\alpha+1)_n\, B_n^{(\alpha+1)}\left(-\frac{z}{2}\right).$$

The generalized Bessel polynomials $B_n^{(c)}(z)$ $(n = 0,1,\dots)$ are orthogonal on a circle centered at the origin of the complex plane, with arbitrary radius $r > 0$, weight function $\Psi(z) = z^{-1}\ {}_1F_1\ (1;c+1;z^{-1})$.

In view of this, the biorthogonality relations can be written as follows:

$$\oint_{|z|=r} Y_n^{(c)}(z;k) Z_m^{(c)}(z;k)\Psi(z)dz \quad \begin{cases} = 0, & m,n = 0,1,\dots;\ m \neq n, \\ \neq 0, & m = n. \end{cases}$$

z is a complex variable, c is real but not equal to a non positive integer, k is an arbitrary but fixed integer ≥ 1.

Finally we consider the sequence $(V_n\ (z))_{n=0}^{\infty}$ $(z \in C)$ of totally positive polynomials introduced by the author in 1964 [13].

These polynomials form an orthogonal system with respect to an (indefinite) inner product based on the sequence $((-1)^n c_n)_{n=0}^{\infty}$ where $(c_n)_{n=0}^{\infty}$ is a strictly totally positive sequence. This means: All minors of finite order with any choice of rows and colums taken from the infinite matrix

$$\begin{pmatrix} c_0 & 0 & 0 & 0 & \cdots \\ c_1 & c_0 & 0 & 0 & \cdots \\ c_2 & c_1 & c_0 & 0 & \cdots \\ \cdot & \cdot & \cdot & \cdot & \cdot \end{pmatrix}$$

are positive. This implies: $(c_n)_{n=0}^{\infty}$ is k-quasi normal for $k = 1,2,\ldots$. Hence

THEOREM 3.5. *The system* $(V_n(z))_{n=0}^{\infty}$ *of totally positive polynomials has biorthogonal companions of order* $2,3,\ldots$. □

The inner product referred to above can be given the following integral representation.

$$\forall\ p,q \in P(z)\ :\ \langle p,q \rangle = \frac{1}{2\pi i} \oint_{|z|=\rho+\varepsilon} p(z)\ \overline{q(\bar{z})}\ \Psi(z)\ dz.$$

where $\Psi(z) = \Sigma_{n=0}^{\infty}(-1)^n c_n\ z^{-n-1}$; $\rho = \limsup \sqrt[n]{c_n}$, $\varepsilon > 0$.

Remark. The definition of strictly totally positive above, is an adaptation of Schoenberg's original definition of totally positive sequences in [14]. He also conjectured an explicit expression of the generating function $\Sigma_{n=0}^{\infty} c_n z^n$ of the totally positive sequence $(c_n)_{n=0}^{\infty}$. This conjecture was proved by Edrei [7] using the Padé table. This Padé table was investigated later by Arms and Edrei [2]. The totally positive polynomials correspond to the Padé denominators in this table, see van Rossum [13].

These polynomials are generalizations of the ordinary Bessel polynomials, since the latter polynomials correspond to the Padé table for e^z and e^z is a (very) special case of the generating function mentioned above.

4. In this section we give de Bruin's proofs of Theorems 3.1 and 3.2. They are based on two lemma's due to J.G. van der Corput [5].

Lemma 4.1 Let y_r $(r = 1,2,\ldots,n+1)$ be complex numbers, where $y_r+1-s \notin Z \setminus N$ $(r,s = 1,2,\ldots,n+1)$. Furthermore let $\Psi_s(y)$ be polynomials in y of degree $\le n+1-s$ $(s = 1,2,\ldots,n+1)$.

Then we have

$$\det \left| \frac{\Psi_s(y_r)}{\Gamma(y_r+1-s)} \right| = \prod_{j=1}^{n+1} \frac{\Psi_j(j)}{\Gamma(y_j)} \cdot \prod_{1 \le s < r \le n+1} (y_r - y_s).$$

Lemma 4.2. Let $|e_{r,s}|$ denote an $((n+1)\times(n+1))$-determinant. The elements satisfy the relations

$$e_{r,s+1} = a_s(y_r - x_s)e_{r,s} \qquad (s = 1,2,\ldots,n;\ r = 1,2,\ldots,n+1).$$

Then we have

$$|e_{r,s}| = \prod_{j=1}^{n} a_j^{n+1-j} \cdot \prod_{1 \le s < r \le n+1} (y_r - y_s) \cdot \prod_{r=1}^{n+1} e_{r,1} \cdot$$

Proof of Theorem 3.1. Interchanging columns in (3.1) in $\Delta_n^{(k)}$ where

$$e_{r,s} = c_{(r-1)k+n+1-s} = \frac{(a)_{(r-1)k+n+1-s}}{(c)_{(r-1)k+n+1-s}} =$$

$$= \frac{\Gamma(c)(a)_{(r-1)k}\,(a+(r-1)k)(a+(r-1)k+1)\ldots(a+(r-1)k+n-s)}{\Gamma(c)(c)_{(r-1)k+n+1-s}} =$$

$$= \Gamma(c)(a)_{(r-1)k}\,\frac{\Psi_s(y_r)}{\Gamma(y_r+1-s)} \qquad (r,s = 1,2,\ldots,n+1).$$

and where Ψ_s is given by (3.1) and $\Psi_{s+1}(y) \equiv 1$. Moreover $y_r = c+(r-1)k+n$ $(r = 1,2,\ldots,n+1)$. Take $\Gamma(c)(a)_{(r-1)k}$ out of the row number r $(r = 1,2,\ldots,n+1)$ and apply Lemma 4.1 to the remaining determinant to obtain

$$\Delta_n^{(k)} = (-1)^{\frac{1}{2}n(n+1)} \prod_{r=1}^{n+1} \Gamma(c)(a)_{(r-1)k} \cdot \prod_{j=1}^{n+1} \frac{\Psi_j(j)}{\Gamma(y_j)} \cdot \prod_{1 \le s < r \le n} (r-s)k =$$

$$= (-1)^{\frac{1}{2}n(n+1)} \prod_{r=1}^{n+1} \frac{\Gamma(c)(a)_{(r-1)k}}{\Gamma(y_r)} \cdot \prod_{j=1}^{n+1} \Psi_j(j)\, k^{\frac{1}{2}n(n+1)} \cdot \prod_{j=1}^{n} j\,!$$

The assertion now follows from observing, $\Psi_{n+1} \equiv 1$ and

$$\frac{\Gamma(c)}{\Gamma(y_r)} = \frac{\Gamma(c)}{\Gamma(c+(r-1)k+n)} = \frac{1}{(c)_{(r-1)k+n}} \qquad \square$$

Proof of Theorem 3.2. Make a change of columns in $\Delta_n^{(k)}$:

$$\Delta_n^{(k)} = (-1)^{\frac{1}{2}n(n+1)} \begin{vmatrix} c_n & c_{n-1} & \cdots & c_0 \\ c_{k+n} & c_{k+n-1} & \cdots & c_k \\ \vdots & \vdots & & \vdots \\ c_{nk+n} & c_{nk+n-1} & \cdots & c_{nk} \end{vmatrix}$$

For this new determinant in the right-hand member we have

$$e_{r,s} = c_{(r-1)k+n+1-s} \qquad (r,s = 1,2,\dots,n+1),$$

With $a_s = 1$, $x_s = s$ $(s = 1,2,\dots,n)$ and $y_r = c+(r-1)k+n$ $(r=1,2,\dots,n+1)$ we get

$$a_s(y_r-x_s)e_{r,s} = [c+(r-1)k+n-s]\ c_{(r-1)k+n+1-s} =$$

$$= \frac{c+(r-1)k+n-s}{c(c+1)\dots(c+(r-1)k+n-s)} =$$

$$= \frac{1}{c(c+1)\dots(c+(r-1)k+n-s-1)} = e_{r,s+1}$$

Applying now Lemma 4.2 and

$$\prod_{1 \le s < r \le n+1} (y_r-y_s) = \prod_{1 \le s < r \le n+1} (r-s)k = k^{\frac{1}{2}n(n+1)} \prod_{j=1}^{n} j!$$

The assertion in Theorem 3.2 follows. □

For related applications of these determinantal properties we refer to a paper by de Bruin [3].

REFERENCES

(Starred items do not deal with biorthogonal polynomials)

1. AGARWAL, A.K. and H.L. MANOCHA, A note on Konhauser sets of biorthogonal polynomials. Ned. Akad. v. Wetensch. , Proc. Ser A 83 = Indag. Math. 42, 113-118 (1980).

2.* ARMS, R.J. and A. EDREI, The Padé tables and continued fractions generated by totally positive sequences. In: Mathematical Essays dedicated to A.J. Macintyre, 1-21, Ohio Univ. Press (1970).

3.* BRUIN, M.G. de, Some classes of Padé tables whose upper halves are normal. Nieuw Archief v. Wisk. (3) XXV, 148-160 (1977).

4. CARLITZ, L., A note on certain biorthogonal polynomials. Pacific J. Math. Vol 24, 425 (1968).

5.* CORPUT, J.G. van der, Over eenige determinanten. Proc. Kon. Akad. v. Wetensch., Amsterdam (1), 14, 1-44 (1930) (Dutch).

6. DAVIS, P. and H. POLAK, Complex biorthogonality for sets of polynomials. Duke Math. J. Vol. 21, 653-666 (1954).

7.* EDREI, A., Proof of a conjecture of Schoenberg on the generating function of a totally positive sequence. Canad. J. Math. 5, 86-94 (1953).

8. KARANDA, B.K. and N.K. THAKARE, Some results for Konhauser biorthogonal polynomials and dual series equations. Indian J. Pure Appl. Math. 7. no 6, 635-646 (1976)

9. KONHAUSER, J.D.E., Some properties of biorthogonal polynomials. Journ. Math. Anal. Appl. 11, 242-260 (1965).

10. KONHAUSER, J.D.E., Biorthogonal polynomials suggested by the Laguerre polynomials. Pacific. J. Math. Vol. 21, no 2, 303-304 (1967).

11. PRABHAKAR, T.R., On the other set of the biorthogonal polynomials suggested by the Laguerre polynomials. Pacific J. Math. 37, 801-804 (1971).

12. PREISER, S., An investigation of biorthogonal polynomials derivable from ordinary differential equations of the third order. J. Math. Anal. Appl. 4, 38-64 (1962).

13.* ROSSUM, H. van, Totally positive polynomials. Ned. Akad. v. Wetensch. Proc. Ser A, 68 no 2 = Indag. Math. 27, no 2, 305-315 (1965).

14.* SCHOENBERG, I.J., Some analytic aspects of the problem of smoothing. In: Studies and Essays presented to R. Courant on his 60th birthday, Jan. 8, Interscience, New York 351-377 (1948)

15. SPENCER, L. and M. FANO, Penetration and diffusion of X-rays. Calculation of spatial distributions by polynomial expansion. Journ. of Research, National Bureau of Standards, 46, 446-461 (1951).

16. SRIVASTAVA, H.M., On the Konhauser sets of biorthogonal polynomials suggested by the Laguerre polynomials. Pacific J. Math. 37, 801-804 (1971).

17. SRIVASTAVA, H.M., Some bilateral generating functions for a certain class of special functions. I, II. Proc. Kon. Akad. v. Wetensch. Proc. Ser A 83 = Indag. Math. 42, 221-246 (1980).

H. VAN ROSSUM
Universiteit van Amsterdam
Instituut voor Propedeutische Wiskunde
Roetersstraat 15
Amsterdam (NETHERLANDS)

THE PADÉ TABLE AND ITS CONNECTION WITH SOME WEAK EXPONENTIAL FUNCTION APPROXIMATIONS TO LAPLACE TRANSFORM INVERSION

by

A. Sidi
Computer Science Department
Technion - Israel Institute of Technology
Haifa, Israel

ABSTRACT

The Padé table to the Laplace transform is considered, the equivalence of the approximate Laplace transform inversion by the use of Padé approximants and some weak exponential function approximations to the inverse transform, is shown, and an oscillation theorem for the error is proved. A generalization to cover the case of multi-point Padé approximants and ordinary rational interpolation to the Laplace transform is also suggested. Prony's method of solving some non-linear equations is generalized.

1. INTRODUCTION

Let $f(t)$ be a real valued improperly Riemann integrable function on any finite subinterval of the semi-infinite interval $0 \leq t < \infty$. $\bar{f}(p)$, the Laplace transform of $f(t)$, is defined by the integral

$$\bar{f}(p) = \int_0^\infty e^{-pt} f(t)\,dt, \tag{1.1}$$

whenever this integral converges.

One way of obtaining approximations to the inverse $f(t)$ of $\bar{f}(p)$ is by approximating $\bar{f}(p)$ by a sequence of rational functions $\bar{f}_n(p)$, $n = 1,2,\ldots$, and then inverting the $\bar{f}_n(p)$ exactly to obtain the sequence $f_n(t)$, $n = 1,2,\ldots$. The hope is that if the sequence $\{\bar{f}_n(p)\}$ converges to $\bar{f}(p)$ quickly, then so will the sequence $\{f_n(t)\}$ converge to $f(t)$ quickly. There are several ways of obtaining rational approximations to a given function, one of them being by expanding this function in a Taylor series and forming the Padé table associated with the Taylor series. For a detailed discussion of this subject and references to various applications, the reader is referred to Longman (1973).

In Section 2 of this work we review briefly some algebraic properties of the Padé approximants. In Section 3 we show that the approximate Laplace transform inversion by the use of the Padé approximants is equivalent to the approximation of the inverse transform by a linear combination of exponential functions in some weak sense and also give an oscillation theorem for the error in the approximation to the inverse transform. In Section 4 we extend the results of Section 3 to cover the case of multi-point Padé approximants and ordinary rational interpolation to Laplace transform. In Section 5 we deal with the problem of interpolation by a sum of exponential functions and generalize Prony's method of solution and Weiss and McDonough's result concerning this problem.

2. SOME ALGEBRAIC ASPECTS OF THE PADÉ APPROXIMANTS

Let the function $h(z)$ have a formal power series expansion of the form

$$h(z) = \sum_{i=0}^{\infty} c_i z^i. \tag{2.1}$$

The (m,n) entry in the Padé table of (2.1), if it exists, is defined as the rational function

$$h_{m,n}(z) = \frac{P_m(z)}{Q_n(z)} = \frac{\sum_{i=0}^{m} a_i z^i}{\sum_{j=0}^{n} b_j z^j}, \quad b_0 = 1, \tag{2.2}$$

such that the Maclaurin series expansion of $h_{m,n}(z)$ in (2.2) agrees with the formal power series in (2.1) up to and including the term z^{m+n}; i.e.,

$$h(z) - h_{m,n}(z) = O(z^{m+n+1}). \tag{2.3}$$

It is possible to express (2.3) also in the form

$$\sum_{i=0}^{m} a_i z^i - \left(\sum_{j=0}^{n} b_j z^j\right)\left(\sum_{i=0}^{\infty} c_i z^i\right) = O(z^{m+n+1}), \tag{2.4}$$

from which, by setting the coefficients of the powers z^i, $i = 0,1,\ldots,m+n$, on the left hand side, equal to zero, we obtain the two sets of linear equations

$$\sum_{j=0}^{\min(i,n)} c_{i-j} b_j = a_i, \quad i = 0,1,\ldots,m \tag{2.5a}$$

$$\sum_{j=0}^{\min(i,n)} c_{i-j} b_j = 0, \quad i = m+1,\ldots,m+n, \tag{2.5b}$$

which, together with the condition $b_0 = 1$, completely determine $h_{m,n}(z)$. As is clear from equations (2.5), if the c_i are real, then the a_i, b_i and $h_{m,n}(z)$ (for real z) are all real. For the subject of the Padé table as defined above, see Baker (1975, Chapters 1 and 2).

For the purpose of the approximate inversion of the Laplace transform, it turns out that the Padé approximants $h_{N+n-1,n}(z)$, where N is a non-negative integer, i.e., those $h_{m,n}(z)$ for which $m \geq n-1$, are of interest, therefore, we shall concentrate on these approximations.

Let us assume that in $h_{N+n-1,n}(z)$ all the common factors in the numerator $P_{N+n-1}(z)$ and in the denominator $Q_n(z)$, if there are any, have been cancelled out and that the numerator is of degree less than or equal to $N+n'-1$, and the denominator is of degree exactly n' $(n' \leq n)$; i.e., $h_{N+n-1,n}(z) \equiv h_{N+n'-1,n'}(z) = P_{N+n'-1}(z)/Q_{n'}(z)$. Dividing the numerator $P_{N+n'-1}(z)$ by the denominator $Q_{n'}(z)$, we can express $h_{N+n-1,n}(z)$ as

$$h_{N+n-1,n}(z) = R_{N-1}(z) + {}_N h_n(z) , \tag{2.6}$$

where

$$ {}_N h_n(z) = \frac{\bar{P}_{n'-1}(z)}{Q_{n'}(z)} , \tag{2.7}$$

such that $R_{N-1}(z)$ and $\bar{P}_{n'-1}(z)$ are polynomials of degree at most $N-1$ and $n'-1$ respectively, which are uniquely defined. Needless to say ${}_N h_{n'}(z) \equiv {}_N h_n(z)$.

Lemma 2.1 The Maclaurin series expansion of the rational function ${}_N h_n(z)$ agrees with the formal power series expansion in (2.1) through the terms z^i, $i = N, N+1,\ldots,N+2n-1$.

Proof. Let ${}_N h_n(z)$ have the Maclaurin series expansion $\sum_{i=0}^{\infty} d_i z^i$ and let $R_{N-1}(z) = \sum_{i=0}^{N-1} e_i z^i$. Using these in (2.6) and substituting (2.6) in (2.3) we obtain

$$c_i = e_i + d_i , \quad i = 0,1,\ldots,N-1 \tag{2.8a}$$

$$c_i = d_i , \quad i = N,N+1,\ldots,N+2n-1, \tag{2.8b}$$

which is the desired result.

From this the following can easily be obtained:

Corollary. The Maclaurin series expansion of the N-th derivative of ${}_N h_n(z)$, agrees with the formal N-th derivative of the series in (2.1), obtained by differentiating (2.1) term by term N times, up to and including the term z^{2n-1}.

We note that the rational function ${}_N h_n(z)$ is determined by the $2n$ equations given in (2.8b).

Since we assume that the denominator $Q_{n'}(z)$ of ${}_N h_n(z)$ is exactly of degree n', we let $z_1,z_2,\ldots,z_s$, be all the zeros of $Q_{n'}(z)$, of multiplicities $\mu_1,\mu_2,\ldots,\mu_s$, respectively, such that $\sum_{j=1}^{s} \mu_j = n'$. Then it is possible to express ${}_N h_n(z)$ in partial fractions as

$$ {}_N h_n(z) = \sum_{j=1}^{s} \sum_{k=1}^{\mu_j} \frac{A_{j,k}}{(z-z_j)^k} . \tag{2.9}$$

From (2.9), the Maclaurin series expansion of ${}_Nh_n(z)$ is

$$(2.10)\qquad {}_Nh_n(z) = \sum_{j=1}^{s}\sum_{k=1}^{\mu_j} \frac{A_{j,k}}{(-z_j)^k} \sum_{i=0}^{\infty} \binom{-k}{i}\left(-\frac{z}{z_j}\right)^i ,$$

which, upon using the fact that $\binom{-k}{i} = (-1)^i \binom{k+i-1}{i}$, becomes

$$(2.11)\qquad {}_Nh_n(z) = \sum_{i=0}^{\infty} \left\{ \sum_{j=1}^{s}\sum_{k=1}^{\mu_j} (-1)^k \binom{k+i-1}{i} \frac{A_{j,k}}{z_j^{k+i}} \right\} z^i .$$

If we now use the result of Lemma 2.1, i.e., equations (2.8b), we obtain the following result:

<u>Lemma 2.2</u> The parameters $A_{j,k}$ and z_j of the partial fraction decomposition of ${}_Nh_n(z)$ satisfy the $2n$ non-linear equations

$$(2.12)\qquad c_{N+i} = \sum_{j=1}^{s}\sum_{k=1}^{\mu_j} (-1)^k \binom{N+k+i-1}{N+i} \frac{A_{j,k}}{z_j^{N+k+i}} , \quad i = 0,1,\dots,2n-1 .$$

Consider now the power series expansion $\bar{h}(z) = \sum_{i=0}^{\infty} \bar{c}_i z^i$, $\bar{c}_i = c_{N+i}$, $i = 0,1,\dots$, and let $\bar{h}_{m,n}(z)$ be its Padé approximants. As can be seen by equations (2.5b), the coefficients of the denominator of $\bar{h}_{n-1,n}(z)$ and those of $h_{N+n-1,n}(z)$ satisfy the same equations. Assuming that the Padé table of $h(z)$ and hence that of $\bar{h}(z)$ are normal (see Baker (1975), Chapter 2), we know that the $h_{N+n-1,n}(z)$ and $\bar{h}_{n-1,n}(z)$ are irreducible and have denominators of degree exactly n and furthermore these denominators are identical. Being identical, the denominators of $h_{N+n-1,n}(z)$ and $\bar{h}_{n-1,n}(z)$ have the same zeros which we now assume are simple. Then ${}_Nh_n(z) = \sum_{j=1}^{n} A_j/(z-z_j)$ and $\bar{h}_{n-1,n}(z) = \sum_{j=1}^{n} \bar{A}_j/(z-z_j)$. Using these expansions, for ${}_Nh_n(z)$ equations (2.12) become

$$(2.13)\qquad c_{N+i} = \sum_{j=1}^{n} -\frac{A_j}{z_j^{N+i+1}} \quad i = 0,1,\dots,2n-1 ,$$

and recalling that $\bar{h}_{n-1,n}(z) \equiv {}_0\bar{h}_n(z)$, for $\bar{h}_{n-1,n}(z)$ equations (2.12) become

$$(2.14)\qquad \bar{c}_i = \sum_{j=1}^{n} -\frac{\bar{A}_j}{z_j^{i+1}} \qquad i = 0,1,\dots,2n-1 .$$

Remembering that $\bar{c}_i = c_{N+i}$, $i = 0,1,\dots,$ and comparing equations (2.13) and (2.14) we see that

$$(2.15)\qquad A_j = \bar{A}_j z_j^N, \qquad j = 1,2,\dots,n .$$

Thus, we have shown that if $h(z)$ has a normal Padé table, then the partial fraction expansion of ${}_N h_n(z)$ can be obtained very easily from that of $\bar{h}_{n-1,n}(z)$, provided $\bar{h}_{n-1,n}(z)$ has simple poles. This way we also avoid the problem of the division of $P_{N+n-1}(z)$ by $Q_n(z)$.

3. APPROXIMATE LAPLACE TRANSFORM INVERSION BY THE USE OF THE PADÉ TABLE

Let f(t) be as described in Section 1 and let $\bar{f}(p)$ be its Laplace transform as defined in (1.1). One property of $\bar{f}(p)$ is that it is an analytic function of p whenever Re $p > \gamma$, for some γ. Then $\bar{f}(p)$ is analytic at $p = w$ for w real and $w > \gamma$, and hence can be expanded in a Taylor series as

$$(3.1) \qquad F(z) \equiv \bar{f}(p) = \sum_{i=0}^{\infty} \frac{\bar{f}^{(i)}(w)}{i!} z^i, \quad z = p-w .$$

Another important property of the Laplace transform $\bar{f}(p)$ is that it goes to zero as Re $p \to \infty$, and this implies that not every analytic function is a Laplace transform. This property, therefore, puts a restriction on the functions that can be used as approximations to $\bar{f}(p)$ for the purpose of inversion. For example, among the Padé approximations $F_{m,n}(z)$ to $F(z)$ (or equivalently $\bar{f}(p)$), only those with $m < n$ can be used for obtaining approximations to the inverse transform f(t), whereas the others can not. In particular, among the $F_{N+n-1,n}(z)$, only those with $N = 0$, i.e. the $F_{n-1,n}(z)$, can be used for this purpose, and they have indeed been used with success. (See the references given in Longman (1973).) A question then is : Is it possible, somehow, to make use of the $F_{N+n-1,n}(z)$, for $N > 0$, for the purpose of obtaining approximations to f(t)? Now we know the mathematical relationship between $\bar{f}(p)$ and the Padé approximations to it. Another even more interesting question then is: What is the mathematical relationship between f(t) and the approximations to it obtained by inverting the Padé approximations $F_{n-1,n}(z)$, and what is the relationship between f(t) and the approximations to it obtained by using the $F_{N+n-1,n}(z)$, for $N > 0$, if the answer to the first question is in the affirmative? The following theorem answers both of these questions simultaneously.

<u>Theorem 3.1</u> Define the sets G_n as follows:

$$(3.2) \qquad G_n = \{g(t) = \sum_{j=1}^{r} \sum_{k=1}^{\sigma_j} B_{j,k} t^{k-1} e^{\alpha_j t} \mid \alpha_j \text{ distinct}, \ \sum_{j=1}^{r} \sigma_j = n' \leq n\} .$$

(It is clear from (3.2) that $G_1 \subset G_2 \subset G_3 \ldots$.)

Let now $g_n(t)$ be that function, if it exists, belonging to G_n, which approximates f(t) on $[0,\infty)$ in the following weak sense:

$$(3.3) \qquad \int_0^{\infty} t^N e^{-wt} [f(t) - g_n(t)] t^i dt = 0, \quad i = 0,1,\ldots,2n-1.$$

Then $\bar{g}_n(p)$, the Laplace transform of $g_n(t)$ is simply ${}_N F_n(p-w)$, furthermore $g_n(t)$ is a real function of t.

Proof. If the function $g_n(t)$ exists, it is then of the form

$$g_n(t) = \sum_{j=1}^{s} \sum_{k=1}^{\mu_j} \frac{A_{j,k}}{(k-1)!} t^{k-1} e^{\alpha_j t}, \quad \sum_{j=1}^{s} \mu_j = n' \leq n. \tag{3.4}$$

Substituting (3.4) in (3.3) and using the relations

$$\int_0^\infty t^\ell e^{-pt} f(t)\,dt = (-1)^\ell \bar{f}^{(\ell)}(p), \tag{3.5a}$$

$$\int_0^\infty t^\nu e^{-pt}\,dt = \frac{\nu!}{p^{\nu+1}}, \qquad \nu > -1, \tag{3.5b}$$

we obtain

$$(-1)^{N+i} \bar{f}^{(N+i)}(w) = \sum_{j=1}^{s} \sum_{k=1}^{\mu_j} \frac{(N+i+k-1)!}{(k-1)!} \frac{A_{j,k}}{(w-\alpha_j)^{N+i+k}}, \quad i = 0,1,\dots,2n-1. \tag{3.6}$$

These equations can be rewritten as

$$\frac{\bar{f}^{(N+i)}(w)}{(N+i)!} = \sum_{j=1}^{s} \sum_{k=1}^{\mu_j} (-1)^k \binom{N+i+k-1}{N+i} \frac{A_{j,k}}{(\alpha_j - w)^{N+i+k}}, \qquad i = 0,1,\dots,2n-1. \tag{3.7}$$

Recalling from (3.1) that $\bar{f}^{(\ell)}(w)/\ell!$ is the coefficient of the power z^ℓ in the Maclaurin series expansion of $F(z)$, and using Lemma 2.2, we obtain the result

$${}_N F_n(z) = \sum_{j=1}^{s} \sum_{k=1}^{\mu_j} \frac{A_{j,k}}{(z-\alpha_j+w)^k}. \tag{3.8}$$

By using the definition $p = z + w$ we can express (3.8) as

$${}_N F_n(p-w) = \sum_{j=1}^{s} \sum_{k=1}^{\mu_j} \frac{A_{j,k}}{(p-\alpha_j)^k}. \tag{3.9}$$

Now, since the right hand side of equation (3.9) is nothing but $\bar{g}_n(p)$, we have $\bar{g}_n(p) = {}_N F_n(p-w)$. Since $\bar{f}(p)$ is real for real p, the Padé approximants $F_{m,n}(z)$ and hence ${}_N F_n(z)$ and equivalently $\bar{g}_n(p)$ are real for real p, therefore, $g_n(t)$ is a real function of t too. This completes the proof.

Theorem 3.1 tells us then that $F_{N+n-1,n}(z)$ <u>can</u> be used for approximating the inverse transform $f(t)$ provided ${}_N F_n(z)$, i.e., that part of $F_{N+n-1,z}(z)$ which goes to zero as $p \to \infty$, is used as an approximation to $\bar{f}(p)$.

Now, by defining $\psi(t) = t^N e^{-wt}$ with N fixed, and $\varphi_i(t) = t^i$, $i = 0,1,\dots,$ equations (3.3) can be written as

$$(3.3)' \qquad \int_0^\infty \psi(t)[f(t) - g_n(t)]\varphi_i(t)dt = 0, \quad i = 0,1,\ldots,2n-1,$$

which looks very much like a Galerkin-type approximation procedure. Therefore, by analogy with Galerkin approximation methods, we would expect the sequence $g_n(t)$, $n = 1,2,\ldots$, ignoring those $g_n(t)$ which do not exist, to converge to $f(t)$. Another justification for this expectation is the following: The sequences of Padé approximants along the diagonals usually converge very quickly, at least numerically. Now the $g_n(t)$ are obtained from the $_N F_n(p-w)$ which in turn are obtained from the Padé approximants $F_{N+n-1,n}(z)$, and these, for N fixed, form a diagonal of the Padé table.

For future reference, we state the following theorem:

<u>Theorem 3.2</u> Let $u_r(x)$, $r = 0,1,2,\ldots$, be a set of polynomials which are orthogonal on an interval $[a,b]$, finite, semi-infinite, or infinite, with weight function $q(x)$, whose integral over any subinterval of $[a,b]$ is positive. If $A(x)$ is any real continuous function on (a,b) and $\int_a^b q(x)A(x)dx$ exists as an improper Riemann integral and if $\int_a^b q(x)A(x)u_r(x)dx = 0$, $r = 0,1,\ldots,k-1$, then $A(x)$ either changes sign at least k times in the interval (a,b) or is identically zero.

The proof of this theorem for $A(x)$ continuous on $[a,b]$ can be found in Cheney (1966, p. 110) and carries over to the case in which $A(x)$ is as described above without any modification.

We now prove an oscillation theorem for the error in the approximations $g_n(t)$.

<u>Theorem 3.3</u> Let $f(t)$ be as described in Section 1 and be continuous on $(0,\infty)$ and let $\bar{f}(p)$, its Laplace transform, be analytic for $\operatorname{Re} p > \gamma$. Let $w > \gamma$ and let $F(z)$ be defined as in (3.1).

Let $\bar{g}_n(p) = {}_N F_n(p-w)$, if it exists, and assume $\bar{g}_n(p)$ has no poles for $\operatorname{Re} p \geqslant w$. Then $D(t) = f(t) - g_n(t)$, where $g_n(t)$ is the inverse Laplace transform of $\bar{g}_n(p)$, changes its sign at least $2n$ times in the interval $(0,\infty)$ if $f \notin G_n$. If $f \in G_n$, then $D(t) \equiv 0$.

<u>Proof.</u> From Theorem 3.1, $g_n(t)$ is real and satisfies the equations

$$(3.10) \qquad \int_0^\infty t^N e^{-wt} D(t) t^i dt = 0, \qquad i = 0,1,\ldots,2n-1.$$

Choosing v such that $\beta = w + v > 0$, we can write equations (3.10) in the form

$$(3.11) \qquad \int_0^\infty t^N e^{-\beta t} \bar{D}(t) t^i dt = 0, \qquad i = 0,1,\ldots,2n-1 ,$$

where $\bar{D}(t) = e^{vt} D(t)$. By taking appropriate linear combinations, equations (3.11) can be expressed as

$$(3.12)\qquad \int_0^\infty t^N e^{-\beta t}\, \bar{D}(t) L_i^{(N)}(\beta t)dt = 0, \qquad i = 0,1,\ldots,2n-1,$$

where $L_i^{(\alpha)}(x)$ are the Laguerre polynomials which are orthogonal on $[0,\infty)$ with weight function $x^\alpha e^{-x}$. It is easy to see that the $L_i^{(N)}(\beta t)$ are orthogonal on $[0,\infty)$ with weight function $t^N e^{-\beta t}$. Now, using Theorem 3.2, we conclude that $\bar{D}(t)$ and hence $D(t)$ change sign at least 2n times on $(0,\infty)$ or that they are identically zero. But $D(t) \equiv 0$ only when $f(t) \equiv g_n(t)$, and this proves the theorem.

4. GENERALIZATION TO MULTI-POINT PADÉ APPROXIMANTS AND RATIONAL INTERPOLATION

The result of Theorem 3.1 can be carried further as follows:

Theorem 4.1 Let $f(t)$ be as in Section 3 and let $g_n(t)$ be that function, if it exists, belonging to G_n, which approximates $f(t)$ on $[0,\infty)$ in the weak sense

$$(4.1)\qquad \int_0^\infty e^{-w_k t}[f(t) - g_n(t)]t^i dt = 0, \quad i = 0,1,\ldots,n_k, \quad k = 1,2,\ldots,\ell,$$

where the w_k are distinct and $\mathrm{Re}\, w_k > \gamma$, and $\sum_{k=1}^{\ell}(n_k+1) = 2n$.
Then $\bar{g}_n(p)$, the Laplace transform of $g_n(t)$, is the ℓ-point Padé approximation to $\bar{f}(p)$, whose numerator is of degree at most n-1 and whose denominator is of degree at most n, and whose Taylor series expansions about the points $p = w_k$ agree with the Taylor series expansions of $\bar{f}(p)$ about the same points up to and including the terms $(p-w_k)^{n_k}$, $k = 1,2,\ldots,\ell$. (For the subject of multi-point Padé approximants see Baker (1975, Chapter 8).) Furthermore, if the w_k are real, then $g_n(t)$ is a real function of t.

Proof. The proof of the first part follows from the fact that equations (4.1), together with the help of equation (3.5a), can be written as

$$(4.2)\qquad \bar{f}^{(i)}(w_k) = \bar{g}_n^{(i)}(w_k), \quad i = 0,1,\ldots,n_k, \quad k = 1,2,\ldots,\ell\ ,$$

and the fact that $\bar{g}_n(p)$ is a rational function with numerator of degree at most n-1 and denominator of degree at most n. The proof of the second part follows from the fact that $\bar{g}_n(p)$ is real for real p, when the w_k are real. This can be seen easily by observing that equations (4.2) when expressed in terms of the coefficients of the numerator and denominator of $\bar{g}_n(p)$, form a linear system of real equations. This, then completes the proof.

Setting $n_k = 0$ in Theorem 4.1, we can now show that the rational interpolation problem to $\bar{f}(p)$ too is simply related with an exponential function approximation to f(t) in some weak sense.

__Corollary.__ Let $g_n(t)$ be that function, if it exists, belonging to G_n, which approximates f(t) on $[0,\infty)$, in the weak sense

$$(4.3) \qquad \int_0^\infty e^{-w_k t}[f(t) - g_n(t)]dt = 0, \qquad k = 1,2,\ldots,2n$$

where the w_k are distinct and $\operatorname{Re} w_k > \gamma$. Then $\bar{g}_n(p)$, the Laplace transform of $g_n(t)$, is the rational function with numerator of degree at most n-1 and denominator of degree at most n, which interpolates $\bar{f}(p)$ at the points $p = w_k$, $k = 1,2,\ldots,2n$. As before, if the w_k are real, then $g_n(t)$ is a real function of t.

When the w_k are chosen to be real and equidistant, we can also prove an oscillation theorem for the error in $g_n(t)$, in the case when $\bar{g}_n(p)$ interpolates $\bar{f}(p)$ at the points $p = w_k$.

__Theorem 4.2__ Let f(t) and $\bar{f}(p)$ be as in Theorem 3.3 and let $\bar{g}_n(p)$ be that rational function, if it exists, with numerator of degree at most n-1 and denominator of degree at most n, which interpolates $\bar{f}(p)$ at the 2n distinct real points $p = w_o + k\delta$, $k = 0,1,\ldots,2n-1$, where $w_o > \gamma$, $\delta > 0$, and assume that $\bar{g}_n(p)$ has no poles for $\operatorname{Re} p \geqslant w_o$. Then, $D(t) = f(t) - g_n(t)$, changes sign at least 2n times in the interval $(0,\infty)$ if $f \notin G_n$. If $f \in G_n$, then $D(t) \equiv 0$.

__Proof.__ From the corollary to Theorem 4.1, $g_n(t)$ satisfies the equations

$$(4.4) \qquad \int_0^\infty e^{-w_o t} D(t)\, e^{-k\delta t}dt = 0, \qquad k = 0,1,\ldots,2n-1,$$

which can also be written as

$$(4.5) \qquad \int_0^\infty e^{-\delta t}\, \bar{D}(t)\, e^{-k\delta t}dt = 0, \qquad k = 0,1,\ldots,2n-1 \;;$$

where $\bar{D}(t) = e^{(\delta - w_o)t} D(t)$. Now taking appropriate linear combinations, equations (4.5) can be written as

$$(4.6) \qquad \int_0^\infty e^{-\delta t}\, \bar{D}(t)\, P_k^*(e^{-\delta t})dt = 0, \qquad k = 0,1,\ldots,2n-1$$

where $P_k^*(x)$ are the shifted Legendre polynomials which are orthogonal on the interval [0,1] with weight function unity. Making the change of variable $x = e^{-\delta t}$ and defining $E(x) \equiv \bar{D}(t)$, we can express equations (4.6), in the new variable x, as

$$(4.7) \qquad \int_0^1 E(x)P_k^*(x)dx = 0, \qquad k = 0,1,\ldots,2n-1.$$

Using now Theorem 3.2, we conclude that E(x) either changes sign at least 2n times on (0,1) or $E(x) \equiv 0$. Going back to the variable t, we see that $\bar{D}(t)$ and hence $D(t)$ either change sign at least 2n times on $(0,\infty)$ or are identically zero. But $D(t) \equiv 0$ only when $f(t) \equiv g_n(t)$ and this proves the theorem.

5. PRONY's METHOD AND THE PADÉ TABLE

Suppose the function c(x) is to be approximated by a sum of exponential functions

$$u(x) = \sum_{j=1}^{n} \alpha_j e^{\sigma_j x}, \tag{5.1}$$

where the α_j and σ_j are to be determined by the interpolation equations $c_i = c(i) = u(i)$, $i = 0,1,\ldots,2n-1$, which, on defining $e^{\sigma_j} = \zeta_j$, $j = 1,\ldots,n$, become

$$c_i = \sum_{j=1}^{n} \alpha_j \zeta_j^i, \quad i = 0,1,\ldots,2n-1. \tag{5.2}$$

The non-linear equations have been solved by Prony (1795) and the relation of Prony's method of solution with the $(n-1,n)$ Padé approximants to the power series expansion $V(z) = \sum_{i=0}^{\infty} c_i z^i$ has been shown by Weiss and McDonough (1963). It turns out that the ζ_j are the inverses of the zeros of the denominator of the $(n-1,n)$ Padé approximants to $V(z)$, whenever this approximant exists and has simple poles.

Now u(x) as given in (5.1) exists if the ζ_j are distinct. But whenever some of the ζ_j are equal, there is no such u(x). This implies that the function u(x) in (5.1) must be modified. The following theorem shows how this modification is to be made and also generalizes the method of Prony and the result of Weiss and McDonough.

Theorem 5.1 Let c(x) be a given function and denote $c_i = c(i)$, $i = 0,1,2,\ldots$. Suppose furthermore, that the Padé approximant $V_{N+n-1,n}(z)$ to $V(z) = \sum_{i=0}^{\infty} c_i z^i$ exists. Then there exists a function u(x) in G_n which interpolates c(x) at the points $x = N,N+1,\ldots,N+2n-1$, and this u(x) is related to ${}_N V_n(z)$.

Proof. If u(x) exists, it is of the form

$$u(x) = \sum_{j=1}^{s} \sum_{k=1}^{\mu_j} (-1)^k \binom{k+x-1}{k-1} A_{j,k} \zeta_j^{k+x}, \tag{5.3}$$

such that ζ_j are distinct and $\sum_{j=1}^{s} \mu_j = n' \leq n$. (It can be shown that any function in G_n can also be written as in (5.3).) Using now the conditions

(5.4) $\qquad c_i = c(i) = u(i), \qquad i = N, N+1, \ldots, N+2n-1,$

we obtain the equations

$$(5.5) \qquad c_{N+i} = \sum_{j=1}^{s} \sum_{k=1}^{\mu_j} (-1)^k \binom{N+i+k-1}{k-1} A_{j,k}\, \zeta_j^{N+i+k}, \qquad i = 0,1,\ldots,2n-1 .$$

Upon setting $z_j = 1/\zeta_j$ and comparing equations (5.5) with equations (2.12), and using Lemma 2.2, we see that the $A_{j,k}$ and z_j are the parameters of the partial fraction decomposition of ${}_NV_n(z)$ provided ${}_NV_n(z)$ exists. But ${}_NV_n(z)$ exists, if $V_{N+n-1,n}(z)$ exists, and this proves the theorem.

As can be seen from the proof of Theorem 5.1, the interpolant u(x) to c(x) can easily be found by determining the parameters in tne partial fraction decomposition of ${}_NV_n(z)$.

ACKNOWLEDGEMENT

The author wishes to thank Professor I.M. Longman for his continued encouragement and support of this work.

REFERENCES

1. G.A. Baker Jr, Essentials of Padé Approximants, Academic Press, 1975.

2. E.W. Cheney, Introduction to Approximation Theory, McGraw Hill, New York, 1966.

3. I.M. Longman, "On the generation of rational function approximations to Laplace transform inversion with an application to viscoelasticity". SIAM J. Appl. Math., 24 (1973), pp. 429-440.

4. R. de Prony, "Essai expérimentale et analytique ... ", J. Ec. Polytech., Paris, 1, (1795), pp. 24-76.

5. L. Weiss and R. McDonough, "Prony's method, Z-transforms, and Padé approximation", SIAM Rev., 5, (1963), pp. 145-149.

ON SOME CONDITIONS FOR CONVERGENCE OF BRANCHED CONTINUED FRACTIONS

Wojciech Siemaszko
I.Łukasiewicz Technical University
Dept.Math.Phys.
W.Pola 2, 35959 Rzeszów, Poland

1. Introduction.

Although branched continued fractions were defined in their most general form long ago /c.f. [2]/ , their properties have not been satisfactorily investigated so far. For practical reasons, especially when numerical computations are concerned, their most convenient forms are those presented in [4],[5] and [6].

In our paper we deal with branched continued fractions defined in [6]. We will investigate some conditions for convergence of such fractions.

2. Definition and properties of branched continued fractions.

By a branched continued fraction we mean an expression of the form

$$K_0 + \left\{ \frac{a_1^1|}{|K_1^1} + \frac{a_2^1|}{|K_2^1} + \dots \right\} + \left\{ \frac{a_1^2|}{|K_1^2} + \frac{a_2^2|}{|K_2^2} + \dots \right\} \qquad /2.1/$$

where

$$K_i^j = \beta_0^{i,j} + \frac{\alpha_1^{i,j}|}{|\beta_1^{i,j}} + \frac{\alpha_2^{i,j}|}{|\beta_2^{i,j}} + \dots \qquad /2.2/$$

$j=1,2,\ i=0,1,\dots$, $K_0^1 = K_0^2 = K_0$ are continued fractions.

The n-th approximant of the branched continued fraction /we will write shortly BCF / /2.1/ is defined as

$$\frac{P_n}{Q_n} = K_0\left(\left[\tfrac{n}{2}\right]\right) + \frac{P_n^1}{Q_n^1} + \frac{P_n^2}{Q_n^2} = K_0\left(\left[\tfrac{n}{2}\right]\right) +$$

$$+ \left\{ \frac{a_1^1|}{|K_1^1\left(\left[\frac{n-1}{2}\right]\right)} + \dots + \frac{a_i^1|}{|K_i^1\left(\left[\frac{n-i}{2}\right]\right)} + \dots + \frac{a_n^1|}{|K_n^1(0)} \right\} +$$

$$+ \left\{ \frac{a_1^2|}{|K_1^2\left(\left[\frac{n-1}{2}\right]\right)} + \dots + \frac{a_i^2|}{|K_i^2\left(\left[\frac{n-i}{2}\right]\right)} + \dots + \frac{a_n^2|}{|K_n^2(0)} \right\}, \qquad /2.3/$$

$n=1,2,\dots$, $P_0/Q_0 = K_0(0) = \beta_0^0$, where $K_i^j(k)$, $j=1,2,\ i,k=0,1,\dots$ are the k-th approximants of /2.2/

$$K_i^j(k) = \frac{A_k^{i,j}}{B_k^{i,j}} = \beta_0^{i,j} + \frac{\alpha_1^{i,j}|}{|\beta_1^{i,j}} + \dots + \frac{\alpha_k^{i,j}|}{|\beta_k^{i,j}}, \qquad /2.4/$$

and $[x]$ is an "entier" function.

Let in the next $P_n^j(i)/Q_n^j(i)$ denote a "subapproximant" of P_n^j/Q_n^j, $j=1,2$, beginning with i-th "level"

$$\frac{P_n^j(i)}{Q_n^j(i)} = K_{i-1}^j\left(\left[\frac{n-i+1}{2}\right]\right) + \frac{a_i^j|}{|K_i^j\left(\left[\frac{n-i}{2}\right]\right)} + \dots + \frac{a_n^j|}{|K_n^j(0)} \qquad /2.5/$$

$i=2,3,\dots,n$, $n=2,3,\dots$, $j=1,2$. Additionaly, let $P_n^j(1)/Q_n^j(1) = 1$, $n=1,2,\dots$, and let $P_n^j(n+1)/Q_n^j(n+1) = K_n^j(0)$, $P_n^j(n+2)/Q_n^j(n+2) = 1$, $n=0,1,\dots$.

The difference $\frac{P_n^j}{Q_n^j} - \frac{P_{n-1}^j}{Q_{n-1}^j}$ can be written for $j=1,2$, $n=2,3,\dots$ as

$$\frac{P_n^j}{Q_n^j} - \frac{P_{n-1}^j}{Q_{n-1}^j} = -a_1^j\left(\frac{P_n^j(2)}{Q_n^j(2)} - \frac{P_{n-1}^j(2)}{Q_{n-1}^j(2)}\right)\frac{Q_n^j(2)\ Q_{n-1}^j(2)}{P_n^j(2)\ P_{n-1}^j(2)} =$$

$$= -a_1^j\left\{K_1\left(\left[\frac{n-1}{2}\right]\right) - K_1\left(\left[\frac{n-2}{2}\right]\right)\right\}\frac{Q_n^j(2)\,Q_{n-1}^j(2)}{P_n^j(2)\,P_{n-1}^j(2)} +$$

$$+ a_1^j\ a_2^j\left(\frac{P_n^j(3)}{Q_n^j(3)} - \frac{P_{n-1}^j(3)}{Q_{n-1}^j(3)}\right)\frac{Q_n^j(2)\,Q_n^j(3)\,Q_{n-1}^j(2)\,Q_{n-1}^j(3)}{P_n^j(2)\,P_n^j(3)\,P_{n-1}^j(2)\,P_{n-1}^j(3)}. \qquad /2.6/$$

Therefore, denoting as

$$R_{i,n-1}^j = -a_i^j\ \frac{Q_{n-1}^j(i+1)\ Q_n^j(i+1)}{P_{n-1}^j(i+1)\ P_n^j(i+1)} \qquad /2.7/$$

$j=1,2$, $i=1,2,\dots,n+1$, $n=1,2,\dots$,

$$\frac{P_n^j}{Q_n^j} - \frac{P_{n-1}^j}{Q_{n-1}^j} = \sum_{i=1}^{n}\left\{K_i^j\left(\left[\frac{n-i}{2}\right]\right) - K_i^j\left(\left[\frac{n-i-1}{2}\right]\right)\right\}\prod_{p=1}^{i} R_{p,n-1}^j \qquad /2.8/$$

where $K_i^j(-1) = P_0^j/Q_0^j = 0$.

Further, we have

$$K_i^j(k) - K_i^j(k-1) = \frac{A_k^{i,j}}{B_k^{i,j}} - \frac{A_{k-1}^{i,j}}{B_{k-1}^{i,j}} = (-1)^{k+1}\frac{\alpha_0^{i,j}\ \alpha_1^{i,j}\ \dots\ \alpha_k^{i,j}}{B_{k-1}^{i,j}\ B_k^{i,j}} =$$

$$= \prod_{p=0}^{k}\rho_p^{i,j} \qquad /2.9/$$

$j=1,2$, $k=1,2,\dots$, $\alpha_0^{i,j} = 1$, $i=0,1,\dots$, where

$$\varrho_p^{i,j} = -\frac{\alpha_p^{i,j} B_{p-2}^{i,j}}{B_p^{i,j}} \qquad /2.10/$$

$j=1,2,\ p=0,1,\ldots,\ B_0^{i,j} = B_{-1}^{i,j} = B_{-2}^{i,j} = 1,\ i=0,1,\ldots$.

Since for some indices n and i, $K_i\left(\left[\frac{n-i}{2}\right]\right)$ is equal to $K_i\left(\left[\frac{n-i-1}{2}\right]\right)$, we can obtain the following formula

$$\frac{P_n^j}{Q_n^j} - \frac{P_{n-1}^j}{Q_{n-1}^j} = \sum_{i=0}^{\left[\frac{n-1}{2}\right]} \left(\prod_{p=1}^{n-2i} R_{p,n-1}^j \right) \left(\prod_{p=0}^{i} \varrho_p^{n-2i,j} \right) \qquad /2.11/$$

$n=1,2,\ldots,\ j=1,2.$

It leads to the following theorem.

<u>Theorem 2.1.</u>

The value of BCF /2.3/ with $\beta_0^0 = 1$ is equal to the sum of a double series $\sum_{k,l=0}^{\infty} c_{k,l}$, where

$$c_{00} = 1 \qquad /2.12/$$

$$c_{k,k} = \prod_{j=0}^{k} \varrho_j^0, \quad k=1,2,\ldots$$

$$c_{n-i,i} = \left(\prod_{j=1}^{n-2i} R_{j,n-1}^1 \right) \left(\prod_{j=0}^{i} \varrho_j^{n-2i,1} \right) \qquad /2.13/$$

$$c_{i,n-i} = \left(\prod_{j=1}^{n-2i} R_{j,n-1}^2 \right) \left(\prod_{j=0}^{i} \varrho_j^{n-2i,2} \right) \qquad /2.14/$$

$i=0,1,\ldots,\left[\frac{n-1}{2}\right],\ n=1,2,\ldots$.

Proof. The n-th approximant of /2.3/ can be written as

$$\frac{P_n}{Q_n} = 1 + \sum_{k=1}^{n} \left(\frac{P_k}{Q_k} - \frac{P_{k-1}}{Q_{k-1}} \right), \qquad /2.15/$$

so from /2.9/ and /2.11/

$$\frac{P_k}{Q_k} - \frac{P_{k-1}}{Q_{k-1}} = \left(K_0\left(\left[\frac{k}{2}\right]\right) - K_0\left(\left[\frac{k-1}{2}\right]\right) \right) + \left(\frac{P_k^1}{Q_k^1} - \frac{P_{k-1}^1}{Q_{k-1}^1} \right) + \left(\frac{P_k^2}{Q_k^2} - \frac{P_{k-1}^2}{Q_{k-1}^2} \right) =$$

$$= \varepsilon_k \prod_{p=0}^{\left[\frac{k}{2}\right]} \varrho_p^0 + \sum_{i=0}^{\left[\frac{k-1}{2}\right]} \left(\prod_{p=1}^{k-2i} R_{p,k-1}^1 \right) \cdot \left(\prod_{p=0}^{i} \varrho_p^{k-2i,1} \right) +$$

$$+ \sum_{i=0}^{\left[\frac{k-1}{2}\right]} \left(\prod_{p=1}^{k-2i} R_{p,k-1}^2 \right) \left(\prod_{p=0}^{i} \varrho_p^{k-2i,2} \right) = \sum_{\alpha+\beta=k} c_{\alpha,\beta}, \qquad /2.16/$$

where $k=1,2,\dots$, $\varepsilon_k = \begin{cases} 1 & \text{for } k=2s \text{ , } s=1,2,\dots \\ 0 & \text{for } k=2s+1 \text{ , } s=0,1,\dots \end{cases}$.

Now BCF /2.3/ has the value equal to the limit

$$\lim_{n\to\infty} \frac{P_n}{Q_n} = \sum_{k,l=0}^{\infty} c_{k,l} \quad .$$

3. Worpitzky's type of sufficient condition for convergence of BCF.

In some cases it is easy to find a majorant series for the series stated in theorem 2.1.

Let now /2.3/ be a BCF for which

$$\beta_k^{i,j} = 1 \quad , \ i,k=0,1,\dots \ , \ j=1,2 \ . \qquad /3.1/$$

Theorem 3.1.

If for BCF /2.3/ with coefficients β defined in /3.1/ we have

$$|\alpha_1^{i,j}| \leqslant 1/8 \quad , \ i=0,1,\dots \qquad /3.2/$$

$$|a_k^j| \leqslant 1/8 \quad , \ k=1,2,\dots \qquad /3.3/$$

$$|\alpha_k^{i,j}| \leqslant 1/4 \quad , \ k=2,3,\dots \ , \ i=0,1,\dots \qquad /3.4/$$

$j=1,2$, then this BCF is convergent.

Proof. It is easy to see that for a continued fraction of the form

$$1 - \frac{\frac{1}{4}|}{|1} - \frac{\frac{1}{4}|}{|1} - \dots \qquad /3.5/$$

we have

$$\frac{A_1}{B_1} > \frac{A_2}{B_2} > \quad \dots > \frac{A_n}{B_n} > \dots > 1/2 \qquad /3.6/$$

where A_n/B_n denotes the n-th approximant of /3.5/. The latter continued fraction is convergent and its value is equal to 1/2.

By mathematical induction we can prove, that

$$\left|\frac{P_n^j(i)}{Q_n^j(i)}\right| \geqslant \frac{A_{n-i+1}}{B_{n-i+1}} > 1/2 \qquad /3.7/$$

$i=2,\dots,n$, $n=2,3,\dots$, $j=1,2$.

Therefore from /3.3/ and /3.7/

$$|R_{i,n}^j| = |a_i^j| \left|\frac{Q_n^j(i+1)}{P_n^j(i+1)}\right| \left|\frac{Q_{n+1}^j(i+1)}{P_{n+1}^j(i+1)}\right| \leqslant 4|a_i^j| \leqslant 1/2 \qquad /3.8/$$

for $j=1,2$, $i=1,2,\dots,\left[\frac{n-1}{2}\right]$, $n=3,4,\dots$. Obviously

$$|R^j_{1,0}| = |R^j_{1,1}| = |a^j_1| < 1/2$$
$$|R^j_{2,1}| = |a^j_2| < 1/2 \quad . \qquad /3.9/$$

On the other hand

$$\varrho^{k,j}_p \, \varrho^{k,j}_{p+2} = -\frac{\alpha_{p+2}}{\alpha_{p+1}} - \frac{1 + \alpha^{k,j}_{p+1} + \alpha^{k,j}_{p+2}}{\alpha^{k,j}_{p+1}} \, \varrho^{k,j}_{p+2} \qquad /3.10/$$

for $j=1,2,\ k=0,1,\dots,\ p=3,4,\dots$. It gives that

$$\varrho^{k,j}_{p+2} = - \frac{\alpha_{p+2}}{1 + \alpha^{k,j}_{p+2} + \alpha^{k,j}_{p+1}\left(1 + \varrho^{k,j}_p\right)} \quad . \qquad /3.11/$$

Therefore $|\varrho^{k,j}_0| = 1$, $|\varrho^{k,j}_1| = |\alpha^{k,j}_1| \leqslant 1/8 = \frac{1}{4} \cdot \frac{1}{2}$, $|\varrho^{k,j}_2| =$

$= \left| \frac{\alpha^{k,j}_2}{1 + \alpha^{k,j}_2} \right| \leqslant 1/3$, $|\varrho^{k,j}_3| = \left| \frac{\alpha^{k,j}_3}{1 + \alpha^{k,j}_2 + \alpha^{k,j}_3} \right| \leqslant 2/4$, so if

we take $|\varrho^{k,j}_p| \leqslant \frac{p-1}{p+1}$, $p=2,3,\dots,n+1$, then from /3.11/

$$|\varrho^{k,j}_{n+2}| \leqslant \frac{|\alpha^{k,j}_{n+2}|}{\left|1 + \alpha^{k,j}_{n+2} + \alpha^{k,j}_{n+1}\left(1 + \varrho^{k,j}_n\right)\right|} \leqslant$$

$$\leqslant \frac{\frac{1}{4}}{1 - \frac{1}{4} - \frac{1}{4}\left(1 + \frac{n-1}{n+1}\right)} = \frac{n+1}{n+3} \quad .$$

From the above arguments we obtain

$$|\varrho^{i,j}_k| \leqslant \frac{k-1}{k+1} \qquad /3.12/$$

$j=1,2,\ i=0,1,\dots,\ k=2,3,\dots$.

It leads us to the following estimations for the coefficients $c_{k,l}$

$$|c_{k,k}| = \prod_{j=0}^{k} |\varrho^0_j| \leqslant \frac{1}{4} \cdot \frac{1}{k(k+1)} \qquad /3.13/$$

$$|c_{k+p,p}| = \left(\prod_{l=1}^{k} |R^1_{1,k+2p-1}|\right)\left(\prod_{l=0}^{p} |\varrho^{k,1}_l|\right) \leqslant 2^{-k} \frac{1}{4p(p+1)} \qquad /3.14/$$

for $p=1,2,\dots$, $k=1,2,\dots$, and

$$|c_{k,0}| \leqslant 2^{-k} \quad . \qquad /3.15/$$

The same inequalities are true for $c_{p,k+p}$.

It gives that the series $\sum_{k,l=0}^{\infty} c_{k,l}$ has the majorant series which can be written as a product of the series $\left\{1 + 2\sum_{k=1}^{\infty}\left(\frac{x}{2}\right)^k\right\}$ and

$\left\{1+\frac{1}{4}\sum_{k=1}^{\infty}\frac{(xy)^k}{k(k+1)}\right\}$ in the point $\langle x,y\rangle = \langle 1,1\rangle$. It proves our theorem.

4. Necessary condition for convergence of BCF .

The convergence of BCF need not lead to a convergence of K_i "branches". For example, let in BCF /2.1/ $K_0 \equiv 0$, $a_1^2 = 0$, and $\alpha_1^i = 0$ for $i=2,3,\dots$. If now

$$\lim_{n\to\infty}|K_1^1(n)| = \infty$$

and if the continued fraction

$$\frac{a_2^1|}{|\beta_0^2} + \frac{a_3^1|}{|\beta_0^3} + \dots$$

is convergent, then our BCF is convergent to the limit equal to zero.

Until now, we do not know general conditions which will connect the convergence of BCF with a convergence of K_i "branches", except for fractions with real positive coefficients [1] .

In the case when all the continued fractions K_i are convergent and moreover their approximants are uniformly bounded

$$|K_i(n)| \leqslant b_i \quad , n=0,1,\dots , \ i=1,2,\dots \qquad /4.1/$$

we can prove an analogue of von Koch [3] sufficient condition for convergence of continued fractions.

Let us consider a branched continued fraction of the form /2.1/, for which, as above, $K_0 \equiv 0$, $a_1^2 = 0$, and $a_i^1 = 1$, $\alpha_k^i = 1$, $i=1,2,\dots$, $k=1,2,\dots$

$$\frac{1|}{|K_1} + \frac{1|}{|K_2} + \dots \quad , \qquad /4.2/$$

$$K_i = \beta_0^i + \frac{1|}{|\beta_1^i} + \frac{1|}{|\beta_2^i} + \dots \ . \qquad /4.3/$$

Theorem 4.1.

If for BCF /4.2-3/ all the continued fractions K_i $i=1,2,\dots$ are convergent, their approximants fulfill the condition /4.1/ and the series $\sum_{i=1}^{\infty} b_i$ is convergent, them this BCF is divergent.

Proof. Let $P_{n,k}/Q_{n,k}$ be an approximant of BCF /4.2-3./ which has the form

$$\frac{P_{n,k}}{Q_{n,k}} = \frac{1|}{|K_1\left(\left[\frac{n-1}{2}\right]\right)} + \frac{1|}{|K_2\left(\left[\frac{n-2}{2}\right]\right)} + \dots + \frac{1|}{|K_k\left(\left[\frac{n-k}{2}\right]\right)} \qquad /4.4/$$

$k=1,2,\dots,n-1$, $P_{n,n}/Q_{n,n} = P_n/Q_n$.

From the uniform boundedness of K_i we obtain, using mathematical induction, that the numerators of the n-th approximant of /4.2/ are satisfying the following inequalities

$$|P_n| \leqslant \prod_{i=1}^{n}(1 + b_i) \ , \ n=1,2,\dots \ . \qquad /4.5/$$

The same will be true for $P_{n,k}$

$$|P_{n,k}| \leqslant \prod_{i=1}^{k}(1 + b_i) \ , \ k=1,2,\dots,n-1 \ , \ n=2,3,\dots \qquad /4.6/$$

as well as for Q_n and $Q_{n,k}$.

Further, we have

$$P_n = K_n(0)P_{n,n-1} + P_{n,n-2} \ . \qquad /4.7/$$

Thus for $n=2,3,\dots$

$$P_{2n} = \sum_{k=1}^{n} K_{2k}(n-k)P_{2n,2k-1} \ . \qquad /4.8/$$

Let $\varepsilon > 0$ be arbitrarily chosen. Since the series $\sum_{k=1}^{\infty} b_k$ is convergent we have that the sequence $\{P_{2n}\}$ is bounded

$$|P_{2n}| \leqslant C_2 < \infty \ , \qquad /4.9/$$

where $C_1 = \sum_{k=1}^{\infty} b_k$, $C_2 = \prod_{k=1}^{\infty}(1 + b_k)$. In addition it gives that there exists an integer $N_1 > 0$, such that for $m,n > 2N_1$

$$\sum_{k=m}^{n} b_k < \frac{\varepsilon}{6C_2} \ . \qquad /4.10/$$

On the other hand, the convergence of K_i continued fractions gives us that there exists an integer N_2 , $N_2 > 2N_1$, and such that for $m,n > N_2$

$$|K_i(m) - K_i(n)| \leqslant \frac{\varepsilon}{3C_2N_1} \ , \qquad /4.11/$$

$i=1,2,\dots,2N_1$.

For every $k=1,2,\dots,2N_1$, the numerators $P_{n,k}$, $n > 2N_1$ are convergent as combinations of $K_i(p)$, $i=1,2,\dots,2N_1$. Therefore, there exists another integer N_3, such that for $m,n > N_3 > 2N_1$

$$|P_{m,k} - P_{n,k}| \leqslant \frac{\varepsilon}{3C_1N_1} \ . \qquad /4.12/$$

Thus for $m,n > \max\{N_2,N_3\} = N$, $m > n$

$$|P_{2m} - P_{2n}| \leqslant \sum_{k=1}^{n} |P_{2m,2k-1} K_{2k}(m-k) - P_{2n,2k-1} K_{2k}(n-k)| + \sum_{k=n+1}^{m} |P_{2m,2k-1} K_{2k}(m-k)| \leqslant$$

$$\leqslant \sum_{k=1}^{N_1} \{ |P_{2m,2k-1}||K_{2k}(m-k) - K_{2k}(n-k)| + |K_{2k}(n-k)||P_{2m,2k-1} - P_{2n,2k-1}| \} + 2C_2 \sum_{k=N_1+1}^{m} b_{2k} \leqslant N_1C_1 \frac{\varepsilon}{3C_1N_1} + N_1C_2 \frac{\varepsilon}{3C_2N_1} + 2 \cdot C_2 \cdot \frac{\varepsilon}{6 \cdot C_2} = \varepsilon \ . \qquad /4.13/$$

It gives that the sequence $\{P_{2n}\}$ is convergent to a finite limit. In a similiar way we can prove that the same is true for the sequences $\{P_{2n+1}\}$, $\{Q_{2n}\}$, $\{Q_{2n+1}\}$.

Now we can show that

$$\lim_{n\to\infty} \left(P_{2n}Q_{2n-1} - P_{2n-1}Q_{2n}\right) = 1 \ . \qquad /4.14/$$

From the formula /2.8/

$$P_nQ_{n-1} - P_{n-1}Q_n = \sum_{i=1}^{n} \left\{K_i\left(\left[\frac{n-i}{2}\right]\right) - K_i\left(\left[\frac{n-i-1}{2}\right]\right)\right\}\prod_{p=1}^{i} R_{p,n-1}\, Q_{n-1}Q_n \qquad /4.15/$$

$n=2,3,\ldots$.

From the fact that $Q_n(i) = P_n(i+1)$, $n=2,3,\ldots$, $i=1,2,\ldots,n$, we can write /4.15/ in the form

$$P_nQ_{n-1} - P_{n-1}Q_n = 1 + \sum_{i=1}^{n-1} \left\{K_i\left(\left[\frac{n-i}{2}\right]\right) - K_i\left(\left[\frac{n-i-1}{2}\right]\right)\right\}(-1)^i\, Q_{n-1}(i+1)\,Q_n(i+1) \ . \qquad /4.16/$$

Using estimations presented above, we can show that the sum in /4.16/ tends to zero as $n \longrightarrow \infty$.

Therefore the sequences of even and odd approximants of BCF /4.2/ are convergent to two different limits, and BCF /4.2/ is divergent.

Acknowledgements.

I am very grateful to dr. M.G.de Bruin for the very careful correction of this paper.

References.

[1] D.I.Bodnar, in "Continued Fractions and Applications",Kiev, 1976 /in Russian/

[2] P.I.Bodnarcuk,W.J.Skorobogat'ko, Branched Continued Fractions and their Applications, Naukowa Dumka, Kiev, 1974 /in Ukrainian/

[3] H.von Koch, Bull.Soc.Math. de France, vol.23/1895/,23-40

[4] K.J.Kutschminskaja, Dokl.Akad.Nauk USRR,No.7,ser.A./1978/,614-617 /in Ukrainian/

[5] J.A.Murphy,M.R.O'Donohoe, J.Comp.Appl,Math,,vol.4,no.3/1978/,181-190

[6] W.Siemaszko, J.Comp.Appl.Math.,vol.6,no.2 /1980/,121-125

[7] H.S.Wall,Analytic Theory of Continued Fractions, Van Nostrand, New York, 1948 .

RATIONAL INTERPOLATION TO MEROMORPHIC FUNCTIONS

Hans Wallin
Department of Mathematics
University of Umeå
S-90187 Umeå/Sweden

0. INTRODUCTION

Let f be a meromorphic function with ν poles in a subset of the complex plane $\mathbb{C}$. We want to approximate f by interpolation by means of a rational function $R_{n\nu}=P_{n\nu}/Q_{n\nu}$ of type (n,ν), i.e. $P_{n\nu}$ and $Q_{n\nu}$ shall be polynomials of degree $\leq n$ and $\leq \nu$, respectively. We define $R_{n\nu}$ in the following way, if n and ν are non-negative integers and

$$\beta_{jn\nu}, \quad 1 \leq j \leq n+\nu+1, \tag{0.1}$$

are $n+\nu+1$ complex numbers, distinct or not, the interpolation points, where f is analytic. We introduce

$$\omega_{n\nu}(z) = \prod_{j=1}^{n+\nu+1} (z-\beta_{jn\nu}) \tag{0.2}$$

and determine polynomials $P_{n\nu}$ of degree $\leq n$ and $Q_{n\nu}$ of degree $\leq \nu$, $Q_{n\nu} \not\equiv 0$, such that

$$(f(z)Q_{n\nu}(z) - P_{n\nu}(z))/\omega_{n\nu}(z) \text{ is analytic at } \beta_{jn\nu},\ 1 \leq j \leq n+\nu+1. \tag{0.3}$$

In fact, (0.3) is equivalent to a system of $n+\nu+1$ linear equations stating that $fQ_{n\nu} - P_{n\nu}$ is zero at $\beta_{jn\nu}$, $1 \leq j \leq n+\nu+1$; if k of the numbers $\beta_{jn\nu}$, $1 \leq j \leq n+\nu+1$, coincide, the corresponding zero shall have multiplicity k. Since the number of unknown coefficients in $P_{n\nu}$ and $Q_{n\nu}$ is $n+\nu+2$, i.e. larger than the number of equations, we can always find $P_{n\nu}$ and $Q_{n\nu}$, $Q_{n\nu} \not\equiv 0$, satisfying (0.3). Furthermore, it is easy to prove that $R_{n\nu} = P_{n\nu}/Q_{n\nu}$ is unique. We call $R_{n\nu}$ the rational interpolant of type (n,ν) to f and the interpolation points $\beta_{jn\nu}$, $1 \leq j \leq n+\nu+1$.

The problem (0.3) is a linearized form of Hermite's interpolation problem which consists of finding a rational function $r_{n\nu}$ of type (n,ν) such that $(f(z)-r_{n\nu}(z))/\omega_{n\nu}(z)$ is analytic at $\beta_{jn\nu}$, $1 \leq j \leq n+\nu+1$. Hermite's problem is not always solvable when $\nu>0$. However, when $r_{n\nu}$ exists, $r_{n\nu}=R_{n\nu}$. The linearized problem (0.3) is easier to handle than Hermite's problem and it is the reasonable generalization to rational interpolation of polynomial interpolation which is the case $\nu=0$. Polynomial interpolation has been studied in many monographs; we refer to Walsh [16], Smirnov-Lebedev [13], Krylov [10], and Goncharov [6]. In particular, the convergence problem, i.e. the problem to decide when $R_{n\nu}\to f$, as $n\to\infty$, has been studied in great detail when $\nu=0$ by a number of people including Fejer, Fekete, Kalmár, and Walsh; see for instance [16, Ch.VII] or [13, Ch. 1].

The case when $\nu>0$ and $\beta_{jn\nu} = 0$ for $1 \leq j \leq n+\nu+1$ was first studied in some detail a century ago by Padé; see [3] or [5] for a survey of some of the modern theory. In that case the rational interpolants are now known as Padé approximants (we get the Taylor series when $\nu=0$) and because of that the rational interpolants $R_{n\nu}$ introduced by (0.3) are in the general case often called multipoint Padé approximants.

The general interpolation problem (0.3) has been studied only in a rather small number of papers including papers by Saff, Karlsson, Warner, Gončar, Lopes, Gelfgren, and Wallin; see [15] for references.

This paper is a continuation of some of the results of [15] but may be read independently. We study the convergence $R_{n\nu}\to f$, as $n\to\infty$, for functions f which are meromorphic with ν poles in certain open sets. We extend some of the theory of Walsh-Kalmár from the case $\nu=0$ to the case $\nu>0$. We give convergence results (Theorems 1 and 2 in Sections 1 and 2) and a converse result (Theorem 3 in Section 3). In particular, we study by means of logarithmic potential theory the influence on the convergence of the choice of interpolation points $\beta_{jn\nu}$. This leads to the concepts of majorizing sequence and (μ,E)-regular sequence (Definitions 1 and 2 in Section 1) which are investigated in Proposition 1 in Section 1, Proposition 2 in Section 2, and Propositions 3-6 in Section 4. As far as I know these concepts lead to weaker assumptions on the set of interpolation points than considered earlier even in the case $\nu=0$. It also allows a unified treatment of cases which have so far been considered separately (see Section 1.2).

Finally, we would like to remark that in the rational interpolation problem (0.3) the poles of $R_{n\nu}=P_{n\nu}/Q_{n\nu}$ are free in the sense that they are determined by the interpolation conditions (0.3). Another rational interpolation problem exists in the literature where the poles of the interpolant are prescribed; see [1] and [2].

1. A CONVERGENCE RESULT

1.1. Some notation and definitions. Let $R_{n\nu}=P_{n\nu}/Q_{n\nu}$ be the rational interpolant of type (n,ν) to f and the interpolation points $\beta_{jn\nu}$, $1 \leq j \leq n+\nu+1$, introduced by (0.3). We assume throughout the paper that all the interpolation points $\beta_{jn\nu}\in E$, where E is a compact subset of the complex plane $\mathbb{C}$ with complement $\complement E = \mathbb{C}\setminus E$, and that f is analytic at $\beta_{jn\nu}$ for $1 \leq j \leq n+\nu+1$, $n = 1, 2, \ldots$; $\nu\geq 0$ is a fixed integer. Our results and proofs are, however, valid also if $\beta_{jn\nu}$ are allowed to belong to $\complement E$ but have no limit point in the complement of E with respect to the extended complex plane.

We introduce the associated measure $\mu_n = \mu_{n\nu}$ to $\beta_{jn\nu}$, $1 \le j \le n+\nu+1$, as the probability measure on E, i.e. non-negative measure with total mass 1 supported by E, which assigns the point mass $1/(n+\nu+1)$ to each of the interpolation points $\beta_{jn\nu}$, $1 \le j \le n+\nu+1$ (compare [10, Chapter 12] and [18]). Let μ throughout denote a probability measure on E. We shall use the logarithmic potentials of μ and μ_n which we denote by $u(z;\mu)$ and $u(z;\mu_n)$, i.e.

$$u(z;\mu) = \int \log \frac{1}{|z-t|} \, d\mu(t)$$

and analogously for $u(z;\mu_n)$. The fundamental relationship between the associated measures μ_n and $\omega_{n\nu}(z)$ defined by (0.2) is

$$u(z;\mu_n) = \frac{-1}{n+\nu+1} \log|\omega_{n\nu}(z)|.$$

We write $\mu_n \to \mu$ if $\int g \, d\mu_n \to \int g \, d\mu$ for all continuous functions on E (convergence of μ_n to μ in the w*-topology or vague convergence).

Two properties of the set of interpolation points $\beta_{jn\nu}$, $1 \le j \le n+\nu+1$, which are relevant for the convergence $R_{n\nu} \to f$, as $n \to \infty$, are formulated in Definitions 1 and 2.

DEFINITION 1. The sequence $\{\mu_n\}$ of associated measures majorizes μ if, for every real number α and every compact set $K \subset \mathbb{C}$, there exists a constant $n(\alpha,K)$ such that

$$u(z;\mu) > \alpha \text{ on } K \Rightarrow u(z;\mu_n) > \alpha \text{ on } K \text{ for } n > n(\alpha,K). \tag{1.1}$$

An equivalent characterization is given in Proposition 3 in Section 4.1.

DEFINITION 2. $\{\mu_n\}$ is (μ,E)-regular if $\{\mu_n\}$ majorizes μ and

$$u(z;\mu_n) \to u(z;\mu), \text{ as } n \to \infty, \text{ for } z \in \complement E. \tag{1.2}$$

The conditions (1.1) and (1.2) are closely related but not identical (see Section 4, Examples 2 and 3 and Propositions 3, 5, and 6). Examples of sequences $\{\mu_n\}$ of associated measures which are (μ,E)-regular are given in Propositions 1 and 2 below. In these cases the convergence problem $R_{n\nu} \to f$ has been studied before. The case corresponding to Proposition 2 and $\nu=0$ is the classical situation (see Theorem 2 in Section 2.2 and, for instance, [16]). For the case corresponding to Proposition 1 and Theorem 1, see [18] and [15]. One advantage to study (μ,E)-regular sequences is that it includes both cases (corresponding to Propositions 1 and 2) in the same treatment.

1.2. A sufficient criterion for (μ,E)-regularity. As an important example of (μ,E)-regularity we prove Proposition 1; compare also Proposition 2 in Section 2 and the discussion in Section 4 which shows, among other things, that the converse of Proposition 1 holds if and only if E has empty interior (Proposition 5).

PROPOSITION 1. <u>Let</u> μ_n <u>be the associated measure to</u> $\beta_{jn\nu} \in E$, $1 \leq j \leq n+\nu+1$. <u>If</u> $\mu_n \to \mu$, <u>then</u> $\{\mu_n\}$ <u>is</u> (μ,E)-<u>regular</u>.

A simple example when $\mu_n \to \mu$ is if E is an interval $[a,b]$ of the real axis, the interpolation points $\beta_{jn\nu}$, $1 \leq j \leq n+\nu+1$, are uniformly distributed on $[a,b]$, and μ is uniformly distributed on $[a,b]$. For Proposition 1 see [15, Lemma 1] and [18, Theorem 4]. We shall give a simple proof using compactness.

<u>Proof</u>. (1.2) follows since $\mu_n \to \mu$ and μ_n and μ are supported by E. We prove (1.1) in two steps.

1) For $N>0$ we introduce the continuous function

$$u_N(z;\mu) = \int \min(N, \log \frac{1}{|z-t|})d\mu(t)$$

and $u_N(z;\mu_n)$ defined analogously. Assume that $u(z;\mu)>\alpha$ on K. Since $u_N(z;\mu) \uparrow u(z;\mu)$, as $N \uparrow \infty$, there exists, for any $z_0 \in K$, an N_0 such that $u_{N_0}(z;\mu)>\alpha$ in a neighbourhood of z_0. From the compactness of K we get the existence of an N such that $u_N(z;\mu)>\alpha$ on K.

2) For a fixed such N we study the sequence of functions $u_N(z;\mu_n)$, $n = 1,2,\ldots$. This sequence is clearly an equicontinuous family of functions on $\mathbb{C}$ which converges pointwise to $u_N(z;\mu)$, since $\mu_n \to \mu$. By a variant of the Arzela-Ascoli theorem the sequence converges uniformly on K. Combined with step 1 this gives $u(z;\mu_n) \geq u_N(z;\mu_n) > \alpha$ on K, if $n > n(\alpha,K)$, and Proposition 1 is proved.

<u>Remark</u> 1. The argument in the proof which was based on the Arzela-Ascoli theorem also gives the well-known fact that (1.2) implies that $u(z;\mu_n) \to u(z;\mu)$ uniformly in compact parts of $\complement E$.

1.3. <u>The convergence result</u>. We now state

THEOREM 1. <u>Let</u> ν <u>be a fixed non-negative integer and</u> μ_n, $n = 1,2,\ldots$, <u>the associated measure to</u> $\beta_{jn\nu} \in E$, $1 \leq j \leq n+\nu+1$, <u>where</u> E <u>is a compact subset of</u> $\mathbb{C}$. <u>Assume that</u> μ <u>is a probability measure on</u> E <u>such that</u> $\{\mu_n\}$ <u>is</u> (μ,E)-<u>regular and introduce the set</u>

$$E_{\mu,\rho} = E_\rho = \{z \in \mathbb{C}: u(z;\mu) > \log \frac{1}{\rho}\} \quad \text{for} \quad \rho>0.$$

<u>Let</u> f <u>be a function which, for some</u> $\rho>0$, <u>is meromorphic in an open set containing</u> $\bar{E}_\rho \cup E$ <u>with exactly</u> ν <u>poles</u> $z_1, \ldots, z_\nu$ $(\neq \beta_{jn\nu})$, <u>counted with their multiplicities, such that</u> $z_j \in E_\rho$ <u>for</u> $1 \leq j \leq \nu$. <u>Let</u> $R_{n\nu} = P_{n\nu}/Q_{n\nu}$ <u>be the rational interpolant of type</u> (n,ν) <u>to</u> f <u>and the interpolation points</u> $\beta_{jn\nu}$, $1 \leq j \leq n+\nu+1$. <u>Then</u> $R_{n\nu} \to f$, <u>as</u> $n \to \infty$, <u>uniformly on compact parts of</u> $E_\rho \setminus \{z_j\}_1^\nu$ <u>with geometric degree of convergence, i.e. we have, for each compact set</u> $K \subset E_\rho \setminus \{z_j\}_1^\nu$

$$\limsup_{n\to\infty}\{\sup_{z\in K} |f(z) - R_{n\nu}(z)|\}^{1/n} < 1.$$

Furthermore, $R_{n\nu}$ has exactly ν poles in $\mathbb{C}$, if n is large, and these converge to the poles z_j, $1\leq j\leq\nu$, of f, as $n\to\infty$.

This theorem has a long history. When $\nu=0$ and $\beta_{jn\nu}=0$ for all j and n we may choose μ as a point mass at 0 which means that E_ρ becomes a disk and we get the classical result that the Taylor series of f around 0 converges in the largest open disk around 0 where f is analytic, uniformly in compact parts. In the multipoint case, i.e. when the interpolation points $\beta_{jn\nu}$ are not all the same it was given when $\nu=0$ (polynomial interpolation) in a comparatively general form by Walsh in 1933 (see [16, Ch. VII, Th 2]); see also Section 2, Theorem 2 for a discussion of the result by Walsh. When $\nu>0$ and $\beta_{jn\nu}=0$ for all j and n (the Padé approximation case) Theorem 1 is a result by Montessus de Ballore from 1902 proved by him by means of Hadamard's theory of polar singularities of functions defined by means of power series. In the multipoint case the theorem was given when $\nu>0$, with simpler proofs than the one by Montessus de Ballore and different degrees of generality, in [12] (see the discussion below after Theorem 2), [18], and [15, Theorem 1].

Proof. The proofs in [18] and [15] were given with the stronger assumption that $\mu_n\to\mu$ (compare Prop. 1). However, an examination of these proofs shows that it is enough that $\{\mu_n\}$ is (μ,E)-regular (compare Lemma 1 of [15]). This gives Theorem 1.

Remark 2. From the proof in [15] we see that it is enough with a somewhat weaker assumption than (μ,E)-regularity. Instead of (1.2) it is enough to assume that $u(z;\mu_n)\to u(z;\mu)$ in $\complement(E\cup\overline{E}_\rho)$ and we need (1.1) for $\alpha>\log(1/\rho)$ only. It is also possible to formulate the convergence theorem without the concept of (μ,E)-regularity. In fact, from the proof in [15] we see that $R_{n\nu}\to f$ at $z\in E_\rho$ if, using the notation of [15],

$$u(z;\mu_n) - u(\xi;\mu_n) \geq \delta > 0 \quad \text{for} \quad \xi\in\Gamma \quad \text{and} \quad n\geq n_0, \tag{1.3}$$

and that $R_{n\nu}\to f$ uniformly on $K\subset E_\rho$ if (1.3) holds uniformly for $z\in K$.

2. A BASIC SPECIAL CASE

2.1. The role of the equilibrium distribution. Of special importance is the case when $\{\mu_n\}$ is (τ,E)-regular where τ is the equilibrium distribution of E for the logarithmic potential, i.e. the unique propability measure on E such that, for some constant $V(E)$, $u(z;\tau)\leq V(E)$ on $\mathbb{C}$ and $u(z;\tau) = V(E)$ on E except on a subset of E having logarithmic capacity 0; here $V(E) = \log(1/\text{cap } E)$ where cap E stands for the logarithmic capacity of E and we assume that cap $E>0$

(see for instance [14, Ch. 3], [8, Ch. 16], or [7, Ch. 5] for the definition of logarithmic capacity, for the existence of a unique equilibrium distribution, and for other basic facts about potentials). We start by proving that when $\mu=\tau$ (1.1) follows from (1.2) in the case when E is a regular set in the sense that the equilibrium potential $u(z;\tau)$ equals its maximum $V(E)$ for all $z\in E$ (for a converse result see Proposition 6 in Section 4); this regularity concept has an equivalent formulation stating that the unbounded component of $\complement E$ has a classical Green function with pole at infinity (see [14]).

PROPOSITION 2. Let $E\subset\mathbb{C}$ be a compact set which has positive logarithmic capacity and is regular. Let τ be the equilibrium distribution of E. Suppose that $u(z;\mu_n)\to u(z;\tau)$, as $n\to\infty$, for $z\in\complement E$. Then $\{\mu_n\}$ is (τ,E)-regular.

A simple example when $\{\mu_n\}$ is (τ,E)-regular is if E is the interval $[-1,1]$ of the real axis and μ_n the associated measure to the points $\beta_{jn\nu}$, $1\le j\le n+\nu+1$, which are chosen as the zeros of the Chebyshev polynomial of degree $n+\nu+1$. In this case τ is absolutely continuous on $[-1,1]$ with density $1/\pi(1-x^2)^{1/2}$ (see for instance [10, Ch. 12]). See also the discussion after Theorem 2.

Proof. Assume that $u(z;\tau)>\alpha$ on the compact set $K\neq\emptyset$. Since $u(z;\tau)\le V(E)$ on $\mathbb{C}$ and $\equiv V(E)$ on the regular set E we conclude that $\alpha<V(E)$ and $K\cup E\subset\{z\in\mathbb{C}: u(z;\tau)>\alpha\}$. Choose a compact set K_1 with boundary $\partial K_1\subset\complement E$ such that $(K\cup E)\subset K_1\subset\{z: u(z;\tau)>\alpha\}$. On ∂K_1 $u(z;\mu_n)\to u(z;\tau)$ uniformly by the assumption and Remark 1. Hence, $\alpha-u(z;\mu_n)<0$ on ∂K_1 if n is large and since $\alpha-u(z;\mu_n)$ is subharmonic in $\mathbb{C}$ we conclude by the maximum principle [7,Th.2.3] that for n large $\alpha-u(z;\mu_n)<0$ in each bounded component of $\complement(\partial K_1)$. Consequently, $u(z;\mu_n)>\alpha$ on K_1 and thus on K, if n is large. This means that $\{\mu_n\}$ majorizes τ and that $\{\mu_n\}$ is (τ,E)-regular.

2.2. The basic special case. As a corollary of Theorem 1 and Proposition 2 we now prove the following result which gives a condition for the convergence of $R_{n\nu}$ to f on $E\setminus\{z_j\}_1^\nu$.

THEOREM 2. Let $E\subset\mathbb{C}$ be a compact set of positive logarithmic capacity such that E is regular and $\complement E$ connected. Let τ be the equilibrium distribution of E and assume that

$$u(z;\mu_n)\to u(z;\tau),\quad \text{as } n\to\infty,\quad \text{for } z\in\complement E, \tag{2.1}$$

where μ_n is the associated measure to $\beta_{jn\nu}\in E$, $1\le j\le n+\nu+1$, for $n=1,2,\ldots$; ν is a fixed non-negative integer. Let f be analytic at the interpolation points $\beta_{jn\nu}$ and meromorphic in an open set containing E with exactly ν poles $z_j\in E$, $1\le j\le\nu$. Let $R_{n\nu}=P_{n\nu}/Q_{n\nu}$ be the rational interpolant of type (n,ν) to f and the interpolation points $\beta_{jn\nu}$, $1\le j\le n+\nu+1$.

<u>Then $R_{n\nu}\to f$, as $n\to\infty$, uniformly on compact parts of $E\setminus\{z_j\}_1^\nu$ with geometric degree of convergence. Furthermore, $R_{n\nu}$ has exactly ν poles in $\mathbb{C}$, if n is large, and these converge to the poles z_j, $1\leq j\leq \nu$ of f, as $n\to\infty$.</u>

This is the convergence result given by Walsh in the case $\nu=0$ in 1933 (see [17] or [16, § 7.2]). Walsh stated his assumption in a different way. If $w=\phi(z)$ is the function which maps $\complement E$ conformally on $|w|>1$ so that the points at infinity correspond to each other, and $\omega_n=\omega_{n0}$ is defined by (0.2), he assumed that

$$\lim_{n\to\infty} |\omega_n(z)|^{1/(n+1)} = \text{cap}\, E \cdot |\phi(z)|$$

uniformly on compact parts of $\complement E$. However, $\phi = \exp(G+iH)$ where G is Green's function for $\complement E$ with pole at infinity and H is the conjugate to G in $\complement E$ [16, § 4.1] and $V(E) - G(z) = u(z;\tau)$, where $V(E) = -\log \text{cap}\, E$ [14, Th. III.37]. This gives $u(z;\tau) = V(E) - \log|\phi(z)|$ in $\complement E$ and hence, as is well-known, the condition used by Walsh may be written in the form (2.1) (compare Remark 1). It should be observed that (2.1) does not mean that $\mu_n\to\tau$ except when E has empty interior (see § 4, Example 1 and Proposition 5). However, note that by Propositions 6 and 3 in Section 4 the condition (2.1) may be replaced by the assumption that $\{\mu_n\}$ majorizes τ. For the classical examples when (2.1) is satisfied see [13, Section 1.3] or [16, Ch. VII]. For $\nu>0$ Theorem 2 was given by Saff in 1972 [12] in a different form and with a different proof.

<u>Proof</u>. Let $d(z,E)$ be the distance from z to E and put, for $a>0$, $O_a = \{z\in\mathbb{C}: d(z,E) < a\}$. We put $b = \max\{u(z;\tau): z\in \complement O_a\}$ and observe that the maximum exists since $u(z;\tau)$ is continuous in $\complement O_a \subset \complement E$ and minus infinity at infinity. This means that $b < V(E)$ since, by the maximum principle, $u(z;\tau)<V(E)$ in the unbounded component of $\complement E$ and hence in the connected set $\complement E \supset \complement O_a$. Put $\rho = \exp(-b)$ and

$$E_\rho = \{z\in\mathbb{C}: u(z;\tau) > \log \tfrac{1}{\rho}\} = \{z: u(z;\tau) > b\}.$$

Then $E\subset E_\rho$, since $u(z;\tau) = V(E) > b$ on E. Also, $\overline{E}_\rho\subset\overline{O}_a$ and f is meromorphic with ν poles in a neighbourhood of $\overline{O}_a$, if a is small, and, consequently, also in $E\cup\overline{E}_\rho$. By Proposition 2 $\{\mu_n\}$ is (τ,E)-regular and Theorem 2 follows from Theorem 1.

3. A CONVERSE THEOREM

Theorem 2 gives a sufficient condition on the set of interpolation points for the convergence $R_{n\nu}\to f$ on E. That this condition is in some sense also necessary when $\nu=0$ is a result by Kalmár from 1926 for special sets E and by Walsh for general regular sets E with connected complement (see [16, § 7.3] or [13, § 1.2]). We extend this result to the case $\nu>0$ and non-regular sets and give a proof based on properties of logarithmic potentials.

THEOREM 3. *Let* E *be a compact set of positive logarithmic capacity such that* $\complement E$ *is connected. Let* $\nu\geq 0$ *be a fixed integer and* $\beta_{jn\nu}\in E$, $1\leq j\leq n+\nu+1$, *given points with associated measure* μ_n, *for* $n = 1, 2, \ldots$. *Assume that the following conditions* a) *and* b) *hold for every function* f *which is meromorphic with exactly* ν *poles in a neighbourhood of* E *with poles* $z_1, \ldots, z_\nu$ (*which may depend on* f) *belonging to* $E\setminus\{\beta_{jn\nu}\}$: *If* $R_{n\nu} = P_{n\nu}/Q_{n\nu}$ *is the rational interpolant of type* (n,ν) *to* f *and the interpolation points* $\beta_{jn\nu}$, $1\leq j\leq n+\nu+1$, *then*

a) $R_{n\nu}\to f$, *as* $n\to\infty$, *on* E *except on a subset of* E *of logarithmic capacity zero*, *and*

b) $R_{n\nu}$ *has, when* n *is large*, ν *poles in* $\mathbb{C}$ *and these converge to the poles of* f.

Under these assumptions $u(z;\mu_n)\to u(z;\tau)$, *as* $n\to\infty$, for $z\in \complement E$, *where* τ is the *equilibrium distribution of* E.

Proof. By choosing a subsequence if necessary, we may assume that $\mu_n\to\mu$ and that $u(z;\mu)\not\equiv u(z;\tau)$ on $\complement E$, since otherwise there is nothing to prove. The proof now proceeds in three steps.

Step 1. We claim that there exists a point t_0 such that

$$u(t_0;\mu) > V(E) = \log(1/\text{cap } E),\quad t_0\in \complement E. \tag{3.1}$$

In fact, suppose that $u(z;\mu)\leq V(E)$ in $\complement E$. This means that

$$\limsup_{z\to\xi}(u(z;\mu) - u(z;\tau)) \leq 0,\quad \text{for all } \xi\in\partial E\setminus e,$$

where e is the set of points of E which are irregular in the sense that $u(z;\tau)<V(E)$ at these points, because then $u(z;\tau)\to V(E)$ as $z\to\xi\in\partial E\setminus e$. We shall use the maximum principle [14, Th. III. 28] on $u(z;\mu) - u(z;\tau)$ and note that $\text{cap } e = 0$ and that $u(z;\mu) - u(z;\tau)$ is harmonic and bounded above in $\complement E$. Hence, $u(z;\mu) - u(z;\tau)\leq 0$ in $\complement E$, including at infinity where it is zero, and, accordingly, $u(z;\mu) - u(z;\tau) = 0$ in the unbounded component of $\complement E$ and, consequently, in $\complement E$ since $\complement E$ is connected. This means that $u(z;\mu) = u(z;\tau)$ in $\complement E$ which contradicts our assumption and so (3.1) is proved.

Step 2. We choose ν points $z_j\in E$, $1\leq j\leq\nu$, put $h_\nu(z) = (z-z_1)\ldots(z-z_\nu)$ and

$$f(z) = \frac{1}{(z-t_0)h_\nu(z)},$$

where t_0 is the point in (3.1). Let $R_{n\nu} = P_{n\nu}/Q_{n\nu}$ be the rational interpolant of type (n,ν) to f and the interpolation points $\beta_{jn\nu}$, $1\leq j\leq n+\nu+1$. Then, by Cauchy's integral formula, for $z\in E$,

$$\frac{h_\nu(z)(fQ_{n\nu}-P_{n\nu})(z)}{\omega_{n\nu}(z)} = \frac{1}{2\pi i}\int_\Gamma \frac{h_\nu(t)(fQ_{n\nu}-P_{n\nu})(t)}{\omega_{n\nu}(t)(t-z)}\,dt,$$

if Γ is a cycle with index (winding number) $\mathrm{ind}_\Gamma(a) = 1$ for $a \in E$ and 0 for $a=t_0$. By deforming Γ into infinity we find that the term in the right member which contains $P_{n\nu}$ is zero since the degree of $h_\nu(t)P_{n\nu}(t)$ is $\leq n+\nu$ and the degree of $\omega_{n\nu}(t)(t-z)$ is $n+\nu+2$. But $h_\nu f$ has a simple pole at t_0 and the calculus of residues gives, for $z \in E\setminus\{z_j\}_1^\nu$,

$$f(z) - R_{n\nu}(z) = \frac{\omega_{n\nu}(z)}{\omega_{n\nu}(t_0)}\cdot\frac{Q_{n\nu}(t_0)}{Q_{n\nu}(z)}\cdot\frac{1}{(z-t_0)h_\nu(z)}, \tag{3.2}$$

if $Q_{n\nu}(z) \neq 0$, which, by the assumption b) in Theorem 3, is satisfied if n is large. By this assumption $Q_{n\nu}(t_0)/Q_{n\nu}(z) \to h_\nu(t_0)/h_\nu(z) \neq 0$, as $n\to\infty$, and by the assumption a), $R_{n\nu}(z) \to f(z)$, as $n\to\infty$, on E except on a subset of capacity zero. From (3.2) we consequently conclude that

$$\exp\{(u(t_0;\mu_n) - u(z;\mu_n))(n+\nu+1)\} = \left|\frac{\omega_{n\nu}(z)}{\omega_{n\nu}(t_0)}\right| \to 0, \quad \text{as } n\to\infty, \tag{3.3}$$

for $z \in E$ except on a subset of capacity zero.

Step 3. We shall now choose a suitable point $z \in E$. We claim that there exists a point $t_1 \in E$ such that

$$u(t_1;\mu) \leq V(E) \quad \text{and} \quad \liminf_{n\to\infty} u(t_1;\mu_n) = u(t_1;\mu), \tag{3.4}$$

and such that (3.3) holds for $z=t_1$. In fact, $u(z;\mu) \leq V(E)$ on $E_0 \subset E$ where $\mathrm{cap}\,E_0 > 0$, because otherwise we would have $u(z;\mu) > V(E)$ on E except on a subset of capacity zero and this would give

$$V(E) \geq \int u(z;\tau)d\mu(z) = \int u(z;\mu)d\tau > V(E),$$

which is a contradiction. Since $\liminf u(z;\mu_n) = u(z;\mu)$ everywhere except on a set of capacity zero (see [11, Th. 3.8]) this means that there exists a point $t_1 \in E_0$ satisfying (3.4) such that (3.3) holds for $z=t_1$. If we combine (3.1) and (3.4) we get $\limsup(u(t_0;\mu_n) - u(t_1;\mu_n)) = u(t_0;\mu) - u(t_1;\mu) > 0$. This contradicts the fact that (3.3) holds for $z=t_1$ and Theorem 3 is proved.

Remark. As is seen from the proof it is in a) in Theorem 3 enough to have one set of points $z_1, \ldots, z_\nu$ and functions f of the form considered in step 2 of the proof. Condition b) may also be replaced by a weaker condition.

4. SOME PROPERTIES OF MAJORIZING SEQUENCES

4.1. Let μ_n be the associated measure to the points $\beta_{jn\nu} \in E$, $1\leq j\leq n+\nu+1$, E compact. We start by giving an alternative characterization of majorizing sequences.

PROPOSITION 3. $\{\mu_n\}$ majorizes μ if and only if

$$\liminf u(z;\mu_n) \geq u(z;\mu), \quad \text{for every } z \in \mathbb{C}. \tag{4.1}$$

Proof. The only if part follows immediately from Definition 1. Assume that the if part is wrong, i.e. that (4.1) holds and that there exist a real number α and a compact set K such that $u(z;\mu) > \alpha$ on K but that, for certain arbitrarily large n, $u(z_n;\mu_n) \leq \alpha$ for some $z_n \in K$. Choose a subsequence of these n so that $\mu_{n_j} \to \mu'$ for some measure μ'. Then

$$\liminf_{j \to \infty} u(z;\mu_{n_j}) = u(z;\mu')$$

everywhere in $\mathbb{C}$ except on a subset of logarithmic capacity zero [11, Th. 3.8], and, in particular, a.e. with respect to twodimensional Lebesgue measure. By combining this with (4.1) we get that $u(z;\mu') \geq u(z;\mu)$ a.e. and, consequently, everywhere since the mean of a potential over a disk with radius r is a continuous function tending pointwise to the potential, as $r \to 0$ (see for instance [11, Th. 1.11] or [7, Th. 2.12]). Hence, by our assumption, $u(z;\mu') > \alpha$ on K and if we combine this with Proposition 1, remembering that $\mu_{n_j} \to \mu'$, we get a contradiction to our assumption that $u(z_n;\mu_n) \leq \alpha$. This proves the if part.

Remark. A different treatment of Propositions 1 and 3 was given by U. Cegrell (unpublished) after a discussion with the author. Compare also a lemma going back to Hartogs (see for instance [9, Th. 1.6.13]).

4.2. The condition that $\{\mu_n\}$ is (μ,E)-regular does not imply that $\{\mu_n\}$ converges in the vague sense when E has non-empty interior or that μ is uniquely determined by $\{\mu_n\}$ as the following simple example shows.

Example 1. Let $E \subset \mathbb{C}$ be any compact set with non-empty interior and $B \subset E$ a closed disk with radius $r>0$ and, center z_0. Let μ be the unit mass at z_0 and τ the equilibrium distribution of B, i.e. the mass 1 uniformly distributed on ∂B. Then it is easy to check that $u(z;\mu) = u(z;\tau) = u(z_0;\tau) = \log(1/r)$ for $z \in \partial B$ and hence, by the maximum principle, $u(z;\mu) = u(z;\tau)$ on $\complement B$ and, consequently, on $\complement E$. Let μ_n be the associated measure to $\beta_{jn\nu}$, $1 \leq j \leq n+\nu+1$, where $\beta_{jn\nu} = z_0$ for all j and n. Then $\mu_n = \mu$ and $\{\mu_n\}$ is both (μ,E)-regular and (τ,E)-regular. In this case $\mu_n \to \mu$ but if we instead choose $\beta_{jn\nu} = 0$ for $1 \leq j \leq n+\nu+1$, for some n, and $\beta_{jn\nu}$, $1 \leq j \leq n+\nu+1$, uniformly distributed on ∂B, for some n, then $\{\mu_n\}$ is (τ,E)-regular but $\{\mu_n\}$ does not converge in the vague sense.

4.3. The behaviour in Example 1 is caused by the fact that there are different propability measures on E having identical logarithmic potentials on $\complement E$. If E has empty interior this can no longer occur.

PROPOSITION 4. Let $E \subset \mathbb{C}$ be a compact set with empty interior. If μ and μ' are two probability measures on E such that $u(z;\mu) = u(z;\mu')$ in $\complement E$, then $\mu=\mu'$.

Proof. If $\sigma=\mu-\mu'$, then $u(z;\sigma) = 0$ in $\complement E$ and hence [4, Lemma 1] a.e. in $\partial(\complement E) = \partial(E) = E$, i.e. $u(z;\sigma) = 0$ a.e. in $\mathbb{C}$. This implies that $\sigma=0$ (see [11, Th.1.12] or [4, Lemma 2]), i.e. $\mu=\mu'$.

Remark. An instructive proof of Proposition 4 follows from Mergelyan's approximation theorem when $\complement E$ is connected. In fact, if $\sigma=\mu-\mu'$ and $u(z;\sigma) = 0$ in $\complement E$, then we may define the logarithm so that $\int \log \frac{1}{z-t}\, d\sigma(t)$ becomes an analytic function which is constant in some open part D of $\complement E$ since the function has real part $u(z;\sigma) = 0$ in $\complement E$. By expanding the logarithm in power series we conclude that

$$\sum_{n=1}^{\infty} \frac{1}{nz^n} \int_E t^n \, d\sigma(t) = \text{constant in } D,$$

and so, since $\int d\sigma(t) = \sigma(E) = 0$,

$$\int_E t^n \, d\sigma(t) = 0 \quad \text{for } n = 0, 1, \ldots .$$

By Mergelyan's theorem [4], $\int g(t)d\sigma(t) = 0$ for all continuous functions g on E which are analytic in the interior of E. Since the interior of E is empty, the last integral is zero for all continuous g on E and hence $\sigma=0$ [8, p.278], i.e. $\mu=\mu'$.

Remark. As is seen by Exemple 1, the conclusion in Proposition 4 is not true when E has interior points. It is interesting to compare this fact with the corresponding proposition for Riesz potentials which holds for all compact sets E [8, Th. 16.4.1].

4.4. As a consequence of Proposition 4 we get the following fact which gives a converse of Proposition 1 for sets E with empty interior and shows that when E has empty interior (1.2) implies (1.1), i.e. $\{\mu_n\}$ is (μ,E)-regular if $u(z;\mu_n)\to u(z;\mu)$ in $\complement E$.

PROPOSITION 5. Let E be a compact set with empty interior, μ_n the associated measure to $\beta_{jn\nu} \in E$, $1 \le j \le n+\nu+1$, and μ a probability measure on E. Assume that $u(z;\mu_n)\to u(z;\mu)$ on $\complement E$. Then $\mu_n \to \mu$ and $\liminf u(z;\mu_n) \ge u(z;\mu)$ on $\mathbb{C}$.

Proof. Choose a subsequence of $\{\mu_n\}$ such that $\mu_{n_j}\to\mu'$, as $j\to\infty$. By the assumption $u(z;\mu') = u(z;\mu)$ on $\complement E$ and hence $\mu'=\mu$ by Proposition 4. This means that $\mu_n\to\mu$ and, by Proposition 1, that $\liminf u(z;\mu_n) \ge u(z;\mu)$.

Example 2. If E has non-empty interior, then (1.2) does not imply (1.1) (compare, however, Proposition 2). In fact, if we in Example 1 for all n choose $\beta_{jn\nu}$, $1 \le j \le n+\nu+1$, uniformly distributed on ∂B and remember that μ is the unit mass at the center of B, then we see that $u(z;\mu_n)\to u(z;\mu)$ in $\complement E$ but that $\liminf u(z;\mu_n)$ is not $\geq u(z;\mu)$ in $\mathbb{C}$.

Remark. It follows from Propositions 1 and 5 that the condition that $\{\mu_n\}$ is (μ,E)-regular is equivalent to the simpler condition that $\mu_n\to\mu$, if E has empty interior. A simple such set E is a compact interval of the real axis which is the case considered in Krylov's monograph [10, Ch. 12].

4.5. We finally prove that (1.1), or its equivalent formulation (4.1), implies (1.2) when $\complement E$ is connected.

PROPOSITION 6. *Let* E *be a compact set with connected complement,* μ_n *the associated measure to* $\beta_{jn\nu}\in E$, $1\leq j\leq n+\nu+1$, *and* μ *a probability measure on* E. *Assume that* $\liminf u(z;\mu_n)\geq u(z;\mu)$ *on* $\mathbb{C}$. *Then* $u(z;\mu_n)\to u(z;\mu)$ *on* $\complement E$.

Proof. Choose a subsequence of $\{\mu_n\}$ such that $\mu_{n_j}\to\mu'$, as $j\to\infty$. Then (compare the proof of Proposition 3) $\liminf u(z;\mu_{n_j}) = u(z;\mu')$ a.e. in $\mathbb{C}$ and, by the assumption, $u(z;\mu')\geq u(z;\mu)$ a.e. and hence everywhere in $\mathbb{C}$. We conclude that $u(z;\mu')-u(z;\mu)$ is a non-negative harmonic function in $\complement E$ which is zero at infinity. By the maximum principle and the fact that $\complement E$ is connected, $u(z;\mu') = u(z;\mu)$ in $\complement E$. This means that $u(z;\mu_n)\to u(z;\mu)$ in $\complement E$.

Example 3. The assumption that $\complement E$ is connected is important in Proposition 6. In fact, if $E = \{0\}\cup\{z:|z| = 1\}$ and μ_n is the mass 1 at 0 and τ the equilibrium distribution of E, then $\liminf u(z;\mu_n)\geq u(z;\tau)$ on $\mathbb{C}$ but $u(z;\mu_n)$ does not converge to $u(z;\tau)$ in $|z|<1$ and thus not everywhere in $\complement E$, i.e. (1.1) does not imply (1.2) in this case.

REFERENCES

[1] T. Bagby, On interpolation by rational functions, Duke Math. J. 36, 1969, pp. 95-104.

[2] T. Bagby, Rational interpolation with restricted poles, J. Approx. Theory 7, 1973, pp. 1-7.

[3] G.A. Baker, Jr, Essentials of Padé Approximants, Academic Press, 1975.

[4] L. Carleson, Mergelyan's theorem on uniform polynomial approximation, Math. Scand. 15, 1964, pp. 167-175.

[5] C.K. Chui, Recent results on Padé approximants and related problems, in Approximation Theory II, G.G. Lorentz, C.K. Chui and L.L. Schumaker, Eds., Academic Press, 1976, pp. 79-115.

[6] V.L. Goncharov, The Theory of Interpolation and Approximation of Functions, Moscow, 1954 (in Russian).

[7] W.K. Hayman and P.B. Kennedy, Subharmonic Functions, Vol I, Academic Press, 1976.

[8] E. Hille, Analytic Function Theory, Vol II, Ginn and Co., 1962.

[9] L. Hörmander, An Introduction to Complex Analysis in Several Variables, Van Nostrand, 1966.

[10] V.I. Krylov, Approximate Calculation of Integrals, Macmillan, 1962.

[11] N.S. Landkof, Foundations of Modern Potential Theory, Springer-Verlag, 1972.

[12] E.B. Saff, An extension of Montessus de Ballore's theorem on the convergence of interpolating rational functions, J. Approx. Theory 6, 1972, pp. 63-67.

[13] V.I. Smirnov and N.A. Lebedev, Functions of a Complex Variable, Constructive Theory, Iliffe Books Ltd, London, 1968.

[14] M. Tsuji, Potential Theory in Modern Function Theory, Maruzen, Tokyo, 1959.

[15] H. Wallin, Potential theory and approximation of analytic functions by rational interpolation, in Proc. Coll. Complex Anal., Joensuu, Finland, 1978, Lecture Notes in Math. 747, Springer-Verlag, 1979, pp. 434-450.

[16] J.L. Walsh, Interpolation and Approximation by Rational Functions in the Complex Domain, Colloq. Publ. XX, Amer. Math. Soc., 1969.

[17] J.L. Walsh, Note on polynomial interpolation to analytic functions, Proc. Nat. Acad. Sci. USA 19, 1933, pp. 959-963.

[18] D.D. Warner, An extension of Saff's theorem on the convergence of interpolating rational functions, J. Approx. Theory 18, 1976, pp. 108-118.

Vol. 728: Non-Commutative Harmonic Analysis. Proceedings, 1978. Edited by J. Carmona and M. Vergne. V, 244 pages 1979.

Vol. 729. Ergodic Theory. Proceedings, 1978. Edited by M. Denker and K. Jacobs. XII, 209 pages. 1979.

Vol. 730 Functional Differential Equations and Approximation of Fixed Points. Proceedings, 1978 Edited by H.-O. Peitgen and H.-O. Walther XV, 503 pages. 1979.

Vol. 731: Y Nakagami and M. Takesaki, Duality for Crossed Products of von Neumann Algebras IX, 139 pages 1979.

Vol. 732: Algebraic Geometry. Proceedings, 1978. Edited by K. Lønsted. IV, 658 pages. 1979.

Vol. 733: F. Bloom, Modern Differential Geometric Techniques in the Theory of Continuous Distributions of Dislocations. XII, 206 pages. 1979

Vol. 734 Ring Theory, Waterloo, 1978. Proceedings, 1978. Edited by D. Handelman and J. Lawrence XI, 352 pages 1979.

Vol. 735: B. Aupetit, Propriétés Spectrales des Algèbres de Banach. XII, 192 pages. 1979.

Vol. 736: E. Behrends, M-Structure and the Banach-Stone Theorem. X, 217 pages. 1979.

Vol. 737: Volterra Equations. Proceedings 1978. Edited by S.-O. Londen and O. J. Staffans. VIII, 314 pages. 1979.

Vol. 738: P. E. Conner, Differentiable Periodic Maps. 2nd edition, IV, 181 pages. 1979.

Vol. 739: Analyse Harmonique sur les Groupes de Lie II. Proceedings, 1976-78. Edited by P. Eymard et al. VI, 646 pages. 1979.

Vol. 740: Séminaire d'Algèbre Paul Dubreil. Proceedings, 1977-78. Edited by M.-P. Malliavin. V, 456 pages. 1979

Vol. 741: Algebraic Topology, Waterloo 1978. Proceedings. Edited by P. Hoffman and V. Snaith. XI, 655 pages. 1979.

Vol. 742: K. Clancey, Seminormal Operators. VII, 125 pages 1979.

Vol. 743: Romanian-Finnish Seminar on Complex Analysis. Proceedings, 1976. Edited by C. Andreian Cazacu et al. XVI, 713 pages. 1979.

Vol. 744: I. Reiner and K. W. Roggenkamp, Integral Representations. VIII, 275 pages. 1979.

Vol. 745: D. K. Haley, Equational Compactness in Rings. III, 167 pages. 1979.

Vol. 746 P. Hoffman, τ-Rings and Wreath Product Representations. V, 148 pages. 1979.

Vol. 747: Complex Analysis, Joensuu 1978. Proceedings, 1978 Edited by I. Laine, O. Lehto and T. Sorvali. XV, 450 pages. 1979.

Vol. 748. Combinatorial Mathematics VI. Proceedings, 1978. Edited by A. F. Horadam and W. D. Wallis. IX, 206 pages. 1979

Vol. 749: V. Girault and P.-A. Raviart, Finite Element Approximation of the Navier-Stokes Equations. VII, 200 pages. 1979.

Vol. 750 J. C. Jantzen, Moduln mit einem höchsten Gewicht. III, 195 Seiten. 1979.

Vol 751: Number Theory, Carbondale 1979. Proceedings. Edited by M. B. Nathanson. V, 342 pages. 1979.

Vol 752: M. Barr, *-Autonomous Categories. VI, 140 pages. 1979.

Vol. 753: Applications of Sheaves Proceedings, 1977. Edited by M. Fourman, C. Mulvey and D. Scott. XIV, 779 pages. 1979.

Vol. 754: O. A. Laudal, Formal Moduli of Algebraic Structures. III, 161 pages. 1979.

Vol. 755. Global Analysis. Proceedings, 1978. Edited by M. Grmela and J. E. Marsden. VII, 377 pages. 1979.

Vol. 756: H. O. Cordes, Elliptic Pseudo-Differential Operators – An Abstract Theory. IX, 331 pages 1979.

Vol. 757: Smoothing Techniques for Curve Estimation. Proceedings, 1979. Edited by Th. Gasser and M. Rosenblatt. V, 245 pages. 1979.

Vol. 758 C. Năstăsescu and F. Van Oystaeyen; Graded and Filtered Rings and Modules. X, 148 pages 1979.

Vol. 759: R. L. Epstein, Degrees of Unsolvability: Structure and Theory. XIV, 216 pages. 1979

Vol. 760: H.-O. Georgii, Canonical Gibbs Measures. VIII, 190 pages. 1979.

Vol. 761: K. Johannson, Homotopy Equivalences of 3-Manifolds with Boundaries. 2, 303 pages 1979.

Vol. 762 D. H Sattinger, Group Theoretic Methods in Bifurcation Theory. V, 241 pages. 1979.

Vol. 763: Algebraic Topology, Aarhus 1978. Proceedings, 1978. Edited by J. L. Dupont and H. Madsen. VI, 695 pages 1979.

Vol. 764: B. Srinivasan, Representations of Finite Chevalley Groups. XI, 177 pages 1979

Vol. 765: Padé Approximation and its Applications Proceedings, 1979. Edited by L. Wuytack VI, 392 pages 1979.

Vol. 766. T tom Dieck, Transformation Groups and Representation Theory. VIII, 309 pages. 1979.

Vol. 767. M. Namba, Families of Meromorphic Functions on Compact Riemann Surfaces. XII, 284 pages. 1979.

Vol. 768: R. S. Doran and J. Wichmann, Approximate Identities and Factorization in Banach Modules. X, 305 pages. 1979.

Vol. 769: J. Flum, M. Ziegler, Topological Model Theory. X, 151 pages. 1980.

Vol. 770: Séminaire Bourbaki vol. 1978/79 Exposés 525-542. IV, 341 pages. 1980.

Vol. 771: Approximation Methods for Navier-Stokes Problems. Proceedings, 1979. Edited by R. Rautmann. XVI, 581 pages. 1980.

Vol. 772: J. P. Levine, Algebraic Structure of Knot Modules. XI, 104 pages. 1980.

Vol. 773: Numerical Analysis. Proceedings, 1979. Edited by G. A. Watson. X, 184 pages. 1980.

Vol. 774: R. Azencott, Y. Guivarc'h, R F. Gundy, Ecole d'Eté de Probabilités de Saint-Flour VIII-1978. Edited by P L. Hennequin. XIII, 334 pages. 1980.

Vol. 775: Geometric Methods in Mathematical Physics. Proceedings, 1979. Edited by G. Kaiser and J. E. Marsden. VII, 257 pages. 1980.

Vol. 776: B. Gross, Arithmetic on Elliptic Curves with Complex Multiplication. V, 95 pages. 1980.

Vol. 777: Séminaire sur les Singularités des Surfaces. Proceedings, 1976-1977. Edited by M. Demazure, H. Pinkham and B. Teissier. IX, 339 pages. 1980.

Vol. 778: SK1 von Schiefkörpern. Proceedings, 1976. Edited by P. Draxl and M. Kneser. II, 124 pages. 1980.

Vol. 779: Euclidean Harmonic Analysis. Proceedings, 1979. Edited by J. J. Benedetto. III, 177 pages. 1980.

Vol. 780: L. Schwartz, Semi-Martingales sur des Variétés, et Martingales Conformes sur des Variétés Analytiques Complexes. XV, 132 pages. 1980.

Vol. 781: Harmonic Analysis Iraklion 1978. Proceedings 1978. Edited by N. Petridis, S. K. Pichorides and N. Varopoulos V, 213 pages. 1980.

Vol. 782: Bifurcation and Nonlinear Eigenvalue Problems. Proceedings, 1978. Edited by C. Bardos, J M Lasry and M. Schatzman VIII, 296 pages. 1980.

Vol. 783: A. Dinghas, Wertverteilung meromorpher Funktionen in ein- und mehrfach zusammenhängenden Gebieten. Edited by R. Nevanlinna and C Andreian Cazacu. XIII, 145 pages. 1980

Vol. 784 Séminaire de Probabilités XIV. Proceedings, 1978/79. Edited by J. Azéma and M. Yor. VIII, 546 pages. 1980.

Vol. 785: W. M. Schmidt, Diophantine Approximation. X, 299 pages. 1980.